THE LATELY TORTURED EARTH

Alfred de Grazia

THE LATELY TORTURED EARTH

EXOTERRESTRIAL FORCES & QUANTAVOLUTIONS IN THE EARTH SCIENCES

Metron Publications

THE LATELY TORTURED EARTH
Exoterrestrial Force and Quantavolutions in the Earth Sciences
ISBN: 978-1-60377-0941
Library of Congress Control Number: 2016949778 Copyright ©
1983, by Alfred de Grazia; 2016, by Anne-Marie de Grazia
All rights reserved

METRON PUBLICATIONS P.O. Box 1205
Princeton, N.J. 08542-1205 USA
http://www.metron-publications.com

Nihil difficile naturae est,
Utique ubi in finem sui properat*

Seneca,
De Quaestiones Naturae

*Nothing is difficult for Nature, especially if she
hastens to destroy herself

CONTENTS

FOREWORD

The title of this book, *The Lately Tortured Earth,* may appear flamboyant, but such is the fault of reality. The author objects to the pedantic sweet view of the world of nature that insists, often insidiously, upon its tranquility and beauty, and sees in the ravaged countenance of the Earth the smooth cheeks of a baby; in same view, when attention is called to evidence of harrowing experiences, the Earth becomes old alma mater, whose blemishes are to be expected in a being of some billions of years of age. By way of contrast, here we stress the devastations of the Earth and the short time span in which they have occurred.

'Tortured' means to be acted upon violently, strained, twisted, distorted, burned, choked, immersed in liquids, and electrically shocked. And, because the word 'recent' in geology goes back millions of years, we choose the word 'lately, ' signifying the past dozen or so thousands of years, approximately the Holocene period. I aim to be theoretical. I do not wish to pile up horrific details, nor do I.

By 'Earth' of course is meant our familiar globe, cracked and slightly irregular, with its high speed motions and far stretching atmosphere. We are interested in the geophysical column that extends radially from a presumably iron and nickel center, through an enormous mantle of molten rock and water, to where the biosphere dwells, then outwards by means of gases and electric charges to beyond the Moon. We seek to know what has happened lately to this Earth system.

The amateur has his place in scientific revolutions, as in civil wars and politics generally. G. Grinnell, in an historical lecture on the revolt against early catastrophism, which prevailed at the London Geological Society from 1807 to 1832, had this to say: "What is extraordinary about the London Geological Society is that none of the original members were geologists. 'The little talking dinner club, ' as Davy put it, was a club for gentlemen given to talk, not to hammering rocks." Now of course, we all believe in hammering rocks, but prior to revolutionary occasions, a great deal of talk must be heard, and this book contains some of it.

I am addressing the book to amateurs, whose numbers will include, I suppose, liberally educated gentlepersons, bomb throwing intruders, and the specialist who seeks interdisciplinary enlightenment. Who, nowadays, among scientists and scholars, need not broaden his scopes anxiously and hopefully beyond his strict area of competence? Specialists in geophysics are unlikely to know archaeoastronomy or the sociology of science. An expert on radio-chronometry may be at a loss in archaeological chronology. Everyone must become an amateur to enter the lists of cosmogony, where the theory of quantavolution seeks to establish itself.

It hardly needs be said that I am myself an amateur, and could be nothing else, even if I had won my spurs in electrical engineering instead of political theory and behavior. This will become quite clear as the panorama of scientific materials and methods begins to unfold in the following pages. I hope to be regarded as an honest amateur, although I am professionally aware of the tricks that the unconscious underground mind can play upon an otherwise sincere scientist. There is no Piltdown Man fraud here, probably no wrong-headed Yale dinosaur, perhaps just plain errors, inadvertent omissions and foolhardiness, which I hope will be promptly discovered and publicized.

Isaac Newton, says a careful student of his work, fudged the members of one equation to improve its numbers for his proposition on the precession of the equinoxes; he manipulated averages in using the Moon's distance from the Earth to better correlate gravity with the Moon's motions; and "his use of the 'crassitude' of the air particles to raise the calculated velocity [of sound] by more than 10 percent was nothing short of deliberate fraud."[1] He then devoted some years to proving Biblical chronology correct, allowing catastrophes to rule the natural history of the universe until the Hebrew genesis put it into a sacred clocklike order. Aside from this, he became abnormally mad.

I have tried, on the problems that I set for myself here, not to fudge the facts, to select perforce my materials from the inordinate mass while letting the reader recognize the manipulations necessarily entailing, and to let no fraud enter my meagre calculations. I intend to prove correct no sacred scriptures, beyond recognizing the contributions which they may make to scientific historiography. In all of this, I doubt that I am more than normally mad, unless it be in the presumption that this work will be useful to science.

ALFRED DE GRAZIA

---

[1] Richard S. Westfall, « Newton and the Fudge Factor, » 179 *Science* 4075 (23 February 1973), 751-8.

CHAPTER ONE

QUANTAVOLUTIONS

Clarence King was the first Director of the United States Geological Survey. He was liberally educated at Yale University and spent years in field work thereafter. According to the historian Bancroft, he "had acquired a reputation and a position second to no scientist in America." When he returned to lecture at the Sheffield Scientific School of Yale in 1877, he argued against the prevailing opinions in geology and evolution, insisting on the basis of his experiences and visions as a surveyor that the Earth had been lately devastated. The belief in catastrophism, he said, in surprising pre-Jungian language, was a true grasp of what had happened to the World. "Catastrophism is therefore the survival of a terrible impression burned in upon the very substance of human memory."[1]

Because catastrophism is a word that excites emotion and connotes only destruction, the present work and the series to which it belongs prefers the more general idea implied in the word quantavolution. The concept allows a more peaceful invasion of the realms of gradualism, uniformitarianism, evolution, and anthropology.

I do not mean this book to be violent and bloodcurdling. We have far too much of such stimulus today on television, in movies and in other books and magazines. I even go so far as to

---

[1] *Scientific America*, Supplement N°80, 14 July 1877, 1276.

say that the Earth system has been settling down - this without
conclusive evidence. But facts must be faced. The Earth has been
severely traumatized in the memory of mankind. In words that I
have used before, any place on earth can be viewed as a
Quantavolutionary Column:

> **Any tube of one-kilometer diameter circumscribed
> anywhere on the surface of the Earth, which reaches as
> high as the end of the magnetosphere hundreds of miles
> upwards, and as low as the upper mantle some thirty
> kilometers down, will have endured within the past 14,000
> years radical changes in its absolute and relative
> orientations, its atmosphere, its rocks and its biosphere,
> including any long-lived human cultures.**

Several   principles   characterize   the   theories   of
quantavolution:

> Every major feature of the Earth's surface is an effect of
> quantavolution;   hence   every   feature   figured   in
> evolutionary theory is translated more realistically into
> quantavolutionary theory.

> The dominant shape of the most determining events in
> natural history is a logarithmic or exponential curve
> where, from a pre-existing state, sharp change occurs,
> followed by a steep exponential decline in the effect. After
> a time, the curve of the effect flattens out, and an illusion
> may arise that the processes under scrutiny have always
> been as they are now.

> The several descriptive spheres of natural activity:
> atmosphere, lithosphere, hydrosphere and biosphere,
> transact regularly, but most emphatically and completely
> under catastrophic impulsion.

Partly because of the greater force of inanimate being and partly because its own basic nature is identical with the inanimate, the biosphere is as subject to quantavolutionary experience and interpretation as the physical spheres.

The theory of quantavolution depends upon the evidence that catastrophes really happened, for it is upon such abrupt, large-scale natural events that the quick leaping changes of quantavolution in the holosphere depend. By the same token, a quantavolutionary theory must show either that large spans of assigned time in natural history are fictitious, or, if they occurred, little of the natural world changed during their passage. Every chapter of our book is dedicated to these tasks, but several general comments may be offered in advance.

If our minds were still strapped to the ideological framework of the seventeenth century, there would be less of a problem in these regards. For we should normally believe that great floods, fires and earthquakes had happened in ancient times, and operated on such a vast scale that many "miracles" were associated with them. By miracles, I mean such phenomena as the falling of edible material, manna or ambrosia, from the sky, and the specters of enormous brilliant comets to which the Earth around us responded like a giant animal coming alive.

No mental gymnastics would be required to see in the Earth's behavior an abundance of evidence of at least the one great Flood of Noah in which the whole world was deluged and inundated. Indeed, we should see so many marks of catastrophe that we would have to invent several such floods and conflagrations, and comets to explain the complex piling up of ruin upon ruin, fossil upon fossil, and their bizarre collection and combinations. Practically every extensive ancient document and legend known to us from around the world would repeat the same kind of catastrophic history and lend support to the testimony of our eyes and the voice of religious and social authorities.

We might have been granted different, or additional, heroes of science, too: the brave Spanish priests who rescued from certain destruction the iconography and writings of the original inhabitants of the Americas; astronomers like William Whiston who perceived an exoterrestrial cause for the Noachian deluge; anthropologists like Nicholas-Antoine Boulanger who recognized the symptoms of catastrophic fear in the history of religion; palaeontologists such as Cuvier who discovered the layerings of catastrophe; anthropological-biological explorers like Humboldt who accorded respect to aboriginal accounts. Charles Lyell and his supporters thereupon might have had less success in dominating natural history --even allowing that they were riding on the crest of English world power, political power always being consciously or unconsciously imperialistic in the dissemination of ideas.

Admittedly there is a world history of science to be written from the standpoint of the sociology of knowledge as a first step in the opening-up of thought upon quantavolution. We must nevertheless still provide in the here and now the evidence of catastrophes called for earlier. Fortunately, and yet unfortunately, the here and now is prejudicial to quantavolution. Fortunate it is that mankind up to the time of the atom bomb has had a respite from cosmic catastrophes for over two thousand years. However, the respite has permitted a thoroughgoing sublimation of memories of general disaster even in religion, all of which are rooted in proto-historic disaster, not excluding the Judaeo-Christian-Islamic faiths. The greatest secret of religion today is the ostensible fact, too obvious for continuous attention, that religion is originally founded upon the terrifying behaviors of its founding gods. Jesus and Mahomet originate in the Books of Moses, in the frightful times of Exodus when Yahweh became God of the Jews. The history of religion as the history of catastrophes is also to be written.

Once more we return to the quantavolutionary evidence in the here and now. If science, politics, and religion are using the relatively peaceful natural world of today to cover up ancient catastrophes, how are the catastrophes to be uncovered? So far as research goes, one must read between the lines of natural science and politico-religious arguments, picking up here and there bits of knowledge and threats of argument. Ultimately, these can amount to many thousands of pieces and a strong line of argument.

The mills of conventional science, originally churning out milk and honey, are beginning to grind stones and salt, as in the ancient Scandinavian myth of the end of the world. This trend is faster than generally believed. I would guess that the leading scientific magazines such as *Nature, Science,* and *Sky and Telescope* have carried since 1945 an ever increasing number of quantavolution-oriented articles, minute proportion to the total, to be sure. But this number has been increasing exponentially in the past several years and by the year 1993, I would expect that fully a quarter of all publications in natural history will treat of quantavolutions.

Going farther, in geology and geophysics a number of scientists are deliberately hypothesizing catastrophes at the boundaries of several geological ages and adducing old and new evidence, especially by chemical examination of sediments, to prove that they occurred. The space programs of U. S. A. and U. S. S. R. have naively reported ancient catastrophes and on-going explosiveness wherever their vehicles have gone -Venus, Moon, Mars, Mercury, Jupiter and Saturn. Astrophysicists and astronomers are edging into catastrophic explanations of the surfaces of the inner planets and the asteroidal belt between Mars and Jupiter. Whereas in Charles Darwin's youth many scientists disbelieved in meteors striking the Earth, today certain scientists are advancing serious proposals for a space project aimed at exploding meteoroids that might appear to be on collision courses with the Earth. Where once the evolution of coal beds was supposed to have occupied millions of years in the ample time depots of natural history, today at least one authoritative textbook adopts great fires and floods as the most possible explanation of the origin of coal.[2] Biology is moving swiftly, but biology (and in the case of man -anthropology) as the history of life moves much more slowly, moves even in reverse motion, sucking up ever greater draughts of time.

Still, Walter Sullivan, dean of science reporters, could declare in the *New York Times* in December 1981 that serious challenges to the conventional tempo and mode of evolution were arising; they came out of proof concerning links between catastrophe and extinction/genesis of species, out of the capacities of genetic engineering for modelling new life forms, and from the growing tendency to interpret the rarity of so-called missing links or transitional types as the non-existence of said types, introducing therefore the alternative presumption that macroevolution (quantavolution) introduced distinctly new forms suddenly. What Lyell wrote a hundred and fifty years ago, "that no causes whatever have changed the earth except those that still do so under the eyes of man," can be easily updated: today man's eyes are wider; they can see more and can see into themselves.

The surface of the Earth that appears before our mind's eye is largely a crystallized image, a set of snapshots of a whole too large to be embraced by a single thought - valleys, plains, deserts, seas, mountains, clouds, jungles, islands, cities and more - ten, twenty, thirty, until the mind tires and says 'enough' and that is our Earth image. And, if we were quickly to call out words that we associate with each snapshot, we should probably begin with a couple of descriptive terms like, 'tall' or 'dry' or 'water' or 'trees', but then somewhere in the early words of each list there would perhaps be words like 'slow', 'long', 'evolving', 'the same', and 'old' that hint at 'long, slow processes in Nature. ' Without conscious awareness, we perceive and recite the ideology of the prevailing science. Yet only when we imagine the cities of the Earth are we describing a surface feature that is surely known to be very recent, because these are manmade.

---

[2] Wilfred Francis, *Coal, Its Formation and Composition*, 2nd ed., London: Arnold, 1961, 625.

We mostly come from western countries whose dominating perspective on the Earth and its history has been shaped by the victorious currents of scientific thought of the past two centuries. Other peoples, and our own peoples in other times, and many of our own peoples who do not participate in this phase of our culture, would not exhibit the same responses. As they imagine the Earth's scenery, they would think in terms of 'creation' and often use the very word. This would mean to them an animate god, the creative force. And when they say 'long ago' they mean 'very lately' in geological terms, and the same if they were to say 'in the beginning. '

Between the gradualists and the creationists are those whose outlook is quantavolutionary, thinking that the Earth here and now presenting itself is both natural and young. To them this Earth is a setting recently arranged by disasters. Quantavolution has had a foot in both camps. Insofar as it claims the methods of science and the empirical positivism of science, it is in the evolutionist camp. Insofar as it adheres to facts and theories resembling the earliest stories of the great and small religions, it is in the creationists camp.

The combination of ideas has never been given a full trial. When, in the early nineteenth century, a few quantavolutionists were active, they were known as catastrophists, or revolutionists, or saltationists. They were soon identified with the enemy by the uniformitarian and gradual evolutionists and crushed in the same battles that saw the defeat of the creationists.

Let us identify ourselves as quantavolutionist and, confronting the Earth's features, ask "How and when did what make what?" For instance, "In the 1980's exploding and erupting magma rising under high pressure fashioned the top of Mt. St. Helens as it appears today. "This is not much of an answer but it suffices to introduce the complicated subject of this volcano.

If "what is made" has to be thought of as the whole surface of the Earth, large categories are needed. So we adopt several arenas or spheres of activity, and place this volcano under volcanoes in general, and volcanoes in general are part of the lithosphere, inasmuch as what remains on the spot is now frozen into rock. Much of what emerged from the Earth rose as ashes, and gases, as electric discharges, too, and water, in a veritable cyclone.

For some purposes, then, Mt. St Helens could find a place under a second category, the atmosphere, which was much affected locally by the eruption. The clouds of water vapor ultimately fell upon the ground and the seas and circulated widely in the hydrosphere, another principle arena for geophysical activity. Except for a few insects and plants, the close-in biosphere was wiped out by the disaster. Some biosphere specimens of homo sapiens cleverly moved to a safe distance and observed the events; a few persons were killed. So in the instance, forces typical of the lithosphere changed a feature of the lithosphere and affected the atmosphere, hydrosphere and biosphere to a noticeable extent.

There are not so many different crustal forms of the Earth that they cannot be encompassed by the mind and by this book. The splendid and fascinating variety of nature is in its details. We hope to treat the major features in a general way: volcanoes, rifts, mountain ranges, ocean basins, etc. in the lithosphere; gases and electric charges etc., in the atmosphere, but too, exoterrestrial intrusions by meteoroids, electricity, gases and dust; further, the waters acting in the oceans, floods, tides, rocks and rivers; and the biosphere of the plant and animal kingdoms. These spheres are the general answer to the question: where does change on Earth occur?

The features or forms are the "what is made." As to "what makes them," we have to settle upon a classification of forces or energies. Here we prefer a pragmatic approximation which is close to the phenomena as experienced, so most of the terms are straight from the newspapers: the volcano, though a feature, becomes also a force. Meteoroids as well, and others, too. Most of the chapter titles convey an impression both of cause and effect. Atmospherics are the workings of and in the atmosphere; hydrospherics of and in the hydrosphere; and so on.

Had it seemed more useful, a highly abstract nomination of forces might have been attempted; electromagnetic, inertial, 'weak' force, and the whole Earth described as built from the working of forces beginning at the level of particle physics. Something like this procedure is followed in an accompanying book (Solaria *Binaria*). But as matters stand, here we have already enough abstraction for our needs and perhaps even too much for the tastes of the reader.

The forceful phenomena that landscape the Earth and impress mankind go by a score of names. Some surprising consequences attend even the seemingly ephemeral noises and sights that attend natural operations; they are, to be sure, powerless effects in one sense, but in another sense, as we shall see, they are forces in their own right. The "music of the spheres" and "the wheels within wheels" are but ancient inherited words fossilizing for us ancient phenomena of sound and sight. They help make man what he is and this can be regarded as a criterion of a natural force; thus, what concerns us about the atmosphere is partly that the air we breathe and the food we eat are governed by atmospheric processes. Such are the homocentric beginnings of ideology, that which inspires our curiosity about nature in the first place.

Otherwise, the categories of forces are commonplace enough and group themselves fairly readily in the several spheres of natural operations. We name them as winds, hurricanes, cyclones, lightning and other electrical flows; as meteoroids and fallouts of all kinds, terrestrial and exoterrestrial in origin, including especially radiation. We call up as forces too, the downpours of rain or cataclysms, the floods, tides, tsunamis, accretions of ice, the ocean currents and chemical 'baths. ' And of the land we speak of continental drift or rafting, of seismism, volcanism, the folding and thrusting of mountains, erosion both fast and slow, the rising and sinking of land, the electrical processes in the land as well as air. And, so far as concerns the biosphere, we are interested in the mutational forces that speciate life forms and the human work that can often transform the landscape and affect the atmosphere and oceans.

We may become most general in our language and conceive of a holosphere, all spheres transacting among themselves. As in the case of Mt. St. Helens, effects of a natural force are likely to be experienced in all spheres, immediately or with the passage of time. An earth tremor will divert a stream, gather and discharge electricity, send the animals fleeing in all directions, and set humans to praying. Seismism is neatly numbered by intensity nowadays, and it is easy to test the holospheric principle by observing effects in all spheres produced in association with a Richter scale 1 and, say, 9, but allowing that this reading of 9 may have, in times before measurement and, more, before conscious memory, reached hypothetical reading of 12 or 20. What would the Richter-scale reading have been when the Indian sub-continent split off East Africa? Or when the fabled island continent of Atlantis "sank in a day of furious trembling," according to Plato?

Now a criticism can be launched against quantavolutionism. India split from Africa, not in a day, but by an exceedingly numerous series of a centimeters a year, as Arabia is pulling away from Africa today - so it is argued. This might be measurable on the ordinary reaches of the scale. So the event, as grand as it appears on maps, was not a catastrophe; besides, the argument goes, it happened a hundred million years ago.

This kind of argument is bound to brew trouble. The "when" problem occurs in conjunction with the "how" problem. The "when and how" are answered together. First, an up-strain from below works gradually along a weak line of rock and slowly insinuates a crack which lengthens and widens until India is separated from Africa and, impelled by mantle-located forces of the same type, is slowly pushed towards Asia. Millions of years were consumed in accomplishing the clear break, many millions more in rafting to Asia. In such circumstances, the hydrosphere, biosphere, and atmosphere would be hardly affected; even the lithosphere would not be severely disturbed; there are always a few crumbs falling when a slice is cut from a cake and slid across the table. All to the tune of numbers 1 to 9 on the Richter scale.

Adversely, a catastrophe is asserted. India's separation from Africa was part of a worldwide fracturing of the globe. It happened quickly, with a hard blow impacting somewhere. Within hours, India was cut off and moving rapidly through watery wastes lately occupied by other lands that, too, were dispersed and moving eastwards. Not only was the event consummated suddenly, but it happened lately -thousands, not many millions of years in the past. So goes the quantavolutionary argument. We shall join the argument again and again in the chapters to come.

A classic case of holospherics is the much-studied and well-discussed theory of world-disaster befalling about the year 1450 B.C. at the instigation of a great comet. Here I shall repeat only the hypotheses, as I have stated them elsewhere, suggesting that the reader may resort to my *Chaos and Creation* and *God's Fire: Moses and the Management of Exodus* for a fuller account, or to the famed book of Velikovsky called *Worlds in Collision* and the debates surrounding it.[3] In regard to that fateful year, and throughout the world, the quantavolutionary hypotheses may be stated as follows:

(a)   **No geophysical feature or process that manifested a sensible form then, and which is capable of exhibiting the effects of discontinuous stress when examined by current geophysical techniques, will fail to show that such stress occurred.**

(b)   **No record of astronomical events available for the period around that year will present astral, planetary, or solar movements as unchanged or uniformly changing from before to after the year.**

(c)   **No retroactive calculation or index (such as of carbon 14 levels) or historical reference will fail to show atmospheric turbulence and atmospherically implicated irregularities.**

---

[3] (a) Princeton, N.J.: Metron Publications, 1981; (b) *ibid.* : 1982, New York : MacMillan, 1950 ; and see the files of the *Society for Interdisciplinary Studies Review,* hereafter *SISR, Kronos,* and *Pensée* magazine, *passim.*

(d)   No survey of biological history around this year can deny
        highly unusual animal and human behavior and
        widespread destruction in the plant and animal kingdom,
        including agriculture.
(e)   No graphic, legendary, or archaeological account will
        produce a human settlement in the world that escaped
        heavy destruction from natural causes.
(f)   No religious temple that was constructed anywhere
        beforehand and rebuilt thereafter shows the same
        astronomical orientation before and after.
(g)   No god passed through this year without change of status,
        rites, family relations, and serious personal incident, and,
        correspondingly, all religions changed.
(h)   No culture complex can be shown to have avoided, with
        or without detectable hiatus, significant changes in
        institutions, ruler ship, and artifacts.
(i)   No institution, behavioral pattern, and natural setting
        existing today, if its history is complete, will fail to recall
        the effects of the events of these times.

In brief, no sphere of existence escaped intense
experiences and transactions with other spheres in the
quantavolution of the times. All quantavolutions imply heavy
holospheric events. For periods before human race had
quantavoluted (the subject of my work, *Homo Schizo I)*,
anthropological spheres of existence would, of course, be
excluded.

It will be appreciated that, under evolutionary theory,
holospherics tend to be less stressed. When large effects are
reduced by time to minute causes, the side-effects are
proportionately and even exponentially reduced. The more
intense and sudden the event, the more spheres will be transacting.
The larger the scale of an event, too, the more spheres will enter
the action.

Suppose the Earth's rotational speed were to be slowed.
This is a mighty event and takes a mighty force; Earth's rotational
energy is calculated at 10^{36} ergs. Yet it has been observed (by
Danton) to happen recently, if only for a millisecond. No account

of effect has yet been rendered; perhaps the effects were immeasurably small, or perhaps the reactions of scientists were too slow. If large solar flares caused the retardation, as seems to have been the case, worse flares or other causes might produce a larger rotational lapse, perhaps a second of time would be lost; perhaps then a minute; why not an hour? -Hypotheticals are cheap. The effects of lengthening the slowdown would be heavy. Every sphere of Earth, every force, would be activated in using up the energy surrendered by Earth in the deceleration. One would have holospherics on a grand scale. Ordinary language, the most archaic religious language, and scientific language could each provide the description required.

Now the quantavolutionist reverses the logic as well. We say, "the more affected the holosphere, the greater the force to be sought." The effects are proportional to the original force. When the effects exceed (or are theoretically calculated as having exceeded) a certain intensity, we must even go beyond the Earth into cosmic forces drastically simplifying. Only in the supra-terrestrial arena, the planetary and galactic systems, are to be found forces large enough to do the Earth what appears to have been done. Only cosmically can truly great holospheric transactions be generated.

One can realize, then, the importance of the "when" and "how long." To say "speedy reactions" is to invite ultimately the cosmos in to explain our terrestrial phenomena. To say "slow reactions" is to keep the Earth within its cocoon in space, traveling evenly and safely. If the Alps tower above Europe, some force must have pushed then up. If the Alps are to arise suddenly, then something besides earthly forces are behind the event. We move into the cosmic realm. If the Alps are to arise over a great many millennia, then the force might be generated in energy measures conceivable from some mysterious, but still earthly, internal force.

PART ONE

ATMOSPHERICS

Here the atmosphere is pictured as a heavy-working transmitter and transformer of the holosphere. Once it was part of a vastly larger gaseous plenum of the solar system. In quantavolutionary episodes, it was repeatedly destabilized and altered. Much of the crust and its deformations are exoterrestrial effects, which passed through the atmosphere. The Earth's magnetic field also functions in the atmosphere, and electricity is prominent there. Gases, electricity, and fire have combined with winds – all on a quantavolutionary scale – to help mold Earth and life forms.

CHAPTER TWO

THE GASEOUS COMPLEX

The atmosphere of Earth is so delicate that most sudden and violent transactions in space or on Earth transform its constituents and their behavior. Considering what is to come in this book by way of demonstrating terrestrial catastrophes, one may wonder how it happens that life has survived five thousand, much less five billion years. The very fragility of the aura around us bespeaks the recency of the atmosphere as we know it.

For example, in-coming cosmic particles collide with atoms of the atmosphere, giving off neutrons that interact with nitrogen to make carbon 14. Then C14 couples with oxygen to form carbon dioxide, and is often ingested by plants and passed along to animals through the plants. When any plant or animal (living from plants) dies, it ceases to acquire C14 and the C14, which is radiogenic, decays at a constant rate into nitrogen. In the short term, the process is fairly regular. The ratio, in a specimen, of C14 to C12, a non-decaying type of carbon, can be used to date its decease. But lightning, smoke, dust, explosions, vapors and cosmic particle flux can alter the density of C14 in the atmosphere, hence in organic material.

Soviet investigators found C14 deviations in connection with galactic supernovas of the years 1054 and 1700.[1] Judging by

---

[1] B. K. Konstantinov and G. E. Kocharov, "Astrophysical Phenomena and Radiocarbon," 10 Sov. Physics 11 (May, 1966), 1043-4.

the C14/C12 ratio in annual tree rings in or about the year 1908, when the Exoterrestrial Tunguska body exploded with heavy local effect in Siberia, 1% less of the C14 was available in that year by comparison with the year before and after.[2] In another case, during a period called the Maunder Minimum, 1645 to 1715, when the Sun exhibited no sunspots and the Earth was gripped by a "Little Ice Age," the C14 found in tree rings of the period averaged 20% more than before and after.[3] Grave events disturbed the atmosphere on other occasions. Between 3200 and 3700 B.C. and in the eighth and fifteenth centuries B.C. the quantity of C14 in the air fluctuated heavily.[4]

A theoretical calculation by Cook that retrogressively computed the presence of C14 in the atmosphere, basing itself on a presently observed slight built-up of the gas, concluded that today's volume of C14 would have had to originate from a zero point 13,000 years ago. Why the rate would decrease to zero around that date has been interpreted as an indication of an extremely short Earth history; we here regard the hypothetical absence of C14 around that time as owing to several factors, most importantly (a) the presence of a plenum of gases incomparably more impenetrable by cosmic radiation that the present atmosphere, (b) a stronger geomagnetic shielding produced by a stronger geomagnetic field than exists today, and (c) exoterrestrially produced turbulence in the Earth's gaseous

---

[2] C. Cowan, C. R. Atluri, and W. F. Libby, 206 *Nature* (1965), 861.

[3] *Science News*, March 6, 1976; *Astronomy* (March 1979), 58; J. A. Eddy, P. A. Gilman, and D. E. Trotter, "Solar Rotation During the Maunder Minimum," 46 *Solar Physics* (1976), 3-14.

[4] A. F. M. de Jong, W. G. Mook and B. Becher, "Confirmation of the Suess Wriggles: 3200-3700 B.C." 2180 *Nature #5717* (July 5,1979) 48-9; I. U. Olsson, ed. "Radiocarbon Variations and Absolute Chronology," (12th Nobel Symposium, 1969; Alqvist and Wiksell, Stockholm and New York: Wiley, 1970) esp, H. E. Suess; Alfred de Grazia, *Chaos and Creation*, 48-52.

complex.[5] The inference here would be that major events before that time might have reconstituted the atmosphere, at which time C14 would have begun to accumulate.

Obviously C14's history indicates that other atmospheric components would not have escaped turbulent experiences. Carbon dioxide in the air fluctuates with industrial and domestic combustion. The amount in the air is increasing (it is some .03% of the atmospheric mass) and concern is expressed that the Earth's climate may change so as put much of the biosphere in jeopardy.[6] So also it has been surmised by students of the ozone (O_3) constituent of the upper atmosphere that its destruction as a particle shield by aerosol discharges on Earth would engender high risks of biosphere damage.[7] All of this may happen within the next century or two.

Very similar types of blue-green algae live under the skins of rocks in the frigid Antartic desert and in the heat of the Sahara.[8] Abyssal organisms live beyond the reach of light. The limits of humans and their predecessors are much more narrow, whether we speak of oxygen or a dozen other basic requirements. (Later we shall examine the claim that simple organisms can traverse and inhabit space-conveyed meteoritic vehicles even "on their own.") Humans have been known to acclimatize themselves to high altitudes with low oxygen and low barometric pressure.[9] But

---

[5] Melvin Cook, "Carbon 14 and the Age of the Atmosphere," *Creation Res. Soc. Q.*, June 1970. Reuven Ramaty (U. C. L. A., Calif) has studied extensively geomagnetic effects.

[6] Gilbert N. Plass "Carbon Dioxide and Climate," Sci. Amer. (July 1959), 3

[7] S. W. Tromp, *Biometeorology* (Philadelphia: Heyden, 1980), 12, 16-17, 19.

[8] George W. Gray, "Life at High Altitudes," 193, *Sci. American* (Dec. 1955), 58; "Respiration and Respiratory Systems," *Ency. Britannica* (1974), 763.

[9] 9. E. I. Friedmann and R. Ocampo, "Endolithic Blue-Green Algae in the Dry Valleys" (Antarctica), 193 *Sci.* (24 Sep. 1976), 1247.

beyond 20,000 feet, the human dies. Pure oxygen is, of course, a poison and an explosive.

There is little certainty about the history of the atmosphere, even during human times.[10] The primeval air must have contained some molecular oxygen (O_2) for the lung-breathers. Not too much lest the air catch fire. Legends do report "world-burnings," that Donnelly and Velikovsky, for instance, attribute to hydrogen gas pockets of exoterrestrial origin. Nitrogen might not be needed but the air must then also have held much other gas; for terrestrial life forms are constructed to deal with outside pressures. The diaphragm and chest muscles are made to operate as a bellows sucking the oxygenated air into the lungs and exhaling it with carbon dioxide. A pressure gradient must be accommodated between the external air and the internal metabolism. Yet if the air had been too dense, creatures such as humans would be too burdened by it to move about.

Considerable leeway is permitted for the amounts of inhalable oxygen, the mixes of gases inhaled (barring poisonous gases such as carbon monoxide), the atmospheric density (pressure) and degree of vaporization, the kinds and amounts of radiation such as ultraviolet rays, temperature (from 40 to 100 Fahrenheit as a milieu), and luminosity of the environment. Dew will suffice in place of other freshwater sources. Edible plants or animals, including one's own species *in extremis,* must be available, and these, of course, are atmosphere-dependent too. The present human cannot survive in the highest mountain altitudes or underwater without artifices.

Given the prolific potential of human reproduction, the atmosphere might have been severely ravaged and changed

---

[10] L. V. Berkner and L. C. Marshall, "A History of Major Atmospheric Components," 63 *Proc Nat'l Acad Sci* 6(1965) 1215; John A. Eddy, "The Sun Since the Bronze Age," *Int. Sym. on Solar-Terres, Phy.*, June, 17, 1976; J. S. Sawyer, ed., *Proceedings Intl Sym on World Climate:* 8000 to B.C. (London: Royal Meterological Soc., 1966; Donald W. Patten, *The Biblical Flood and the Ice Epoch* (Seattle: Pacific, Meridian, 1966), Chapter. 9.

without destroying utterly the species. The human body is built upon and functions with the basic elements of nature. It is catastrophized and by the very fact catastrophe-proofed to some degree. Its incubating young are deeply encased and easily transportable. What it cannot cope with internally it seeks to escape by rapid mobility and exponential rates of reproduction.

The atmosphere presently consists of a changing mix of gases and vapors that moves from surface levels upwards to where the magnetosphere ends at any moment of measurement. What is beyond may be called outer space, where space plasmas, solar winds, cosmic particles, and meteoritic material play about in some disorder.

The atmosphere itself is a model of disorder. It is continuously moving and reorganizing. Everyday its pressure goes up and down. About 99% of its mass blankets the globe at under 19 miles of altitude. This consists of the gases, molecular nitrogen (78%), molecular oxygen (21%), argon (1%) and carbon dioxide (0.03%). Water vapors rarely reach 1% of the total: normally, half of the globe is covered by clouds, which form, reform, and discharge their vapors almost entirely within six miles of the surface.

Below the clouds hang most of the "pollutants" of industry, consumption, war, and transportation. But some of this may rise so high as to threaten the layer of ozone, a poisonous triple-atom oxygen molecule ($O3$), which, so long as it stays out of the animal system, performs a vital function in stopping solar ultraviolet rays from reaching the animals.

As one moves up the atmospheric column from ground-zero one passes successively through "belts." These are statistical entities, not the usually discontinuous strata of the lithosphere. The sixty-mile homosphere is divided into troposphere, stratosphere, and mesosphere. Then occurs a heterosphere, and, at around 300 miles, an exosphere. The homosphere is a molecular region where nitrogen and oxygen are the principal actors; but at bottom are cloud and pollutant behaviors and at the top occur some vigorous radiation, dissociation of molecules, formation of hydrogen compounds, and ionization.

In the heterosphere, atomic oxygen, helium and hydrogen are the abundant elements. Some of the helium and hydrogen is on its way into farther space, but is replaced, it is believed, to produce an equilibrium. However, Melvin Cook, a quantavolutionary geophysicist, has asked, "Where is the Earth's Radiogenic Helium?".[11] Cosmic-ray sources are alleged to generate helium at 3×10^9 g/year. The same amount is estimated to be generated from the uranium and thorium in rocks of the lithosphere. With an Earth age of 5×10^9 years, about 10^{20} grams of helium should have passed into the atmosphere by now. The atmosphere contains 3.5×10^{15} grams of helium-4; if a steady state, it must have passed out through the exosphere the equivalent of the aforesaid 10^{20} grams.

However, helium-4 does not concentrate in the upper atmosphere significantly and "at the escape temperature of 1500°K at the base of exosphere, the rate of escape of helium-4 would be only about 600 g/year, or only about 10^{-7} as great as the replenishment rate from the lithosphere." Only by raising temperatures at the base of the exosphere by thousands of degrees could the helium be allowed to escape in sufficient quantities to permit equilibrium. This can be conceived as possible only by means of a number of immense solar storms that would wreak havoc on Earth or, worse, by large-body encounters wrecking the atmosphere. Cook suggests that the helium-4 is still increasing; the atmosphere is not in equilibrium; and if retro calculated, a recent beginning or reconstruction of the atmosphere must be confronted.

Geophysicists and meteorologists nevertheless retain the concept of the atmosphere as a whole being in equilibrium. This is probably not so, even in the short run of a thousand years. The idea is difficult as commonsense, considering that all the way from sea level into outer space the atmospheric column is in continuous

---

11. 179 *Nature* (26 Jan. 1957) 213.

flux. It is agitated and fed from the bowels of the Earth with heat, vapor, etc. and bombarded topside by elemental particles of all kinds. Motion is continuous, too, up and down the column and then horizontally with winds produced by thermal changes, such as the seasons produce, and rotational effect that, for instance, disturb the atmosphere via surface irregularities such as mountains and basins.

Indeed, equilibrium of the atmosphere is probably more of a hope than a fact. What makes the hope into a "fact" is, not surprisingly, the uniformitarian conviction that today's actors and roles are unchanged from eons ago. Given hundreds of millions of years when animals and plants have been surviving, then the mix of vapors, nitrogen, oxygen, carbon dioxide, argon, ozone, and radiation must have been what they are today. And that spells equilibrium.

The belief becomes so strong that meteorologists, possessed of the "fact" of atmospheric equilibrium, can even take their turns at guarding the portals of uniformitarianism, assuring other scientists that meteorology, too, proves the long-enduring stability of present-day conditions. At the same time, ironically, meteorologists are leaders in the campaign to save the world from the atmospheric ravages produced by a few years of industrialism, atom bombs, and aerosol discharges.

A quantavolutionist may share heartily the meteorologists' fear of the poisoning of our present atmosphere. The quantavolutionist would at the same time point out the extreme improbability of the atmosphere's having been preserved intact - free from radical changes and poisons over long periods of time. Unless, of course, there were, before the present atmospheric system came about, some ancestral system that in its nature involved a true long-term equilibrium.

It is generally admitted that the sources of nitrogen and oxygen of the air are uncertain and disputed. Further, the sources of water and salt are unknown. Too, all of the minor gases of the atmosphere are of mysterious origin: neon, helium, methane, krypton, hydrogen, nitrous oxide, and xenon. And some has mysteriously "disappeared;" neon "should be" far more abundant, for example.

Oxygen is supposed to have been exhaled from plants, permitting thus the beginnings of animal life. Orthodoxy puts this "happening" at over a billion years ago. Perhaps the only "hard" evidence for the event is the discovery of a non-oxidized core of uranium and sulphur in Kenya, the presumption being that there was little or no molecular oxygen with which the elements could react when the rock was formed. Yet by this kind of reckoning, it is hard, too, to explain fossils of 3.1 billion-year-old bacteria.[12]

It has long been permissible to speculate that the components of the air came from the "primordial melt," a fiction of science performing very much the same role as the fiction of "the end of the Ice Ages." One may as well speculate that they came from space, since practically every element has been identified within the magnetosphere of Earth.

There are indications that the Earth may have evolved in a binary system such as I have described in *Chaos and Creation* and, with Earl R. Milton, in *Solaria Binaria*. An electrical axis, carrying an arc or current between the Sun and its small and less radiant binary partner, would be a more durable and gently changing source of radiation and chemical energies than the direct glare of the sun today. A magnetic gaseous tube rotating around the axis would provide a full complement of chemical elements, again in a highly stable medium that so minor a product as aerosol sprays could not disrupt. It would be making large quantities of all the substances whose manufacture in the small atmospheric and petrological economy of "Spaceship Earth" has been hard to explain. Atmospheric pressures, too, would be stable. Winds would be largely absent, illumination fairly constant.

It should be permissible to speculate that the magnetic gas tube stretching between the binary's two principals was the source of the Earth's atmosphere. Most of the binary tube gases

---

[12] 12. I. S. Shklovskii and Carl Sagan, *Intelligent Life in the Universe* (New York: Dell, 1966), 223-4.

would have escaped into space with the decline and disappearance of the axial current. The Earth then may be surviving upon the fragment of the gases that its electric-gravitational field retained. The atmosphere now may be only a remanent halo.

The variety and abundance of the atmospheric gases are what would be expected according to the gas tube model. A long-time continuity of the atmosphere and biosphere would have been possible; life could have begun long ages ago (or recently) and enjoyed the same relationships it now enjoys with oxygen, carbon dioxide, water, and salts. The fragile ozone layer was entirely missing, without ill effects, because the Sun and galaxy were not striking directly upon the Earth. Indeed, there would be little need for a stratified, local Earth atmosphere. The Earth could change position along the central axis without losing its atmospheric and thermal equilibrium. In the early declining period of the axial current, the pollutant of meteoroid or large-body contacts could be dissipated into the gas tube environment, and important losses replaced from the same source. Even the effects of an eruption of the Moon from the Pacific Basin would be cushioned by the binary atmosphere.

The postulated magnetic tube would be randomly composed. Its gases would be arranged with the lighter elements nearer the axial current, heavier elements in the middle; simple compounds would occur toward the boundary of the tube, where the planets were rotating. The heavy-bodied planets would accrete their special atmospheres within the tube, even while rotating magnetically around the axial current. But the difference between the terrestrial atmosphere and the tube atmosphere would be far less than between the Earth's atmosphere and its heterosphere or outer space today.

It is understandable, under these postulates, how the Earth's atmosphere, so fragile, might have existed for a considerable period of time. Given the evidences of catastrophes on Earth, I do not see how the atmosphere could have survived without large external atmospheric background. Still the Earth was lucky to escape the fate of Mars, Mercury and possibly other inner planets, whose atmospheres were almost entirely stripped; Venus, with an infernally hot and turbulent atmosphere, was an exception, but a recent arrival. All of this is possible, and dealt with in *Chaos and Creation* and *Solaria Binaria*. Scientific opinion has slowly liberalized in respect to new models. By 1972 a scientist might write offhandedly in *Nature* magazine that "major reorganizations of the solar system are no longer regarded as ridiculous."_[13]

Recently, dendrochronologists, historians, meteorologists, radiocarbon dating specialists, and astronomers combined in a most unusual enterprise. They delivered a blow to the theory of the constant Sun. John A. Eddy of the National Center for Atmospheric Research conveyed the message: "We've shattered the Principle of Uniformitarianism for the Sun."[14] He presented evidence mentioned earlier, showing that for 70 years between 1645 and 1715 A.D. sunspots were almost entirely absent. It proved to be a period of bitter prolonged winters, when Londoners walked across an iced-over Thames River, when the Northern Lights hardly displayed themselves, and when the 11-year sun-spot cycle was absent. Lapses of the same kind were uncovered in other historical periods.

Other conditions may be expected to vary with sunspots - solar flares, ozone density, radiation diminution, precipitation, magnetic fields, atmospheric turbulence, famines and perhaps even human energy and inventiveness. No doubt the last will be among the most difficult to prove. No simple search of the annals of culture will reveal a closely related trend.

Stretching the uniformitarian thesis, more severe storms may be conjectured for pre-historic times, in an attempt to keep the planetary bodies in place, eliminate cometary encounters and still explain catastrophes upon Earth. Thus Harlow Shapley, who led some scientists in an attack upon Velikovsky's catastrophism

---

[13] 13. A. G. W. Cameron, 240 *Nature* (1 Dec. 1972), 229
[14] 13. A. G. W. Cameron, 240 *Nature* (1 Dec. 1972), 229

in 1950, himself had in 1935 proposed a solar nova as the explosive generator of space X-rays.

Hurricanes, volcanism, interrupted rotation, ozone destruction, ice ages, geomagnetic field reversals, biological extinctions and even explosions of cometary and meteorological material on Earth can be rationalized up to a point as effects of solar misbehavior. Such a theory is possible, but it would be like hiring a thief to catch a thief. For the Sun would then become sole factor in quantavolutions, in the effort to exclude other bodies from trespassing upon Earth. As we shall see, there is too much evidence of other operative factors to assign the whole job of quantavolutions to the Sun, even though, as a matter of fact, the Sun is the original sire of quantavolution in the solar system, according to the model of Solaria Binaria, mentioned above, which begins history with a nova of the Sun.

According to the quantavolutionary theory here presented, solar behavior has exhibited only effects of a moderate kind since its gradual emergence as a distinct bright image some thousands of years ago. Before then, the Sun was hidden or a bright prominence in the cloudy firmament. Its indirect influence was of course always paramount. But should the counter-thesis be proposed that the Sun was responsible directly for earthly catastrophes, it would have to be said that its "uniformitarianism," though spotty, was nevertheless much greater than that of the planetary family descended from the Sun's binary partner, which I have called Super-Uranus, after the Greco-Roman first Heavenly Father.

The sunspots may be a trailing-off effect of the exhaustion of the electrical current and magnetic tube. That is, they may be fairly regular attempts of electricity to jump the gap between the Sun and its binary. In such a case, the sunspots should become less intense and more sporadic with the passage of time, like the plasmoids and bolts of Jupiter.

Climate is the typical behavior of the atmosphere over any geological column during a longish time. Every island, they say in the Caribbean and Aegean Seas, has its own climate; "mini-climate" would be precise. More expansively, we can talk of a regional climate or a global climate. Too, we shall soon have a "cosmic climate," since evidence is fast accumulating of solar-planetary transactions on a continuing climatic basis. Earthquakes, volcanism, winds, precipitation, magnetic fields, temperatures, electric currents and the biosphere transact in climatic affairs.

One does not get this sense of a welter and complex of factors in going far back by conventional chronology. Rather one has the sense that climates have swirled around in multiform changes in the Quaternary period but then somehow climates withdraw into the background while we are presented a broad succession of ages in the tens of millions of years each, when life changed very slowly and conditions of biological survival and adaptation must have been constant over long periods of time. One is privileged to view charts in which paleontological developments occur at the slowest imaginable pace, with only a dozen or so boundary lines where, certainly, it is given that climates changed and new names are provided - Devonian, Carboniferous, and so on. Did climates, with all the factors that engender them, stand still for these long periods in rigid constancy? This would be unbelievable. If in between the major boundaries of epochs, climates changed as they have in the brief recent past of the Quaternary, then the paleontological and geological record is far too short, or contains very little information. In sum, either the world has changed and the recent past speeds up wildly in comparison with the remote past, or else the remote past is still quite unknown despite its diligent study over two centuries by numerous disciplines and thousands of scholars.

Hence climatology lends us a great doubt when we imagine it fitting to the long past ages, and many doubts when we try to use it for the turbulent recent times. A great many works on pre-history try to associate events with climatic changes. Considering that geologists have failed to establish confidence in climatic boundaries and periods, the pre-historian's failure is predictable.

For instance, classicist Rhys-Carpenter has endeavored to explain as a climatic worsening over generations the end of the Mycenaean (Greek) civilization and the subsequent so-called "Dark Ages" (an invented period of several hundred years to evade evidence of catastrophes in the eighth and seventh centuries B.C. and to accommodate Greek to Egyptian chronology, the later itself wrong by centuries).[15] Cities were abandoned in the face of desiccation; new hot, dry prevailing winds made impossible the carrying on of their culture.

To believe him, however, one must have a reason why the flowering of Greek culture occurred under the same climatic conditions later on. One must also discount the many evidences of natural destruction by fire and earthquake of the Mycenaean centers.[16] One must cling to a spurious Egyptian chronology, which gives 500 years to Greek and Mediterranean history that, since nothing happened, are not needed.[17] Further, catastrophic changes in winds and precipitation have a cause; that cause can only be celestial changes, whether by introduction of new Earth motions and land forms, or by solar-system particle-outputs.

If the Alaskan musk contains the swept-in plant and animal life of large areas and the species it contains are modern, then one should suspect that sooner or later, as Hibben has opined, humans, even clothed and deep-frozen, should turn up by accident or deliberate excavation. Already, several pre-" Ice Age"

---

[15] Rhys Carpenter, *Discontinuity in Greek Civilization* (Cambridge: Harvard U., 1966)

[16] Claude F. A. Schaeffer, Stratigraphie Cornparé... (London: Oxford, 1948).

17. I. M. Isaacson (pseud.), "Applying the Revised Chronology." 4 *Pensée 4* (Fall). 5.

18. This has been known since O. Heer in the 1860's. See Velikovsky, *Earth in Upheaval* (New York: Doubleday, 1955), *44 et seq. Cf.* H. H. Lamb, "The Earth's Changing Climate," 180-5 in *Encycl.* Britannica Yrbk, 1975; Frank Hibben, *Treasure in the Dust* (195 1).

[17] I. M. Isaacson (pseud.), "Applying the Revised Chronology." 4 *Pensée 4* (Fall). 5.

settlements have been uncovered within the arctic circle by Americans and Russians. Rodents and mammoths froze quickly while eating warm-weather plants. How abrupt was the climatic change that killed them is unreported, if known. The polar regions were recently near-tropical in climate and ecology.[18]

The bafflement of archaeologists over climate is understandable. They follow the evolutionists. But the attic of climatic evolutionism is stuffed with junk. When a modish dress does not suit the facts, an old-fashioned one is tried on.

For example, the heat of the Earth has been described in numerous ways over the past two hundred years; hence, without ostracism, one may propose that the Earth has an enormous internal heat or is cool - whichever advances one's theory of climates. Too, the ages of the Earth and its geological periods have been estimated with tens and hundreds of millions of years of variance and leeway, so that evidence of climatic shift can often be placed in time wherever it will fit the theory at hand. And the melting of the ice sheets can proceed rapidly or slowly, as needed for a particular job of explanation.

Uniformitarians employ typically six mechanics of climatic change:

(a)   a cooling of the Earth's interior over eons of time. (Since this should have ended long ago, with the Earth's interior stabilized, a radioactivity of deep rocks is now believed to be an incessant source of heat from below.)

(b)   a crawling up and crawling back of ice owing to pronounced cyclical solar activity (which has lately received some support by the aforementioned "Maunder Minimum" and sunspot studies.) (c) a reorientation of prevailing winds due to a manmade or artificial desiccation of lands, or to ice movements or Earth cooling (as above.)

---

[18] This has been known since O. Heer in the 1860's. See Velikovsky, *Earth in Upheaval* (New York: Doubleday, 1955), *44 et seq. Cf.* H. H. Lamb, "The Earth's Changing Climate," 180-5 in *Encycl.* Britannica Yrbk, 1975; Frank Hibben, *Treasure in the Dust* (195 1).

(d)   the "inches-per-century" drift of the continents from cold to hot places or *vice-versa.*

(e)   heavy multiple volcanism, called upon to supply the heat for the vaporizing of waters that then proceed northward and drop upon the polar areas as snow and ice.

(f)   changes in solar activity, whereby a period of diminished or augmented sunspots will produce cold weather or stormy weather.

That all of these are explanations inadequate to explain even holocene climatic change is evident in the controversies and the contradictions continually appearing. Geologist Vita-Finzi practically abandons his search for climatic benchmarks in his authoritative work on the holocene. Lacking the engine of a general theory and a time-table to run it on, freight cars may be switched around at will. In one place he is driven to remark: "On the assumption that every yodel in the Alps had its echo on the coast, pebble bands are equated with glacial episodes, truant beds are eroded away, and the uplift of mountains is delayed to justify the absence of glacial features."[19] He prays that the radiochronometrists will rescue the situation. But I have already concluded in my analysis of tests of time, published in *Chaos and Creation,* that a rescue must come from elsewhere.

Perhaps a quantavolutionary scheme may do better. It is not written in some law that enough time must be allowed to let humans get away, bag and baggage, from the changing air. Every catastrophe which they underwent would demand a climatic response as one of its effects. Hence there may have been a score of global shifts in climate within a 14,000 year holocene period.

Certainly the boundaries of the ages would point to climatic change. The onslaughts of the early holocene mark a paramount boundary. There came destruction of a worldwide greenhouse regime and the beginnings of mountain ranges, huge

---

[19] Claudio Vita-Finzi, *Recent Earth History* (New York: Wiley-Halstead, 1973), 106-7.

deserts, stripped shield rock, high plateaus, oceans and their currents, and biosphere revolution.

This Pleistocene-Holocene boundary climax is euphemistically carried in the logbook of the sciences as "the end of the Ice Ages". I treat it as the Lunarian climax in *Chaos and Creation,* because of its apparent connection with the advent of the Moon. Hundreds of titles from many fields are dedicated to it. In oceanography, Emiliani extracts from Gulf of Mexico bottom cores the information that a fresh water avalanche descended upon the basin some 11,500 years ago and he wonders whether this was from a cataclysm such as sank the legendary continent of Atlantis. Tree pollen changed abruptly in the Great Lakes region about 10,000 years ago, according to J. G. Ogden III. "The only mechanism sufficient to produce a change of the kind described here would therefore appear to be a rapid and dramatic change in temperature and/or precipitation."[20]

Oceanographers Broeoker, Ewing, and Heezing gather ocean-bed "Evidence for an Abrupt Change in Climate Close to 11,000 Years Ago."[21]Vita-Finzi reports that a group of geosols, or weathering profiles, ended their development about 12,000 years ago; the date is proposed as the holocene beginning for the U. S. A.[22] From Israel, paleo-zoologist Joseph Heller writes of the faunal remains of a Kebaran Site on Mount Carmel:[23]

---

[20]20. See below, Chapter 31

[21] 21. 258 *Amer. J. Sci..* 429.

22. *Op. cit.,* 42-3.

23. The Faunal *Remains of Iraq es Zihhan, a Kebaran Site on Mt. Carmel; cf.* Livingstone, 1975 "Late Quaternary Change in Africa," *Ann. Rev. Ecology and Systematics* 6: 249-81; Williams, M. 1975 "Late Pleistocene Tropical Aridity Synchronous in Both Hemispheres," 253 *Nature* 617-18; Hamen, Wunstra, and Zagwin "The Floral Record of the Late Cenozoic of Europe," in Turekian, K, ed. *The Late Cenozoic Glacial Ages* (Yale U. Press); Farraud, "The Floral Record," *Ibid*

[22] *Op. cit.,* 42-3.

[23] The Faunal *Remains of Iraq es Zihhan, a Kebaran Site on Mt. Carmel; cf.* Livingstone, 1975 "Late Quaternary Change in Africa," *Ann. Rev.*

What then was the cause of the post-Natufian size crash? (9000-10,000 B.C.) The fact that the crash occurred in certain carnivores and rodents simultaneously suggests that it was not causally related to phases in the evolution of human cultures. Rather this simultaneous dwarfing favors climatic interpretation. Drastic climatic changes occurred in various parts of the world towards the end of the Pleistocene about 12,000 years ago. In tropical Africa, India, South America and Australia, conditions that were extremely arid before 12,500 B.P. suddenly gave way to increase in humidity.

It is generally accepted by pre-historians of Europe that the end of the Pleistocene Ice Ages brought disaster to human races and cultures. The finding is surprising, considering that the warmer the climate, the more abundant the biosphere should be. But if catastrophes were involved, the reduction and retardation would be understandable, indeed demanded.

Ruins of cultures are found in many a harsh climate of the world, in deserts, on high plateaus, amidst perma-frost, and in steaming jungles. (Let us exclude, under the seas, which, after all, involved a climatic change, one which we shall discuss later on.)

When archaeologists and pre-historians cannot explain the death of a culture by enemy invasion, plague, or economic decline, they are prone to seek out a change of climate. But what they seek out is a uniformitarian or gradual change of prevailing winds, rainfall, and temperature. Centuries, if not millennia, are invoked to pursue the death agonies of a culture.

The quantavolutionist tackles the same problem with a markedly different concept, catastrophic climatic change. With the images in mind of an aboriginal greenhouse world afforded by

Ecology and Systematics 6: 249-81; Williams, M. 1975 "Late Pleistocene Tropical Aridity Synchronous in Both Hemispheres," 253 *Nature* 617-18; Hamen, Wunstra, and Zagwin "The Floral Record of the Late Cenozoic of Europe," in Turekian, K, ed. *The Late Cenozoic Glacial Ages* (Yale U. Press); Farraud, "The Floral Record," *Ibid*

many sources, he sees in every desert a likely disaster, every tall plateau another one, under frozen arctic shores still another.

For the quantavolutionist, too, the mechanisms of explanation are available, they are high-energy forces as provoked possibly by changes in the Earth's motion, a change of its orbital path around the Sun, a shift of its angle of inclination to the plane of the ecliptic (axial tilt), and a movement of its crustal shell (continental displacement). They include, further, a bombardment or discharge of particles, including cosmic electricity, affecting the atmosphere and magnetosphere that stretches even now beyond the Moon. And deluges of salt, oil and other dense material that spoils the land.

With all of this, it would seem that the quantavolutionist would necessarily bungle more than the uniformitarian in describing the natural history of climatic change. He is using, it seems, many more variables, and the more the variables, the more complicated the solution of a problem. However, the quantavolutionist has two sources of encouragement, he can see how futile are the explanations of the conventional climatologists of the natural history of climate. And the evidence appears to fall into the line of this theory with surprising ease.

The uniformitarians, in attempting to explain climate by reducing chances of natural catastrophes to a near-zero constant, become bogged down in a morass of special climates; every way they turn they discover new and different climates. They cannot cope with the possibility that in the sudden prelude and aftermath of disaster, short-term climates by the hundreds are created around the world; deserts are deluged, jungles are desiccated, lands are flooded, lands rise, winds change sharply, soils are turned over, the biosphere is transformed; if late in time, cultures terminate, or spring up, or react eccentrically. Nor can they allow that, if several global catastrophes may have occurred in four billion years, several might have occurred in ten thousand years, each transforming atmosphere and climate.

A Woods Hole Oceanographic Institution team reported in the *Scientific American* of March 1982 a set of discoveries which threatens the prevailing theory that oceanic waters are regionally stable, that regional bottoms reflect this aquatic stability, and that world climates can be determined by fossil and chemical balances of the bottom content. Eddies of the great oceanic currents such as the Gulf Stream occasionally break off from these gigantic oceanic flows and set up columnar rings of water that can reach 300 kilometers in diameter, even in this relatively placid age, and endure for 18 months or more. The ring-waters differ significantly in salinity, oxygen content, and temperature from their surroundings. Biological assemblages follow suit. Sedimentation rates are also a function of current velocity. Under such conditions, given several thousand, let alone several hundred million, years false climates can be expected to be inferred practically everywhere. Misleading strata will be exceedingly numerous. Once more, we must warn against the many theoretical structures of climate, hydrology, chronology and paleontology that interlock in varying degrees of poorness of fit. These findings by the Woods Hole scientists may effectively administer the *coup de grace* to the whole lot of them.

But we must not be carried away with the holistic interplay of factors before we have explained them. We may content ourselves at this point with three tentative, even skeptical, remarks. The atmosphere is not stable and has not been for long in its present state of equilibrium. When subjected to quantavolutionary hypotheses, the history of the atmosphere becomes full of mystery and potentiality. The study of climates has been vigorously pursued, but perhaps with the wrong conceptual instruments. Climates, the benchmarks of atmospheric history, seem to us to disintegrate under analysis into ephemeral signals of catastrophic events.

CHAPTER THREE

HURRICANES AND CYCLONES

An explosion of Mt. St. Helens recently blew down thousands of trees. An exoterrestrial explosion at Tunguska in 1908 blew down millions of trees. The Fens of East Anglia contain millions of felled trees. Here the trees were knocked down facing northeast and were buried. They were sheared off a meter above the ground and their stumps remain rooted. Many were tall and thick trees. No volcano is to be located as the source of the blast. What kind of a wind was this?

Winds find a minor place in textbooks on earth features. They erode rock by polishing and pitting it, by making grooves, by shaping and faceting. They make various alcoves and niches in rock walls. They also form sand dunes in deserts, and blow the sand and silt of stream beds hither and yon. A sand sheet in Libya, over a meter thick, rests on bedrock over many thousands of square kilometers and is supposed to have been laid down by winds of the desert. There are others like it around the world. Such aeolian activity is allotted millions of years to help shape the landscape; the number of millions, one or a hundred, is calculated from estimated past climatic conditions working against various constraints, such as whether landforms exist nearby to provide the material of erosion.

Tornados, cyclones and hurricanes now and then wreak havoc upon soil and settlements. Part of the climatic complex of this age, these storms are localized - the "tornado belt" of the south-central United States, the Japan and China Seas, and so on. Of course, bearing in mind the "many changes of climate over the ages," most places on earth would have suffered such storms in turn. When they occur, part of the biosphere is blown away with some of the natural landscaping. Paleo-anthropology and archaeology debate the relative contributions of the Orient and the Eur-African world to the earliest American cultures, for example, without proper attention to the possibilities afforded travelers by changing winds that come with changing climates, now pushing things one way and then again another way. So that even when the possibilities of cataclysmic changes in early human times are ignored, changing climates would carry culture both East and West.[1]

Tornado effects are discoverable in some places where sedimentary beds are interrupted by poorly sorted mixtures of rock which evidence by their shape, fragmentation, and positions a sudden displacement and replacement. Ager calls these storm deposits "tempestite," after a word that he ascribes to Gilbert Kelling, when he observes them, for instance, on the heights of the Atlas Mountain of Morocco[2]. Similar deposits have been identified in a few other places. Missouri, Virginia, the English Channel, the Paris Basin, in rocks of the Mesozoic and Paleozoic. Carozzi and Gerber consider that "such an early generation of cherts in carbonates is more common than generally assumed."[3]

We cannot figure how often such high energy local events have occurred, until the world is better surveyed with this idea in mind. But one can "think big". With a thousand tornados a year (300 in the U. S. A.) tearing up two thousand square kilometers of sediments and breaking down surface features, an area equal to

---

[1] 1. *Cf.* C. L. Riley *et al, Man Across the Sea: Problems of Pre-Columbian Contacts* (Austin, Tex.: U. of Texas, 1971) 302 *et passim.*

[2] Derek W. Ager, *The Nature of the Stratigraphical Record* (New York: Wiley-Halsted, 1973), 39.

[3] A. V. Carozzi and M. S. Gerber, "Late Paleozoic Tornados and Synsedimentary Brecciation of Chert Nodules."

the total land surface of the world (240 million square kilometers) would be superficially pulverized in about 120,000 years. If a conventional age of 3.6 million years is accorded the Earth's crust, the whole of it would have been scoured, not once, but 30,000 times by cyclonic action. In the short term, not all land would be affected equally, but in the long-term, given changing climates and drifting continents, an assumption of randomized strikes could be tolerated. Where then are the scars of 30,000 tornados in every geological column? Or even in any single one anywhere?

From this we might conclude that we have a great deal of field research to do in geological history so as to obtain a realistic estimate of the number of events. This is also the situation, we may as well say, in respect of meteoroid falls, volcanism, and other high-energy events to be discussed. The quantavolutionary approach to history comes naked as a neonate, without systematic hypotheses, data, or applicable mathematics.

If few such effects are discoverable, it may be because catastrophes acting on a large scale have obliterated almost all localized indications of damage. For instance, if great earthquakes have shattered rock strata, lesser violence to the rock would be hardly visible. The schist dropping deep below the city of Athens is infinitely fractured. Is this tempestite, thermotite, seismotite, hydrotite, turbotite, or what? If the wind god, Aeolus, blew at once all around the world, many sediments would be displaced, losing their local cyclone scars in the process and letting no new strikes penetrate deep into the new strata.

But perhaps the Earth's surface has spent 99.9% of its time in a peaceful state with a quiet atmosphere. Such quiescence contradicts uniformitarianism as much as it does catastrophism; that is, I have used above the present "quiet" state to reconstruct the past, as Hutton and Lyell recommended. Yet even so, estimates resulting therefrom would be much more impressive than present conventional history gives one to understand.

A final possibility is that the sedimentary rocks of the Earth are much too young to have experienced all that is supposed to have happened. That is, if the Earth were 100,000 years old, much of its surface would perhaps not have been scarred by tornados (or meteoroids).

Ancient legends speak of a large role for winds. The sacred book of Buddhism, the Visuddhi-Maggia, says that when worlds collide the winds "turn the ground upside down. Large areas crack and are thrown upwards. The winds pulverize the ground and it disappears into space, never to return. Thus ends a cycle of the ages."[4] It is the extreme catastrophic typhoon.

The ancient Meso-Americans said that the former world was brought to an end by the great wind god, Huracan. Probably the origin of the word "hurricane" is here. Huracan is also a manifestation of the great god Quetzalcoatl, who is also identified with the god and planet Venus.[5] Huracan, the Heart of Heaven, fathered a large number of people, who he then destroyed in the darkness of a storm amidst black rain that fell day and night. So records the Quiche book of *Popul Vuh*. Then animal gods mangled the bodies.[6]

"Air" is rarely missing in the legendary and early scientific classifications such as "earth, air, fire and water." The idea of world destruction by wind is, of course, quite disregarded by modern scholars. One hears the term "marine transgressions" but not "wind transgressions." It is surprising how few pages have been devoted to the winds by catastrophists, too. Again, perhaps the effects of hurricanes and typhoons are quickly concealed by other forces operating. Or the effects may be interpreted as tidal wave deposits.

The splintered bones of some fossil assemblages would indicate aerial rather than water transport. Although he does not follow through, F. Hibben provides a rare passage dealing with the immense deposits of bones that he witnesses. "Throughout

---

[4] Warren, *Buddhism in Translation,* p. 328 quoted by Velikovsky, *Worlds in Collision* 70.

[5] William Mullen, "The Mesoamerican Record," 4 *Pensée* 4 (Fall), 34-44.

[6] *Popul Vuh: The Sacred Book of the Ancient Quiche Maya* (Norman, Okla.: U. of Okla. Press, 1950), 90.

the Alaskan mucks, too, there is evidence of atmospheric disturbances of unparalleled violence."[7] The Cumberland Cavern catastrophic life dump shows no evidence of water transport.[8] Probably as many collections of animals and vegetation have been gathered and flung in heaps by winds as by water. In seeking the origins of some coal deposits, catastrophic winds are a prime suspect, along with rock and water thrusts.

What can create deposits can remove them. Heavy winds, operating tidally or cyclonically, can blow away pre-existing structures. Contemplating the early ages of human settlement, one may wonder at the frequent absence of primordial sites. Here, as everywhere in the mythicized realms of science, there is a vision that is perhaps false, of excavating sites layer upon layer until arrival at bed rock, and thereupon pronouncing the last ruins to be the first settlement. But the god Huracan is able quickly to erase settlements down to bed rock one and more times. The typical absence of human vestiges before the Neolithic age is usually taken to signify that human settlement began with the Neolithic. There is small reason to believe this to be the case. In fact, there is a hint of aeolian morphology in the near absence of Paleolithic remains except in caves and *abris* in the Dordogne of France and elsewhere.

The power of winds to push, pull and lift is great. The Hiroshima nuclear fission-bomb explosion is assigned an energy of 7.9x10 18 ergs. The measured energy release of a one-megaton fusion bomb explosion is in the range of 10^{22} ergs. This is about the same energy as exploded in the Berringer meteoroid crater in Arizona. "In one day a large hurricane releases as much energy as a 13,000 megaton nuclear bomb. Some hurricanes take a week to reach such intensity, others mature in a day or so. And during the time another may be at full blast a thousand miles away."[9] Some hurricanes last three weeks and travel 1,000 miles. (One can bear

---

[7] *Op. cit.*

[8] I. Velikovsky, *Earth in Upheaval* (New York: Doubleday, 1955), 60.

[9] Frank W. Lane, *The Elements Rage* (Philadelphia: Chilton, 1965), 6.

in mind the immediate transport of resilient living species around
the world by such means.)

An ordinary Kansas tornado will approximate 4×10^{18} ergs
of kinetic energy. Its power in kilowatts is 10^{18}, "which is in excess
of the capacity of all the generating stations in the United States."
(ca 1959)[10]. The wind velocity at the center of its funnel
theoretically may achieve 2000 miles per hour. By the Fujita scale,
an F-5 wind, indexed at combined forward and rotating speeds of
261-318 mph causes "incredible damage."

Electrical activity is so vigorous that Peltier's words of
1840 can be used as a model for an electrical cyclone theory.
"Everything proves that the tornado is nothing else than a
conductor formed of the clouds which serves as a passage for a
continual discharge of electricity from above."[11] Observers have
been inside of this "enormous vacuum tube, somewhat similar to
a Geissler, neon or fluorescent light tube, conducting very low
density electric current whenever there is a sufficient accumulation
of electricity in the clouds to make the jump to Earth."[12] Typhon,
the cosmic spectral dragon felled by a thunderbolt from Jupiter,
was anciently described by Apollodorus as "rushing at heaven"
with hissing and screams, spouting a great jet fire from his mouth.
This same Typhon is probably the origin of the word "typhoon."[13]

Cyclones and water spouts (water-bearing cyclones) often
appear in groups. An outbreak of 148 tornados was registered in
the United States and Canada on April 3, 1974. Sometimes
associated with a tornado are a number of downbursts of high-
velocity winds that blow down whatever they strike, whether
groves or houses or aircraft. Ted Fujita of the University of
Chicago compares the downbursts with giant garden hoses aimed

---

[10] *Ibid.*, 45.

[11] 38 *Amer. J. Sci. and Arts* (1840) 73, *cf.* William Corliss, compiler,
Strange Phenomena (Glen Arm, Md.: Corliss), GLD052-G2-105.

[12] *Ibid.*, G2-104-5.

[13] Velikovsky, *World in Collision*, 68-70.

downwards upon circles kilometers in diameters; often they end their work in two minutes.

What might cause a vast number of cyclonic events to appear? A meteoroid bombardment, an interruption of the Earth's motion, a tilt of the Earth's geographic axis, magnetic axis, or sidereal axis: these would do, and also a large meteoroid impact, and a large body passing nearby, the latter, however, being tied almost inevitably to other changes in Earth's motions. Too, a deluge of waters might form into many ribbons, mushrooms, or funnels in descending. The winds and other effects of a heavy meteoroid impact would be simulated if a large number of nuclear missiles were trained upon a single spot and exploded at the same moment.

The atmospheric turbulence accompanying such impacts must include more than a blasting power. Its heat can provide the circulating system for a natural instantaneous chemical factory. The turbulence generates disturbing sounds and sends them over long distances and brings intolerable changes in barometric pressures. Volcanic explosions produce similar effects: whether a crater is a volcanic or meteoric effect is often contested, and both produce tornado and hurricane effects.

During the Krakatoa volcanic explosion of 1883, winds stripped all the surrounding area of its lush vegetation before burning it.[14] People heard noises of anchors being hauled up and dropped, of thunder and beating drums: the winds carried the explosions across the Indian Ocean where they were heard as distant cannonading. The barometer on a ship nearby jumped up and down an inch at a time. The air was sucked up so that people could not breathe. The gases were sulfurous, choking and blinding. The sun was obscured, and slightly so around the world for years. In the pitch-black day, a Dutchman groped for a knife to despatch his family.

---

[14] Rupert Furneaux, *Krakatoa* (Englewood Cliffs, N. J.: Prentice Hall, 1964), 34.

So cyclones darken abruptly the sky, and bring ear-bursting and chest-bursting drops in barometric pressure. They explode houses by creating vacuums into which the inside air must burst. They lift boulders and cows, carrying them off, and they dig up the earth. There is a hint in cyclonic action of what may have happened to some of the mammoths and other large-animals that were exterminated a few thousand years ago: suffocation; lifting and dropping; followed by quick freezing; thence to be discovered in the same position today.

Winds act faster than water and have the same exponential effect upon the bodies which they may encounter as their speed increases. Wind pressure, that is, increases as the square of wind velocity, up to the velocity of sound at least. A 500 km/hr wind exerts 25 (not 5) times the pressure of a 100 km/hr wind; gravel then begins to behave like fusillades of bullets. Kelly and Dachille calculated that the winds created by a large meteoroid impact will move laterally and vertically with the speed of sound.[15] Their effect has to be measured, too, in terms of the amount of debris that they transport. A single such blast, moving horizontally, can strip its area of passage bare down to bed rock, or below, especially if it is loaded with detritus, and may continue its major effects for a thousand kilometers. Only a mountain can stand against it and it, too, will be defaced; an instant ablation corresponding to millions of years of ordinary aeolian erosion will occur. Rivers would be wiped out and set up elsewhere. Valleys would be filled with debris. Great vegetable and animal dumps would be established in many places.

Waterspouts have been known to hoist and drop far away the water and biosphere of large ponds; since these events happen under meteorological conditions ordinary to our age, they must be hundreds of times less powerful than the waterspouts (and land

---

[15] 15. Allan O Kelly and Frank Dachille, *Target: Earth, The Role of Large Meteors in Earth Science* (Carlsbad, Calif.: Box 335, 1953), 203, 66 *et passim*.

spouts) that would arise from large-body impact explosion or related events involving catastrophic energies.[16]

The turbulent atmosphere of the planet Venus rotates in six days as contrasted with the 243 days that the body of the planet takes to rotate. Its normal wind velocities of 10 to 100 meters per second are comparable to those of the jet stream that races through the upper atmosphere of the Earth.[17] The surface heat of Venus is of course in the hundreds of degrees Celsius. The mechanism has not been solved. Several effects of a perpetual firestorm might be considered, granted that free oxygen is absent. One is reminded of the firestorms that were engendered in the Chicago fire, the Tokyo earthquake, the Pestigo forest fire, the firebombing of Dresden, and the atomic bomb-burst over Hiroshima. Large areas can become like giant tornados; perhaps a planet can suffer the same fate.

Winds can operate like tides. Thus, if the Earth's rotation is altered, the atmosphere will be subjected to the same influences that cause the alteration and will in effect act turbulently, that is, out of phase with the lithosphere. They will sweep over the globe like a tide of water. The atmosphere, if electromagnetically affected by a conjunction of planets and Sun, will help to disturb the lithosphere and engender seismism.

Differential atmospheric pressures define the existence of a wind; two clouds of gas, essentially isolated but lacking an effective "bag" to contain their isolation, interact. Electric potentials are established. Electrical forces thereupon flow throughout the transacting systems laterally and vertically. It is perhaps axiomatic that where there is wind there is an electric current and discharges. And where there is an electric current there is bound to be a magnetic field. And, lacking a better container, an electric current is contained by its magnetic field.

---

[16] *Ibid.*, 202; Hans Oersted, 1 *Amer. J. Sci.* 37 (1839) 250-67, quoted in Corliss, *op. cit.*, G2-233.

[17] Andrew and Louise Young, "Venus," 233 *Sci. Amer.* (Sept. 1975), 73.

More than one observer has confirmed the testimony of
a man who was caught in the open as a tornado passed above him
by a few meters. He was beneath a tunnel whose walls were
composed of whirling clouds, in the manner of a magnetic field as
this is pictured in drawings of a textbook. He looked up into the
tunnel for at least half a mile; brilliant lightning flashes illuminated
the tube. Where he crouched, the air seemed calm; the gases stank
suffocatingly; screams and hisses could be heard. The tornado,
having deftly raised itself to pass over him just as gently dropped
down upon his neighbor's house, exploding it and its objects.[18]
This small tornado may function very much on the same
principles as the cyclonic effect of a large meteoroid explosion,
and again like the great tube of gases that envelops a binary star
system, such as I outlined for the solar system in *Chaos and Creation*
and discussed at length with Earl R. Milton in *Solaria Binaria*.

In the Uweinat section of the Great Sand Sea of
Southwestern Egypt, a number of possible meteoric impact sites
have been reported. One, positively identified, is of 4 km diameter;
another is of 14 km diameter. Many extinct volcanos are also
evident in this desolate area of sand and sand dunes, which was
occupied by humans until at least the Neolithic period.[18A] A great
climatic change must then have occurred lately.

The region is part of the Sahara Desert, which is also
marked here and there by human traces. The Gobi Desert, greatest
in Asia, bears human relics as well. So do the Mexican and U. S.
deserts, and the Peruvian. The great deserts of the world are
recent, it appears.

The astroblemes and volcanism of Uweinat may have
been associated with the events ending civilization and creating
deserts. The wind-blown dunes are long, wide, and tall; yet the
same winds have not erased the meteoric or volcanic craters, even
though these are often not so deep as the dunes are high; not

---

[18] Alonzo A. Justice, 50 *Monthly Weather Review* (May 1930) 205-6,
quoted in Corliss, *op. cit.*, G2-105-7.
[18A.] Faraouk El-Baz, 213 *Science* (24 July 1981) 439-40.

enough time may have passed. Aeolian dunes, astroblemes, volcanos, climatic switching, and culture extinction together can entertain a hypothesis of holospheric quantavolution, pending the establishment of a chronology that would prove the hypothesis or temporally sunder apart the events.

The largest deposits accorded to winds are not those of the Libyan peneplain mentioned earlier, nor those of Egypt, but the huge areas of the Earth covered by loess. The term itself was invented for glacier deposits of the Rhine and Danube valleys and elsewhere in Europe. It found itself connected with the "drift", the glacial pebbled clay of North America, where vast stretches of the buff and porous earth, compacted but frangible to the fingers, were found distributed. Here transportation by ice sheets and rivers forming from their melts was imagined. Then, west of Peking, an area larger than France exposed its loess to geological inquiry.

Loess can occur at high elevations as well as on great plains. It breaks down into excellent thick soil in China and its cliffs degrade into natural terraces.[19] Old roads cut through it, sometimes passing through the Chinese countryside thirty meters below the houses and farms on the loess above. In Indiana, the highest lands and ridges in particular have the thickest yellow clay (called drift or loess) and it is free of sand and gravel.[20] The loess is not stratified, nor does it contain marine fossils, and land fossils of shells and mammals are only occasionally found in it.

Sedimentation from lakes and rivers seems to be an impossible explanation. Adequate sources of glaciers and ice are often absent, as for example near the loess that occurs inland from the Gulf of Mexico. The favored theory of loess formation stands

---

[19] Frederick W. Williams, "Loess Deposits of Northern China," 22 *Popular Sci. Mon.* (1882) 243-8, quoted in W. Corliss, compiler, *Strange Planet* (Glen Arm, Md. 21057: Sourcebook Project, 1978), ESL001-E2-161.

[20] J. T. Campbell, 23 *Amer. Naturalist* (1889) 785-92, quoted in Corliss, ESL004-E2-167.

upon the transporting power of winds that would carry the material from distant high places or deserts, operating over long periods of time. But where are the loess heaps on the fringes of great deserts? There are none. And why should stratification and cross-bedding not then have occurred? Nor can the chemical composition of loess be assigned to the mountains of its supposed origins. And the loess grains are not rounded by wind or water but are angular, as if exploded, and are settled in vertical lines through which rain readily percolates.

Ignatius Donnelly, in *Ragnarok* (1882), was already ascribing till, drift and loess to fall-out from a great comet, going so far as to deny the very existence of past ice ages, to which most scientists then and still today ascribe these materials. He read many distinct legendary sources and intercepted many sedimentary strata as stories of great winds that picked up the detritus of Earth, whirling it around wildly and depositing it in "intercalated beds."[21] Donnelly's denial of the ice ages in favor of exoterrestrial deposits by comet does not appear so outrageous today. As we shall see, ice age theory has been used (and abused) to the point of exhaustion of the subject and of the geologists working in the field; it has been made responsible for many geological forms and events that might more readily be assigned to other forces. Velikovsky, in a note of the 1940's, before he had himself been subjected to ridicule, commented that Donnelly had been called "the Prince of Cranks" for his books on several difficult and controversial subjects.[22] Donnelly was in fact a superior writer and lecturer, an intense student with a sensuous affinity for the palpability of the ground, a political and social hero, and a precursor in fundamental ways of later writers such as Velikovsky.

---

[21] I. Donnelly, *Ragnarok: The Age of Fire and Gravel* (New York: Appleton, 1883), 53.
[22] "Precursors," 7 *Kronos* 1 (1981), 53.

Fifty years after Donnelly, Penniston was advocating the thesis of an exoterrestrial origin for loess.[23] Citing Shapley (later a violent critic of Velikovsky) and Belot for having proposed a solar nova as the cause of the ice ages, he reasoned upon this as a possible source of the material, which, experiencing high temperatures for a period of time, had its silicates metamorphosed in part to quartz, thus arriving at the loess. That stony meteorites have differed in composition from loess has stood against his theory. The source of meteorites has probably been mainly from the asteroid belt in contemporary times, however, and cannot be well compared with either the solar or the cometary origins hypothesized. Not unnaturally, geologists faced with a choice of wind or exoterrestrial fall, would prefer the wind. Wherever possible, as in middle America, they introduce " glacial sluiceways." Yet we would prefer to discuss the matter once again when it comes time to ask what can and does fall to Earth from outer space.

Let us rest content here if we have but established several points: The force of wind rises with the square of its velocity, with correspondingly large effects upon the landscape. Hurricanes must be associated with every abrupt and intensive geological event. Cyclones convey major electrical and fire phenomena. In large-scale catastrophic events, a great many typhoons could originate to accommodate changed atmospheric and lithospheric motions or multiple meteoroidal instrusions. Finally, if the sediments of the world do not reflect adequately cyclonic effects, the reason may rest in their continuous erasure by more forceful events which themselves require identification. Furthermore, assigned geological times may be too long; maybe not enough events have happened to flesh out the skeletal ages.

---

[23] J. B. Penniston, 39 *Pop. Astro.* (1931) 429-30 and 51 *Pop. Astro.* (1943), 170-2, quoted in Corliss, ESL-003-E2-165.

CHAPTER FOUR

MAGNETISM AND AXIAL TILTS

The Earth has two axes of concern here, its axis of rotation between the geographical north and south poles, and the warped axis of its magnetic field lying between the north magnetic pole and the south magnetic pole. It is easier to imagine the axis of rotation; the imaginary equator divides the globe into two equal halves and this equator marks a circle around the spinning globe which, every 24 hours, completes a turn.

The magnetic poles are distant by some hundreds of kilometers from their corresponding geographic poles. They are denoted by the behavior of a compass needle which assumes a vertical position when at or near the magnetic pole; the nearly global distance that lies between the north and south magnetic poles witnesses a continuously changing dip of the compass needle which reverses itself as it passes approximately half the globe and again turns to the vertical (in reverse) as it approaches the opposite pole. The magnetic poles are in perpetual motion, seemingly traversing a kind of oval figure. In the north, the pole is just south of King Christian Island (1980, 77°19 N; 101°49W) and is moving north by 24.4 km per year and west by 5.4 km per year.[1]

---

[1] 1. Paul H. Serson, "Tracking the North Magnetic Pole," *Geos* (Winter, 1980)

Apart from a certain usefulness in navigation, its extreme weakness may let one think such magnetism to be quite unimportant. But it indicates the presence of several important processes of the atmosphere, lithosphere, biosphere and cosmosphere. An entertaining book might be written concerning the effects on life of the loss of the magnetic field. How will wild geese navigate? Will there be less heart attacks or more? Cox says that the removal of the dipole magnetic field will reduce the total shielding of the biosphere from cosmic rays by 10 to 12%, no more than is involved in a person's moving from the equator to Alaska. Waddington is of the same opinion "unless it is assumed that these periods are associated with greatly increased particle radiation from some external source."[2] This last point stresses the atmosphere-exosphere relationship, and may serve later on to solve some reversal perplexities.

In 1989, NASA's Magnetic Field Satellite confirmed that the field, already weak, is decreasing in strength. The trend indicates a zero strength in about 1200 years.[3] Relying upon studies begun in 1830 by Gauss, Barnes made the same prediction earlier.[4] Theorists are divided, some saying that the field hits zero, then reverses, and then returns to zero, and so on over great periods of time. A few, the present author among them, say that the field is a once and for all thing: it began at higher intensity, endured for a long time, then began to diminish, meanwhile from time to time reversing its direction.

Assuming a continuously increased strength reading backwards in time, however, implies an enormous intensity eons ago; there is a hint here, to our way of thinking, that the field was created and sustained at a constant level, and then abruptly was

---

[2] II *S. I. S. R.* 2(1978), 45.

[3] "Magsat down: Magnetic Field Declining", 117 *Sci. News* (1980), 407

[4] 4. Thomas G. Barnes, 8 *Creation Res. Soc. Q.* 1(1971) 24-9; 9 *C. R. S. Q.* 4 (1973) 322-30; 18 *C. R. S. Q.* (June 1981) 39-41; II. *Soc. for Interdisciplinry Stud. R.* 2 (1978) 42-5, 4(1978), 110-11.

cut off from its source, and began to decline. Barnes declares, too, that "This magnetic decay phenomenon could not have been going on for more than a few thousand years, as the magnetic field would have been implausibly large for a relatively neutral body such as the earth."[5]

The magnetic field constitutes a magnetosphere which is much larger than the Earth itself;[6] it can be imagined as a kind of giant electric globe enclosing the Earth which is perceptible even as one descends into the deepest rocks and which may only end in some kind of an electric current which may be running through the core of the Earth at about the geographical spinning equator, very roughly perpendicular to the geophysical poles.

It is important, too, to appreciate that these two features, the magnetic electric current and the geographical spinning equator may be largely independent of one another. That is, one can conceive of the magnetic and geographical systems operating even at right angles to one another. We have discovered no natural law that says the two equators and sets of poles must be close together.

This implies, however, that the two sets of poles are not stable, that their present positions are a historical accident. But, then, to think so introduces worrisome possibilities: that the axis of spin of the Earth may be changed, too. Both of these possibilities have increasingly occupied the minds and studies of scholars and explorers. Have there indeed been occasion on which the globe has tilted, geographically and magnetically? The answer today is yes, that the axis of spin has shifted and also the magnetic axis has shifted.

But before we consider these two probabilities, it is well to mention yet a third change in the Earth's behavior that would

---

[5] Allan Cox, "Geomagnetic Reversals," 163 *Sci.* (17 Jan. 1969) 237-45; C. J. Waddington, *Sci.,* 17 Nov. 1967; *Cf.* J. Eberhart, "Of Life and Death and Magnetism," *Sci. News* (Mar. 27, 1976), 9.

[6] R. Juergens, "Reconciling Celestial Mechanics and Velikovskian Catastrophism," 2 *Pensée* 3 (Fall, 1972), 6-12.

possibly occur without magnetic or geographic shift. Suppose that the Earth simply tilted in space.

On this phenomenon, Peter Warlow reports that both Needham and Dodwell found oscillatory change in the obliquity of the ecliptic, on the basis of ancient astronomical records. Dodwell concluded that three factors were operative in the movement, the linear drift conventionally ascribed, a decaying oscillation with a period of 1200 years, and a logarithmic-sine decay. Dodwell saw in the exponential decay (quantavolutionary exponentialism that I mentioned earlier and in *Chaos and Creation*) a drastic occurrence some 4500 years ago.[7]

Could the Earth have even turned over completely without interrupting (interrupting very little) its spin or its magnetic field? The geographic poles would be reversed, and along with them the magnetic field. The Earth could not perform such a movement without an external assist, whether from an upsetting explosion of gases from the Sun or from the attraction or repulsion of a large passing body.

According to Warlow, who has however been challenged by Slabinski, the transaction could be relatively delicate; it would amount to the drawing of a force along the Earth's path that would cause it to tip over while containing its spin, in the manner of a tippe-top, a toy that is weighed on top and set to spinning on the board; the top turns completely over continuing to spin all the while in the same direction, North becomes South and East becomes West.[8] The motion performed is technically a fast precession.

A moment's reflection will rid us of any notion that the action would be harmless. The atmosphere, hydrosphere, and lithosphere would be agitated and produce effects that by any measures would have to be called quantavolutionary. For instance, it appears most likely that the widespread sudden destruction

---

[7] Geoffrey Gammon, "Focus: Catastrophism Old and New," V *SISR* 2 (1980-81), 34.

[8] Peter Warlow, *The Reversing Earth* (London: Dent, 1981).

throughout the northern regions of the mammoths and other large mammals occurred in conjunction with a tilt of the Earth's axis in the presence of the exoterrestrial entity causing the tilt. We can say this because a sudden deep vacuum freeze, asphyxiation, thrusting of masses of gravel and bones, and permanent cold ever thereafter, such that the animals are sometimes found still fleshed-out and diagnosed in certain cases as heart-failures or with blood-clotted lungs, must indicate a holospheric event comprising an atmospheric and aquatic withdrawal, the descent of an extreme coldness, and upon the passing of the body, returning tides of water and wind to accomplish quick burial under muck, ice and tundra.

Yet, according to Warlow's theory, the tilt, which might have been complete to 180° and would change East to West and North to South, would require only thousandths of the energy to be disposed of if, by contrast, the Earth were largely cease or reverse its rotation. If such were to happen, it would be most unlikely that the two bodies, Earth and the intruder, would achieve just the mode of encounter and passage that would avoid direct electrical and material exchanges or that would bring about a full 180° reversal; the Earth, unlike the tippe-top, could cease its tilt at any angle not excluding a full 360° circle with its intruder acting momentarily as its binary, and performing a "loop-the-loop."

Should the intruder collide with the Earth, the Earth might tilt, also, and the damage to it would be much greater. Dachille estimates that a body 320 km in diameter, impacting tangentially at a velocity of 12 km/sec would produce an axis shift of a mere 0°32'.[9] Many forms of energy disposal are available, it appears, besides reorientation of the global axis. One is led to suspect that non-colliding encounters involving heavy electrical differentials might more effectively produce axis tilting than would collisions.

Lest the idea be considered quite fanciful, it should be recalled that several ancient sources refer seriously to a reversal of

---

[9] 198 *Nature* (13 April 1963), 176.

directions. Herodotus and Plato cite Egyptian sources of occasions when the Sun changed directions and arose in the West instead of the East. A ceiling in the tomb of Senmut of Egypt also pictures a reversed sky tableau such as would occur were the Earth turned upside down. In fifteen spectacular pages, [10]Velikovsky searches out and orders rationally other indications in legends and writing of a reversal of directions that could only come with the Earth turning upside down. The contexts scarcely permit the alternative, a cessation and reversal of the Earth's rotation.

Thomas Gold once remarked that, if the Earth were a perfect sphere, an insect alighting upon it might turn it over. In revising Warlow's calculations, Slabinski assumes that the Earth has to be turned over in a single pass-by at two Earth's radii distance in a parabolic approach trajectory. He emerges with a requirement for a body with the mass of 62 Suns. Even if the crust of the Earth is shoved around independently of the underlaying layers, a body of the mass of 68 Jupiters is needed.[11] We expect that such an action will be totally catastrophic." Furthermore, "any appeal to electromagnetic forces that does not give a quantitative analysis of how such forces produce the required torque is equivalent to saying..." a miracle occurs."

Ellenberger, although a stout Velikovsky supporter, agrees: "Since motions occur along the path of least resistance, the possibility that a spin reversal has occurred would appear to be greatly reduced and that interpretation of Senmut's ceiling (and other evidence cited) may be in need of a raison d'être other than evidencing a spin reversal. If a spin reversal is a viable alternative, where are there discussions and quantifications of its

---

[10] *Worlds in Collision*, 105-20. See also, A. W. Perrins, (tr., 3 *S. I. S Workshop* 1 (July 1980) 27-8, on the reversed burials of Pharaohs, the inscription of Horemheb's tomb that the Sun rises in the West, and Rameses II at Abu Simbel facing East rather than the orthodox western way to where, with his false beard, he should be oriented.

[11] VII *Kronos* (Winter 1982), 86-94, 92.

mechanism?"[12] Yet Velikovsky, arguing the case for axis displacement, had earlier discussed a calculation by Weizaecker demonstrating that an Earth transaction with a strong magnetic field would affect its axial inclination much more readily than its rotation.[13]

Presently, the evidence for sidereal tilts is considerable, for geographic tilts also some, for upside down tilts little, for stop-and-reverse rotation very little. There is no way in which astronomical assurances can be lent to geologists on this account. Conversely, there is enough doubt on all scores to let geologists be open to the possibility of several catastrophically effective maneuvers of "Spaceship Earth". A moment's consideration of Slabinski's calculation leads to the suspicion that he may be employing a rate in his formulas that soars to wild heights and casts doubts *prima facie* on his procedures: if it would take the gravitational force for 62 Suns to turn the Earth around at a distance of less than 15,000 km, how does a single Sun lock the Earth into fixed orbit at 150 million kilometers? Also, evidence of a geographical shift of the poles is abundant; if this is not to be denied, then we should have to supply the force to do the job; if not 62 Suns, then how many Suns at 15,000 km distance are needed?

The possible occurrence of reversals in proto-historical times may suggest additional reversals in pre-human ages. However, Milton and I have presented in *Solaria Binaria* (Chap. 8) a theory according to which the Earth was in grip of a huge external magnetic field of the solar binary system until perhaps eight thousand years ago; during almost all of geological time, it could not reverse its field. In fact, it is argued that this same magnetic field and its reciprocal electrical current are the present geomagnetic field and current within the Earth, which have been steadily undergoing decay since the grip of the external magnetic

---

[12] V. J. Slabinski, "A dynamical objection to the inversion of the Earth on its spin axis," 14 *J. Physics A.* (1981) 2503-7.

[13] "Straka: Science or Anti-Science," I *Pensée* (Fall 1972), 16.

field was released. This theory permits us here to explain the principal geological problems connected with terrestrial magnetism.

We would have to assert that the numerous alleged reversals of the Earth's magnetic field in geological history simply did not occur. Obviously there is no evidence to be obtained one way or another by atmospheric testing of the field; any number of reversals (or none at all) might have occurred without leaving discernible evidence.

The geophysicist, however, can search for evidence of the magnetic field in rocks.[14] Igneous rocks have often been imprinted with magnetism when in a molten state; hence they hold myriads of tiny compasses, pointed towards the magnetic pole. If for one set of rocks the compasses point north and for another adjoining set they point south, it is conceivable that the magnetic field had reversed itself on an occasion between the melting and hardening of the first set of rocks and the melting and hardening of the second set.

Magnetic mapping of rocks is almost entirely of this century but has burgeoned swiftly and, some say, chaotically. Persuaded that they can tell the ages of rocks by radiometry, explorers have used time as a reliable indicator of the change in the magnetic field of the Earth. Since the rocks of the world have exhibited a bewildering variety of magnetic directions, many "dated" strata of differing magnetic direction have been assigned to the different magnetic periods, usually forced into a preconceived mold of "normal" and "reversed" magnetic field.

Depending upon the angle of declination, not only have such fields been noted, but they have been asserted to pertain to shifting magnetic poles. Some students have supported the idea that hundreds of field reversals have taken place in the several billions of years allotted to the Earth's history. One catalogue reports 433 paleomagnetic poles for 3 to 4 billion years of Pre-

---

[14] S. Matsushita and W. H. Campbell, eds. *Physics of Geomagnetic Phenomena* (New York: Academic Press, 1967).

Cambrian time, an average of one new pole per 7 to 9 million years.[15] Since the Cretaceous, says Heirtzler, 171 reversals of the magnetic field have been identified.[16] Others have perceived certain intervals of time to elapse between reversals, 700,000 years, fifteen million years, and so on; several studies claim that the farther back in time one goes, the longer the period between reversals.

Some observe much more frequent reversals; they can claim that a reversal occurred 2600 years ago, 3500 years ago, a dozen times during the Pleistocene, and so on. If, they say, we cannot perceive so high a frequency in times more ancient, it is because the reversal is not accompanied by a general melting of rocks and therefore cannot be detected, or it is too faint to be recognized because of disturbances or contamination of the strata. Magnetic reversals may be concealed because sedimentation is too slow to capture its duration, when samples are not closely spaced in time and the reversals are brief, when turbulence and contamination affect samples, when the sediments are dumped or shifted, and when biological activity is high at the level being searched for magnetism.[17] Still indications are strong in favor of heavy magnetic disturbances in the mid-first and mid-second millennia B.C., with ceramic, clay, rock, bio stratigraphic, legendary, and historical contributions.

As early as 1907, P.L. Marcanton, using Folgheraiter's method, demonstrated magnetic reversal and intensity changes by studies of the magnetic inclinations imprinted upon Bavarian and Etruscan vases of the period 600-800 B.C., a period that in *Chaos and Creation* I called "Martia."[18] In 1981, K. Games reported upon a similar investigation of Egyptian pottery over a 3000-year period,

---

[15] P. L. Lapointe *et al.*, "What happened to the High-Latitude Paleomagnetic Poles," 273 *Nature* (22 June 1978), 655.

[16] 16. J. R. Heirtzler, "Seafloor Spreading," 219 *Sci. Amer.* (1968), 60-70.

[17] Thomas McCreery, "Krupp and Velikovsky," VI *Kronos* 3 (1981), 44-5.

[18] 112 *Archives des Sciences Physiques et Naturelles* (1907), 467-82.

concluding: "Clearly, the geomagnetic field in Egypt has varied rapidly and by large amounts. The greatest rate of change, which occurred around the maximum at about 1400 B.C. was about 140 nanoteslas/year... and lasted about 300 years either side of the maximum.[19]" He did not study directional changes of the field; further, his date of 1400 B.C. is more likely to have fallen in the 8th century, since he was using an unreconstructed chronology which is backwards by 500 years.

One important off-shoot of this enthusiastic age of magnetic pole discovery is the belief that the discovery of a new magnetic pole means that a new geographic pole has been discovered. If so, and if what is being discovered are true magnetic reversals, the Earth would have suffered thousands of devastations. A shift in a true geographic pole (as opposed to a purely celestial or sidereal tilt) must involve a shift in the axis of rotation, the worst kind of disaster. Apparently some geologists are runaway catastrophists as long as they can run on free time long past. Munk's title, "Polar Wandering: A Marathon of Errors,"[20] deserves sober thought.

The significance of this chaos of findings also lies in the association of magnetic reversals with atmospheric, biospheric and lithospheric turbulence. The magnetic field or magnetosphere, even though it is remarkably weak in the farthest stretches of the atmosphere, nevertheless blocks and deflects a host of incoming particles. It acts thus like the ozone layer and atmosphere in general, as a protective shield. If it is removed, or temporarily "shut off" because it is shifting, or overwhelmed or shunted aside by great blasts of gases and charged particles, species extinctions may occur. Kennett and Watkins claim, on the basis of deep-sea drilling, that volcanism was at a peak in

---

[19] *New Scientist* (11 June 1981); *cf.* Brian Moore, note in V *S. I. S. Rev.* 2 (1980-1), 38, and 4 *S. I. S. Workshop* 2 p. 17.

[20] W. H. Munk, 177 *Nature* 4508 (24 Mar. 1956).

coincidence with changing geomagnetic polarity.[21] Wollin, Ericson and Ryan have noted by faunal and oxygen indicators at various sedimentary levels that cool climates may be associated with high magnetic intensity.[22] These may be short-term indicators, since at least by the *Solaria Binaria* theory, magnetic intensity was stable and high until recently and has since been declining.

A sampling of Siluro-Devonian sedimentary sections from the Arctic Archipelago of Canada reveals a common magnetic reversal. The magnetic inclinations suggest a low equatorial latitude. The rocks were apparently laid down under equatorial conditions, and they magnetized rapidly. Unfortunately, if the globe's axis rotation has since tilted or the continents have shifted or a plenum of clouds then covered the globe, the findings of such studies must be discounted; all three probably occurred. That is, the Devonian has long been thought to have been a warm world; the arctic rocks, whether drifted by conventional modern theory or by quantavolutionary theory, would give false paleomagnetic readings, and the geographical poles may well have shifted as late as the end of the ice ages.

Also, field reversal is an indicator that worse things may be happening. An incoming giant meteoroid may dislocate the magnetic field in the course of destroying life and blasting rock. Whatever it lays down or heats to melting point will be stamped with a deviant magnetic imprint as it cools, provided the field has not sprung back into its original figure.

The complex picture is liable to so many contradictions and misinterpretations that one is tempted to discard it completely. If the magnetic field is due to an original source of electrical current deep in the Earth, can such a current be so fickle,

---

[21] J. P. Kennett and M. D. Watkins, "Geomagnetic Polarity Change, Volcanic Maxima and Faunal Extinction in the South Pacific," 227 *Nature* (29 Aug. 1970), 930-4.

[22] G. Wollin, D. B. Ericson and W. B. F. Ryan, "Variations in Magnetic Intensity and Climatic Changes," 232 *Nature* (20 Aug. 1971), 549-50.

breaking down and resetting itself in a new pattern time after time, so as to mark new orientation upon the rocks and atmosphere above? Runcorn has written that microsecond daily changes in Earth's rotation (one report gives 1 second slowdown every 600,000 years) may cause variations in the shape and intensity of the current; he adds that sudden changes in rotation would produce radial changes in the currents.[23] Michelson argues that the energy required to interchange the Earth's magnetic poles is about that of a moderately strong geomagnetic storm resulting from an intense solar eruption.[24]

Meteors have pronounced magnetic effects. Studies to this end by Jenkins, Gilmor, Campbell and Green are summarized by Corliss, and Dachille has also insisted upon the phenomenon.[25] Passing cometary trains exhibit strong electrical disturbances and can cause the same in transacting bodies as in the space plasma. A large meteoroid, whether impacting or passing close by, will disorder the Earth's electromagnetic field. Also, were the Earth to change its orbital position, it would behave like a comet, with a flaring electric tail representing electrical transactions with the unaccustomed medium of passage.

The most enthusiastic students of terrestrial magnetic changes are the exponents and developers of continental drift. Prof. Billy Glass once told the author that what convinced him of continental drift was paleomagnetic measurements. These generally are held to correlate positively bands of rock, moving away from the central Atlantic ridge, with time; the older rocks are farther from the ridge. Not only do the magnetic measurements depend upon geochronometry but also upon uniformitarianism, because it is assumed that the lava flood extending from the ridge

---

[23] S. K. Runcorn, "The Earth's Magnetism," 193 *Sci. Amer.* (Sept. 1955). 152.

[24] Irving Michelson, "Mechanics Bear Witness," 4 *Pensée* (Spring 1974), 15-21.

[25] A. W. Jenkins, *et al.,* 65 *J. Geophys. Res.* (May 1960), 1617- 19, and in Corliss, compiler, *op. cit.,* GMM-001 to 4 in G2.

has been of the same volume-to-time ratio for many millions of years. More on this last point will be brought forward later.

To conclude these pages on magnetic and geographical tilts, we can state our position: the geographical figure of the rotating Earth can tilt or reverse north and south, with moderate applied exoterrestrial force and with large holospheric damage. It has done so. The magnetic figure of the Earth will tilt or reverse in general accord with a change of geographical figure, but can also tilt or reverse independently depending upon a large electrical exchange between the Earth and a massive agglomeration in space. It has done so repeatedly. The damage is much less. Both types of change -of geographical and magnetic axes -could not have occurred, by the theory of *Solaria Binaria*, until the binary system was collapsing, which has been placed in time by the present author and again by Milton and myself at less than 14,000 years ago.

There remains a more devastating change, whereby the Earth not only tilts but also emplaces its poles upon a new geographical location. The physical force needed to accomplish such a change is many times greater than that required for the tilt alone, because the rotation of the Earth is both interrupted and altered in orientation. It is known that the Sun changes differentially the rotational speed of its several sections and some sharp movements may occur in connection with solar storms.[26]

Too, on Earth, an interrupted rotation is likely to be ramified latitudinally and stratified internally. T. Gold has given attention to such problems; in one place he has demonstrated that the polar positions will change owing to crustal movements and distortions.[27] In another place, too, he insists upon the alteration of the Earth's shape that must accompany a displacement of the

---

[26] 26. "Solar Rotation," 202 *Science* (8 Dec. 1978), 1079.

[27] "Irregularities in the Earth's Rotation," in two parts, 17 *Sky and Telescope*, (March 1958) 216-8 and (April 1958), 284-6.

geographical poles.[28] He points to the evidence of paleomagnetism as indicating numerous different polar locations over geological time, evidence that we must largely discount.

But hard geophysical evidence, as presented by Hapgood, Velikovsky and Cook, for instance, supports belief in a recent ice-age finale that shifted the north geographical pole from a position presently denominated by Baffin Island, 20° south of its present location. There is a measurable spring-back occurring all the way from Scandinavia to the Hudson Bay area, a rising area that may be due to a new rotating figure of the Earth, involving a new equator, and possibly to collapse and sudden removal of a burden of ice that had been weighing down the region. (Inasmuch as the great global cleavage passes through the center of this region, one has to introduce the probability of a forcing apart and expansion of the area between the two rising elements of continental rock.)

Surely, if the Moon were to have erupted from the Pacific Basin, the Earth's shape would have been altered, the crust would have been half removed, and the conditions Gold sets for a shift of geographical poles would be satisfied. A great force moving southwestwards would have tilted the globe, removed the crust, cleaved the globe, set the continental fragments into motion, slowed the speed of rotation, and established a new figure of spin, with a new equator and new geographical poles.

This occasion may have been the one and only time that the Earth changed its true axis of spin, as opposed to a number of other occasions in which the geographical and magnetic axes tilted. All the historical and legendary allusions to the world "turning like a potter's wheel," to celestial dizziness, to changing constellations, suns standing still, and so on may relate only to tippe-top behavior of the globe. Moderate changes in time, that is, of orbital and rotational motion, are not excluded, involving deceleration of the Earth's rotation, whether momentary (the

---

[28] "Instability of the Earth's Axis of Rotation," 175 Nature (26 Mar. 1955), 526-9.

Gibeon phenomenon),[29] or permanent. Claims of heavy deceleration, even so, are suspect; with a tilt, the sun may be visually retarded but the Earth's rotation very little affected.

The full range of possibilities in tilts has not been completed yet. Two additional ideas remain to be presented. The first concerns crustal slippage. The Earth's shell or crust, contributing about 1% to the Earth's radius, lends about one-thirtieth to the moment of inertia of the whole Earth. Apparently, then, if the shell can slip without an identical movement of the mantle and core, the energy required to change celestial and geographical orientations on the shell would be less than that required for a total reversal or retardation of Earth motions.

There are signs that this stratified slippage has occurred in the overwhelming evidence of crustal destruction around the globe as, for example, in the outpourings of lava found everywhere. Even so, the energy required for total shell slippage (following the attraction of a passing body) is formidably high, and where it would be applied is crucial, so that this idea appears, initially at least, to be as totally destructive as any other means of moving the Earth about.

However, if this crustal slippage were to occur at the moment when over half the crust was being blasted into space, then obviously the problems of slipping and venting would be greatly lessened, especially with the assistance of fracturing, rifting, and expansion. These topics cannot well be delved into here, and are reserved for treatment in later chapters.

Archaeology affords support to the proposition that the Earth has changed position relative to the Sun and the planets in recent antiquity. In connection with the human drive to build settlements according to the prevailing cosmological observations

---

[29] 29. J. Gribbin and S. Plagemann, "Discontinuous Change in Earth's Spin Rate Following Great Solar Storm of August 1972," 243 *Nature* (4 May, 1973), 26-7; André Danjon, *Comptes Rendus des Séances de l'Académie des Sciences,* series B, 250: 1399 (22 Feb. 1960), 254: 2479-82 (2 Apr. 1962), 254: 3058-61 (25 Apr. 1963).

and beliefs, the compass orientation of the constructions presents highly important issues in regard to changes in the Earth and the sky. That the earliest humans felt compelled to address their dwellings and public places to astronomical occurrences is generally granted. No one has yet found an ancient settlement capable of taking some shape that is not sky-oriented.

The mind of today's scientist turns first to the Sun, then the routines of the current Sun -the rising and setting, the solstices and equinoxes -to answer all problems of ancient civilizations. When the ruins do not confirm to these directions, then Polaris, the current fixed star of the north, is assumed to guide the primeval builders. One perplexed writer suggested that the Mesoamerican Olmecs aligned their structures with the Big Dipper. When neither the north-south axis nor the solar behavior nor a constellation fits the orientation, then it is that the ancients could not tell directions well, or that the matter in any case was not important to the builders.

What is absent from such reasoning? First, there is a failure to appreciate that the desire to orient to the skies was an obsession, a compulsion, an inescapable tradition, a sacred obligation, a proud duty. Second, the ancients, as far back as we can discover their humanity, could calculate readily and exactly the course of heavenly bodies and orient themselves thereto. Many examples of this are presented in G. de Santillana and H. von Dechend's book, *Hamlet's Mill*,[30] indeed this is the book's theme.

Third, not only the Sun, the North Star and the constellations, but also and especially the Moon and the planets were often objects of sacred (which is to say, all-important) architecture. This point has been stressed in numerous works on many cultures. The ancient pyramids of several countries, the design of Greek temples, the Hebrew Tabernacle and the Temple

---

[30] Boston: Gambit, 1969.

of Solomon -these and all other ancient masterpieces were like wedding rings uniting Earth and Heaven.

Fourth, when the heavenly bodies deviated from their customary paths or when the Earth shifted its position with respect to them, then the plans of temples, buildings, and settlements were shifted to conform to the new order of the skies. That is, celestial and mundane catastrophes of the past can explain many deviations from present "true" orientations.

Controversy naturally is engendered by any claim that the planets and Earth have shifted their axes in millions of years, if not billions. Still, every oriented edifice or monument built since about 2600 years ago (after the last of the catastrophic shifts, as argued by Velikovsky)[31] seem to have remained fixed in relation to the present skies, while those built before then appear to have moved.

Certain claims of "fixed" structures warrant study. The most famous is the Great Pyramid in Egypt. Recently, the Stonehenge megalithic "astronomical observatory" has also been widely discussed. The age of the Great Pyramid of Ghiza is in question. It has been ascribed to around 3200 B.C. and to other times. But no one suggests that it was built after 687 B.C. or for that matter after 1450 B.C. that is, after the end of the Middle Bronze Age. The West face of the Great Pyramid, which Stecchini believes was drawn first and is the basic face, is oriented 2'30" west of true north.[32]

This slight discrepancy, claims Stecchini, may be attributed to the precession of the equinoxes, which occurred from the time at which the plans were drawn to the commencement of work. He thinks that the Egyptians knew of the precession and deliberately allowed this discrepancy. I doubt this thesis, also, which is based partly upon the work of de

---

[31] 31. See Velikovsky, "The Orientation of the Pyramids," 3 *Pensée* 3 (fall), 20-1.

[32] L. C. Stecchini, Appendix to Peter Tompkins: *Secrets of the Great Pyramid* (New York: Harper and Row, 1971), 380.

Santillana and von Dechend, and ascribe the deviation from true north as an increment of continental drift and other seismic movement of the area.

A more important question concerns whether the almost perfect north-south orientation means that no tilt or change of poles has occurred since the Great Pyramid was constructed. The following possibilities ensue:

1.   The Pyramid was imperfectly oriented to true north.

2.   The Pyramid was perfectly oriented to true north but the continuing drift of the African land mass or at least northeastern Africa has amounted to minute disorientation since the Pyramid was built.[33]

3.   The Pyramid was oriented to a pre-existing true north, marked by another star. The axis of the earth shifted celestially. But an abundance of stars can be used to mark true north; Polaris is the most recent star and naturally the Pyramid points to it.

4.   The Pyramid was oriented to a pre-existing true north, which coincided with the present true north. The Earth's axis tilted on one or more occasions and then tilted back to its former position when it was built.

5.   The Pyramid was oriented to the north-south. Subsequently, the rotation of the Earth changed direction, meaning that a new *geographical* (not celestial) true north was set up, but the rotation was either changed by 180° and therefore south became north, or alternatively, accompanying or subsequent land mass thrusts

---

[33] 33. *Cf.* G. S. Pawley and N. Abrahamson, 179 *Science* (2 Mar. 1973), 892.

coincidentally brought the area around Cairo to rest pointing at the true and original north-south axis.

Of these five possibilities, the third appears most acceptable within the framework of this book. It would permit a number of axial tilts but only a minimum land-mass movement affecting Egypt since the Pyramid was constructed. This seems to be in accord with the theories advanced in *Chaos and Creation* that catastrophes subsequent to the great Pyramids construction did not cause major crustal slippage or a changed axis of rotation even though they caused heavy electrical, flooding, hurricane, and volcanic events. Earlier catastrophes involved the major changes in the geographical existence and location of the Earth's land masses.

At least so far as the Egyptian area is concerned, Velikovsky's descriptions in the Venusian case (ca 1450 B.C.) especially may be exaggerated; any implication that the geographical masses moved, or the Earth's axis of rotation changed, would have to be discounted. His evidence that the Pyramid shows signs of great seismic stress should be recalled, however. The most resistant material ever sculpted and fitted by mankind was affected visibly by earth shocks that must have been beyond the present limits of the Richter seismic scale.

The huge stones placed in circles and lines at Stonehenge, England, can be proven to be only generally oriented to observe solar solstices of the present age. Otherwise they display actual rearrangements of stones, done with immense labor, which can best be accounted for by an axial tilt, that is, by catastrophe.

Here, as at other magnetic settings, the earth scientist needs to take into account human motives, asking oneself: is it likely that the stupendous collective labor required to build these great structures, admittedly astronomical, would have been mobilized if the Earth (and hence the skies) were not exhibiting strange and terrifying changes of motion? Was the human urge to control the sources of his terror implicated?

Attempts have been made at dating Stonehenge by C14 on organic objects found in association with it. MacKie is of the opinion that the dates of Stonehenge and other megalithic astronomical sighting locations would not permit one to claim reorientations of the Sun after 1500 B.C.[34] Hence, in Joshua's time or on later occasions, reports of the Sun altering its route would have to be considered false.

Still, Stonehenge, like the Pyramid, is a catastrophized artifact in the first place, and bears also the marks of catastrophic changes in its settings. The C14 dates are not abundant and consistent, nor generally reliable within the span of centuries.

The Mesoamerican sites magnify the uncertainty. There are many of them. All are thought to have been set up after 1500 B.C. Macgowan, (1945), and now we quote Anthony Aveni extensively,[35]

> ... seems to have been the first person to suggest that the plans of a large number of Mesoamerican cities exhibited an east of north axiality. Among those sites which evidenced some orderly arrangement, he observed that the orientations fell into three groups: true north, about 7° east of north, and about 17° east of north; he noted that few sites were oriented west of north. In the 17° group were Teotihuacan, Cholula, Tenayuca, Mexican period buildings at Chichen Itza, Tula, and the pyramid adjacent to the Zocalo in Mexico City. A number of sites of the Peten District seemed to belong to the 7° group. Macgowan suggested that a historical pattern might emerge in the sense that early structures such as Cuilcuilco possessed a nearly true north axiality while the 17° east of north orientation showed up in the later buildings.

[34] Euan W. Mackie, "Megalithic Astronomy and Catastrophism," 4 *Pensée* 5 (Winter), 5-20; "Megalithic Astronomy: Neolithic Stone Circles," I *S. I. S. R.* 4 (Spring 1976), 2-4.

[35] Anthony Aveni, ed., *Archaeoastronomy in Pre-Columbian America* (Austin, Texas: U. of Texas Press, 1975).

Aveni found by transit that fifty of the fifty-six sites surveyed align east of north; the 17° orientations seem to be prevalent in the valley of Mexico.

Yet Carlson, working on centers carbon-dated between 1000- 1400 B.C. says that "Olmec culture is well-characterized by ceremonial centers, which are generally 7° to 12° west of north...".[36] This would suggest that tilts of different ages are represented in the two regions, or that the Olmecs, who invented the magnetic compass, may have oriented their buildings to a magnetic north. Almost all of them deviate from true north orientation.

According to sacred scripture, the four gods who were born of the creator gods govern the four cardinal points of the Earth's compass, and struggle with each other. It would appear from the chart that, while north-south was the way human construction should be engineered, by present direction lines, frequent changes have occurred.

A few years ago, Mesoamerican civilization was considered recent and crude. Today the view has changed and the same respect is given the early Mesoamerican as is accorded to other world civilizations.

In 1976, a lodestone compass was claimed for the Olmec civilization at 1000 B.C. or earlier, before the earliest demonstrable Chinese compass. In this case, it cannot be argued that the Mesoamerican were incapable of planning their settlements and public buildings with accurate reference to north or any other cardinal point. In a letter describing a study trip to Central America, Patrick Julig writes:[37]

---

[36] John B. Carlson, "Lodestone Compass: Chinese or Olmec Primacy?" 189 *Science* (5 Sept. 1975), 753-60.

[37] To William Mullen, 1974 n. d. Julig studied the famous Nazca earth lines of Peru and concluded that they might represent lines of meteoritic falls from which the (sacred) burnt stones were removed.

... I observed changes in the orientation of the foundations of Mayan buildings between the Archaic and Classical periods. Sometimes there were changes within the same building by as much as 10° in later additions to the structure such as in the Palace at Palanque. This could possibly be a way to date the structures, or at least the foundations, as being pre-687 B.C.

One must tentatively conclude that at least Middle America suffered serious crustal slippages. Or, axial tilts occurred frequently and the Mesoamericans were employing a method of determining true north (the Earth's axis of rotation) by a means not dependent upon a star. And, if this technique existed, the alternative presents itself that the object defining true north itself moved on occasion.

A second study by Aveni leads us also to believe that astronomical settings have altered in proto-historical times. He and his associates traced and surveyed the orientations of "The Peaked Cross Symbol in Ancient Mesoamerica" in many places.[38] These peaked crosses are not monuments of the highest level, but remind us in some ways of the frequent crude religious sculptures that are to be found at crossways in many places on Earth, dating up even to the present day. The cross represents the application of the Sacred Year to the four quarters of the world, the cardinal directions, the highly significant merging of time and space that the ancient Mesoamericans achieved.

Teotihuacan was probably the religious center of ancient Mesoamerica, like Rome of medieval Europe. The fundamental Teotihuacan grid as excavated is oriented 15.5° east of north. Of the some 30 symbols that the Aveni group have assembled from elsewhere in Mexico, the orientations of 19 are given. Of the 19, nine are oriented within 2° of the Teotihuacan grid. Of the nine, all except one (carved on an outcrop) are on a floor. Of the remaining ten with known orientations, all range between 35°42 and 80°24'; all are incised upon outcrops except one that is on a

---

[38] A. Aveni. H. Hartung and B. Buckingham, 202 *Science* (20 Oct. 1978), 286-79.

broken flat stone whose "axis points toward Teotihuacan"(TEP3), and another (TUI) that is "pecked on horizontal floor of lava field." Most of these are considered as pointing towards the summer solstice sunrise, which is rather insulting to the intelligence of the humblest pecker. Are we to believe that they could not find the point of farthest advance of the Sun? And why should outcrops be carved east and floors to the north?

It would appear that either (a) the carvers were inexact amateurs with biases towards the east, or (b) the larger part of Mexico shifted its axis by 15.5° to the west of north in response to seismism and/or a tilt of the Earth's axis in reference to the solar and sidereal system, or a geographical transfer of poles implying a changed axis of rotation, or (c) the axis of Teotihuacan shifted at an early time eastwards from true north and its new position was assigned sacred and ritual meaning, a Holy North to be imitated, just as the 260-day Sacred Year was tenaciously preserved, without a celestial referent, until modern times, alongside the 360 and 365-day calendars. In the case of (b) and (c), the extreme eastern orientations of the peckings might have been memorial, without special orientation, to Teotihuacan's gods upon the occasion of faulting, fracture and exposure of new rocks. The geology and the relative dating of the peckings are important in considering these alternatives. Especially, the hypothesis can be entertained of a deliberate attempt to follow a fault line (especially if an electrical current were running) in the outcroppings. (If the Etruscan priests took possession of and catalogued all aspects of a spot struck by lightning, similar obsessions may be expected among the equally obsessional Mesoamericans.)

So long as north-oriented axes were to Holy North, they would be consistent. But east-oriented axes, if there is no "Holy East", would wander with tilt of the Earth's pole, that is perhaps from 30° to 80°, whether in the wake of the Teotihuacan shift or upon some later occasion. The association of the peaked cross symbols with outcroppings must have some significance. If a desperate speculation may be permitted, new outcroppings might have become thereby "holy" too, just as fallen meteoroids have become holy, and perhaps the outcrop orientations might be attempts at affixing the eastern risings of that vagabond planet, Venus.

A research of deviating Egyptian, Mesoamerican, Mycenaean, Greek and other structural orientations may suggest dates for the construction of earliest Teotihuacan-a subject of some controversy -as well as point to causes of the phenomena of the peaked crosses.

Finally, one may observe that the Teotihuacan orientation 15.5° to the east of north could have indicated a transfer of the geographical north pole of the Earth by that amount at some point of time. This shift is not far from the degree of shift in the north pole from a location at Baffin Island to its present location northwest. A number of students believe such shift to have occurred at the "End of the Ice Ages."

CHAPTER FIVE

ELECTRICITY

Tertullian, an early Christian apologist, came to the attention of a contemporary physicist delving into the occult, and he, J. Ziegler, has supplied us with this quotation which can introduce this chapter and the next:

> The philosophers know the distinction between common and mysterious fire. The First that serves man's use is one thing. The fire that ministers to the judgement of God is another, whether flashing the thunderbolts from the heaven or rushing up from the earth through the mountain tops. For it does not consume what it burns, but, even while it spends it, repairs the loss. So the mountains remain, ever burning; and he who is touched by fire from heaven is safe -no fire shall turn him to ashes.

Lightning expresses only a small fraction of electrical processes. Electricity is everywhere. It presents itself in the smallest particle and, some of us believe, commands the behavior of every remote galaxy of stars. It is part and parcel of every natural transaction. Perhaps it is the hunger of protons for electrons that initiates all natural behavior, whatever the scale or intensity.

Earth scientists have been reluctant to admit electricity to their domain. There is a confined interest in "hard" lightning, taken over by meteorologists now, and geophysics must trespass upon nuclear physics in connection with chemical bonding and radioactivity. Historically, earth scientists have led the parade of debunkers when meteoroids were reported to fall or when lightning took unusual forms. Of course, when geologists stood upon mountain tops and St. Elmo's fire flowed from their beards and hammers, they could not well deny this "god's fire" of the ancients.[1] But one searches in vain for a treatise on St. Elmo's fire, one of the oldest and most fascinating phenomena continuously reported.

In fact, there exists no treatise on the full range of electrical behaviors related to geology. This universal presence of electricity in geological events does not excite systematic attention, no more than it has in astronomical events up to the present. If one seeks a rational explanation for this neglect, it may lie in the unreadiness of the lithosphere, hydrosphere, and atmosphere to display their electrical history, letting the electrical be considered transient and superficial.

If one seeks non-rational explanations of an ideological or psychiatric sort for such avoidance, it may be in the quixotic or miraculous appearances of electrical phenomena. Bordering upon the religious and the occult, these set up psychological resistances among "hard" scientists. As we shall see, even the famous subject of lightning, which can hardly be ignored, is little understood. The latest literature on lightning is still at the state of trying to survey its extent and intensity, and not even its forms are classified.

The ancient Etruscans thought that they could discern eleven different types. So wrote Pliny, but a modern Etruscan expert, Rilli, says that they recognized thirty kinds of lightning.[2] Ancient sources that refer to fire often are speaking of electricity, "god's fire." Applying the modern meaning of "fire" as

---

[1] 1. *Cf.* 44 abstracts of such experiences in Wm. Corliss, *Sourcebook* GLD-001 to 044, GI-81 to 110.

[2] Pliny, *Natural History*, Rockham tr. (Cambridge: Harvard U. Press, 1967), II. LIII; N. Rilli, *Gli Etruschi a Sesto Fiorentino* (Florence: Giuntina, 1964).

combustion and conflagration, one cannot comprehend their outlook. To early theologians and philosophers," fire" meant a set of qualities exhibited by the "aether", loosely translated as "air", and when "air" was considered a basic element of existence, electrical phenomena were deemed to be integral with it.

The large importance given to electrical phenomena in ancient times, drives us to believe that their manifestations were much more in evidence. Furthermore, although there are a few indications that the Egyptians may have employed wire on occasion to transmit electricity, unquestionably they were preoccupied with electrostatics, the exploitation of the generous and ready electrical potentials of the ground atmosphere. This I have discussed in my study of Moses.

Lately, the ionization of the atmosphere has come to be studied. Even the ground beneath our feet has come to be conceived as a conveyor of waves of numerous types, ranging from the gross seismic tremors that topple whole cities to the delicate motions of the wire in the hands of dowsers in search of underground water.[3] Ions are electrified particles; they affect the growth, fibers and nervous system of plants, animals, and humans in ways mostly unknown.[4] Many students think that an abundance of negative ions in the atmosphere produces a sense of well-being, but that "excessive" positive ions provoke depression, irritability, and illness.

The Earth's surface contains a charge; it too is unknown in extent and effects.[5] The charge is called negative originally because it is of the kind that comes from rubbed resin, and conventionally because it comes from the ground. On a clear day an electric potential of about 100 volts per meter of height occurs.

---

[3] Guy Underwood, *The Pattern of the Past* (London: Abacus, 1972) Treats dowsing, electricity, geodetic lines, and cultural associations all together.

[4] Fred Soyka and Alan Edmonds, *The Ion Effect* (Toronto: Seal Books, 1978); S. W. Tromp, *op. cit.*, 112-5.

[5] Fernando Sanford, *Terrestrial Electricity* (Stanford, Calif.: Stanford U. Press, 1931), Chapter 4.

The charge of Earth tends to persist in the absence of exoterrestrial intrusions, employing the lower atmosphere as an insulator. The charge in our opinion, will have varied greatly over the human past. Then its variation, as well as its constancy, must have had significant effects upon human behavior and ecology.

The Earth may have presumed once to have been in the grip of a constant heavy charge, for reasons that will unfold below and are also treated in *Solaria Binaria*. It began to lose this charge, both gradually and in series of catastrophic discharges. Today, solar flares excite large surges in the flow of charge from upper atmosphere to ground. Too, thunderstorms may be principally a method of balancing the atmosphere-lithosphere equation by releasing ground electricity.[6]

There persist certain phenomena that may reflect this decline of charge. All over the world there are pathways that were worked out mysteriously (part instinctively and part deliberately) by ancient men and that are followed today. Michell has sought out the English paths especially. He shows that they are often not the shortest way between two points.[7] Rather they have seemingly pursued geodetic "power lines" which thereupon developed as religious routine, ritually followed. As with many customs, people follow behavior that originally had a perceived and sound meaning.

Waterlines have been explored successfully by following the cues provided by traditional water-dowsers. It may well be that underground water moves along paths which are electrically distinctive. In other cases, it may emulate the course of lightning that once travelled along root networks and also fractures formed by lightning. Seismic fractures also are important conduits of water.

Lightning has been used as a kind of naturally-provided instrument for studying the electrical nature of the ground. Aside

---

[6] "Solar Activity and Terrestrial Thunderstorms," 81 *New Scientists* (1979), 256.

[7] *A View Over Atlantis*, (1969).

from numerous ancient observations along these lines, a few modern studies exist[8] to indicate that soils of high conductivity (e. g. marshes) are lightning-prone; that ironstone outcrops attract lightning; that strata discontinuities attract lightning. So do underground springs; so do areas of high negative ion concentrations. Masts, lightning conductors, and buried metal pipes invite strokes.

Experiments by Stekolnikov showed that soils attracted sparks depending upon their conductivity. Certain trees are stroke-prone, the oak, for example. The variety of effects is scarcely understood -the fancy dendritic patterns sometimes displayed underground, the killing of flocks of sheep, the escape unimpaired of a girl enveloped in lightning flames, the subsequent death of a man seemingly unaffected at the moment of stroke, and so on.

In 1977 an American physicist, J. Ziegler, published a study of the knowledge and uses of electrostatics among the ancient Hebrews and other peoples of the Near East and Greece.[9] His thesis, elaborated shortly thereafter by the present author in a book on the period of Moses, maintained that these ancient peoples possessed devices for inducing and displaying electrical effects in their religious practices. The most spectacular of the devices was Moses' Ark of the Covenant, which G. C. Lichtenberg, a German Electrician of the 18th Century, termed a form of Leyden jar.

The Leyden jar is called an electric capacitor. A metal rod based upon a metal lining within an insulating (e. g. wood) vessel will store a charge from the air. When the outside of the vessel is also lined with metal that is in touch with the ground, an opposite charge is induced. The potential between the two poles may accumulate to a level at which a spark will jump the gap between

---

[8] See the survey of unusual ground effects by B. L. Goodlet, *J. Inst. of Elec. Eng.* 81 (1937), 1-26.

[9] Jerry Ziegler (pseud. Zeromiah II), *YHWH,* Princeton: Metron Publns., 1977.

them. The frequency, brilliance, and power of the spark or arc (Ark = box = Aron in old Hebrew) depend upon the size of the gap and the voltage differential that is generated.

The condition of the atmosphere and ground are critical factors. The higher the box and the wetter its grounding contact, the greater the electrical effects. That is, the effectiveness and potency of the devices depends upon local conditions that can to some degree be manipulated. Aside from this, the general electrical state of the Earth and atmosphere (including exoterrestrial influences affecting these bodies) determines the overall effect.

In an atmosphere where electrical and dust turbulence were prevalent, as in times of Exodus and other periods that I have identified elsewhere, and the Earth was discharging at an effectively higher level than it is today, the incitement of electric displays without motors, pumps, and wires was easy: large potential differences continuously presented themselves for exploitation. Electrical effects became essential to political and religious roles and were subjects of jealous contention within and between governments. A full social analysis is presented in my treatise on Moses; what may be stressed here is that the existence and activity of such devices evidences that the Earth was then in a state of heightened electrical activity relative to modern times.

With the settling of the skies, the intensity of electric phenomena diminished. The divine spark manifested itself less and less; the arks were carried more and more up to the mountain temples (e. g. both the Temple of Solomon and the Temple of Jeroboam). The angels, demons, and mountain gods manifested themselves in electrical demonstrations on high with the aid of crosses, trees, and poles.[10] These, too, could not be maintained. Empedocles, when discussing the four elements, fire, earth, air and water, says that fire has ceased to "travel", and no lower forms of fire remain.[11] Plutarch wrote at the end of the pagan age an

[10] *Ibid.,* 53ff.

[11] Hock, *God in Greek Philosophy,* 99, cited in Ziegler.

essay on why the highly placed Delphic oracle had lost its
influence; he gave the vaguest of references indicating a failure of
electric current, but the question itself is significant.[12] By late
classical times, the knowledge of arks and of the exploitation of
"god's fire" was largely defunct.

Yahweh became "invisible", who before, declares the
Bible, could be seen in flaming display upon the Ark of Moses. So
later philosophers gave new meanings to words: realities became
metaphors and abstractions; thus, the "word" and "presence" of
the divine became thoughts, rather than the noises and signs of
electrical divinity. The profuse electrical references in the Bible, in
ancient Near East documents and Greek Mythology, in the Hindu
Vedas - all were reduced to metaphors, generalized into ordinary
meanings ('fire' becomes 'conflagration'), and metaphysical
abstractions (the commandment to worship no other God nor
image is interpreted philosophically rather than realistically). The
obelisks whose points once lit up as the eyes of the hidden god
Amon (Amen) came to be variously interpreted as giant sundials,
emblems of royal power, phallic symbols, or sign boards for
vainglorious inscriptions. As Ziegler suggests, the Greek word
"obelisk" itself might have meant "ob-el-ish," or "serpent-light-
fire."

Von Fange recounts a century-old report on a Babylonian
ziggurat, which may have been the Tower of Babel. The structure
can be placed several centuries earlier than Moses but also in a
highly electrical epoch.

> It appeared that fire struck the tower and split it down to the
> very foundation. In different parts of the ruins immense brown
> and black masses of brickwork had changed into a vitrified
> state. At a distance the ruins looked like edifices torn apart at
> their foundations. Evidently the fiercest kind of fire created the
> havoc. The most curious of the fragments found several

---

[12] "Why the Oracles Cease to Give Answers," IV, 56. See Ziegler,
Chapter 19.

misshapen masses of brickwork, black, subjected to some kind of heat, and completely molten.

The whole ruin has the appearance of a burnt mountain. On one side, beneath the crowning masonry, lay huge fragments torn from the pile itself. The calcined and vitreous surface of the brick had fused into rock-like masses. It is difficult to explain the cause of the vitrification of the upper building. Great boulders were vitrified, and brickwork had been fused by fire.[13]

Here possibly was cosmic fire. Another effect deserves mention. A major electrical discharge in which a number of humans are stimulated, as in a town on an eminence, may proceed slowly and without killing. It leaks rather than blasts. It might affect people's minds. Today, a fearful side-effect of electroshock therapy, which is used to treat persons suffering from depression, is amnesia; whole sections of the person's store of memories will be erased.

The Tower of Babel was probably erected at a time when electrical perturbations were attributed, if my analysis in *Chaos and Creation* is correct, to movements of the planet Mercury.[14] The arrogance of the builders in attempting to reach the sky was punished, recites the Bible, but in a peculiar way. They who spoke the same language when they began their work were caused to "babel" in many tongues. The Earth shook long beforehand; the tower partly sank into the ground, so say Jewish legends; but also much of the tower was destroyed by fire from the sky. The work had to be abandoned and afterwards the nations spoke different languages.

I offer a scenario for consideration. The Tower of Babel was being built in terror and hope of appeasing sky-bodies, possibly Jupiter-Marduk or Mercury. Conscripts or slaves of many

---

[13] Erich A. von Fange, "Strange Fire on the Earth," 12 *Creat. Res. Soc. Q.* (Dec. 1975), 132.

[14] *Op. cit.,* 210 ff.

countries made up a work force of 50,000 men. They put together a rough *lingua franca* from the language of the area to communicate on the job. The approach of a large body (there were actually many adoring and frightened references to planet Mercury around this time) occasioned the build-up of charge and then a flowing discharge through the structure, creating a confusion in administrative orders and a linguistic amnesia especially in the *lingua franca*. No longer could people understand each other. And then the whole edifice was stuck by immense cosmic bolts, partly fractured, and exploded.

"Slow lightning" is the geologically and biologically effective discharge of terrestrial electricity. A "slow lightning flood" may be conceivable, too. The curious vitrified forts of Scotland may be a case in point.[15] They remind us of the Ziggurat of Babylon. Their stone and mortar are fused solidly with the clifftops to which they adhere.

The forts are much in need of study. The early interpretations of them as cattle pens is uncomplimentary to a people that lived in hovels that experienced no such fusion. The idea of brush being heaped outside the precipitous walls, and then burning them with an intense heat, would require a mobile ceramics oven and vent.

We would argue that the lightning here was not "bolt-thin" and "lightning-quick" but poured upwards over seconds, diffusing through its medium, ferruginously-mortared stone. There would have been an approaching unequally charged great body or gas cloud that had pierced the electrically balanced plasma and drawn away or brushed aside the magnetic space sheath of Earth. The Earth below would have collected on its highest surfaces a charge to meet the incoming charge. This would begin to flow upwards. Heavy leader strokes descending would have collapsed the roofs of houses. A tube of ionized dust would arise

---

[15] James Anderson, 5 *Archaelogia* (1777), 241-66; *ibid.,* (1980), 87-99. and see the materials reprinted in W. R. Corliss, *Strange Artifacts* (Glen Arm, Md.: Sourcebook Project, 1974) vols. M-1, M-2, under "Forts."

and descend, make contact from both ends and set up a fierce heat that would scorch its "vessels." A final flash, and then the body would pass or the cloud dissipate, and a rain of dust and vapors would fall back upon the ground, calcinating it.

It is probable that many thousands of burnt eminences exist around the world whose tops have seen the fusion of rocks, perhaps even Troy IIg, the "Burnt City" so-called.[16] The famous site, whether or not it was the real Troy, is on an eminence. While not high, the city would have had many small reservoirs of water, whereas the ground outside might already have been dried out. In Troy IIg a sulphurous color suffuses all outdoor spaces and passageways of the town. A deposit of lead and copper melted and flowed around the town. (It is possible that this melt had been scavenged after Schliemann reported it in the 1880's and the discoloration was all that was discoverable when the Blegen expedition re-excavated the site in the 1930's).

No human hand could have or would have set such a fire. The heat was fierce. The ash was far too abundant for a deliberate fire from local materials, and carried a red color. In places it was like calcinated rock, a meter or more in depth, perhaps like the vitrified Scottish forts. No one would have wanted to destroy precious metals (not so mention even more precious metal left in abundance in the scorched houses and the Treasure of Priam, found on a Wall). Noteworthy is the absence of human and animal skeletal material in the ruins. Either they turned to dust from the heat, or the electrical build-up was sensed, as it is by animals before earthquakes for example, and they fled from the hill onto the plain where the sensations were absent.

Perhaps a heavily charged cosmic body was approaching or was near the Earth with an opposite charge or inducing one to collect on Earth; this would cause numerous discharges. Every eminence, one might imagine, would offer an exit for lightning,

---

[16] A. de Grazia, "Paleo-Calcinology: Destruction by Fire in Pre-Historic and Ancient Times." I *Kronos* (April 1975), 25-36; II *Kronos* (August, 1975), 63-71.

especially if it held the slightest metallic component, and were not surrounded by damp lowlands. Buildings are not needed.

If settlements seem to have been affected by slow lightning flood, unsettled eminence should often have endured the same experience. I have explored as a candidate a conical hill of Stelida, Naxos, Greece (Alt. 152m).[17] The top is a hard silicate with bits of ferruginous rock in the eroded (burst?) rubble. It nests among loose, hardly consolidated rocks that have fast fallen away from the columnar core. This phenomenon is usually seen as an ancient metamorphosis. Somehow the temperature of water-laden deep limestones and granites mounted and caused them to nearly melt and to rise. Limestone is a common environment of silicification. Silification is abundant around igneous metamorphism. In a hot and fast reaction, siliceous fluid is introduced hydrothermally and replaces the host rock, such as limestone, into which it intrudes.

Such is the case where an electric charge is seeking an exit from far below. With or without water, a hot electric discharge current can assemble and proceed quickly up the core of a hill, heating and silicifying as it moves. On top of the hill, it forms a cap just as caps will form on the sparking end of a discharging rod. The charge, that is, uses the plastically flowing rock as a conductor and then builds a deposit from which it may discharge more easily.

The taller the mountain, the less time and chance for the siliceous fluid to reach and cap its peak. At the same time, electricity of this type may even build mountains. Juergens has suggested that mounds may have been formed on the planet Mars by the same process. An electrical process may also be involved in the vigorously erupting mountains of Io, satellite of Jupiter. These are casting material to heights of several hundred kilometers from caldera-like structures. Unless Io is newly emplaced, all water or carbon dioxide would have long ago been exhausted as propellant media. Spectroscopy evidences no water on Io, moreover. Sulfur

---

[17] The author thanks geologists Dr. Gerd Roesler and Dr. Poul Andriessen, who aided me notwithstanding their scepticism.

would be too heavy to gain the speed of eruption required for such lofty explosions.

Therefore, Thomas Gold turns to the electric current of 5×10^6 amperes that cyclones upwards from the Jovian surface arguing that it is "largely conducted through the body of Io.[18] The current contracts along a narrow tube of passage which is kept hot and therefore more conductive. As it emerges into cold space, the current encounters conductive resistance and, hence, forms heat spots of several thousands of degrees kelvin. "Most current spots are likely to be volcanic calderas, either provided by tectonic events within Io or generated by the current heating itself." The electric volcanism is steadied by the "accurately repeating" electric arc from Jupiter. So now we find here a model for processes that may once have occurred on Earth as well, supposing a sufficiently intense terrestrial discharge were occurring at a weak spot for even a few days.

The "slow lightning" may shape not only eminences but also subterranean cavities. Von Fange writes that "The same phenomenon has been observed in the mounds and barrows of the British Isles. Some have at one time been filled with an intense heat. Their walls are melted and their contents fused. The stones of the innermost cell of a long barrow near Maughold on the Isle of Man have been fused together like the mysterious vitrified towers of Scotland and elsewhere."[19] Many Egyptian tombs and the interiors of pyramids are scarred by intense heat. Caliche ($CaCO_3$) adhering to bones and rock undersides in a California burial cairn provide radiometric dates of 19,000 to 21,000 years, whereas archaeological estimates of the many such cairns give 5,000 B.P. or less.[20]

---

[18] "Electrical Origin of the Outbursts of Io," 206 Sci (30 Nov. 1979), 107 1-3. On sulphur as the medium, cf. Guy J. Consolmagno, "Sulfur Volcanos on Io," 205 *Sci* (27 July 1979), 396-7.

[19] *Op. cit.*, 132.

[20] P. J. Wilke, "Cairn Burials of the California Deserts," 43 *Amer. Antiquity* (1978), 444-8.

The famed caves of Aquitaine (France)[21] whose primeval users carved and sculpted images upon the walls, may surprise the naive visitor. One expects to find a general similarity of the interiors. Not at all. Each interior is unique. Some are serpentine, others like grand ballrooms; some have magnificent silicate columns and startling naturally formed shapes; others are plain and dull, save for the signs of human occupancy. All are of limestone; all are elevated, if only slightly, above the flat river and stream valleys around.

Why are they so different? Caves are said to be formed by the percolation of water through weak stone, cracked stone, or interstices of layers of stone. The filtering drops become trickles, and then streams. The cavity is enlarged. The river deviates or dries up and the interior is prepared for occupancy. Time elapsed may be "millions of years."

However, Worrad reports that limestone caves can be rapidly formed by water -"that in one year a cave of 3ft. x 6ft. cross section x 120ft. long would be formed per square mile of the surface," and opines that the Deluge [not to mention other floods] provided huge amounts of water for limestone solution and cave foundation.[22] Dripstone would be formed rapidly, too. A *National Geographic Magazine* photograph (1953) carried a picture of a bat "entombed" inside a stalagmite, which, therefore, could not have formed at the "0.001 inch per annum or so rates " usually assumed.[23] In Brixham caves (Devonshire), the bones of fossil mammals, of the types drawn in the Caves of Dordogne, are stuck

---

[21] *Inter alia cf.* J. P. Rigaud and B. Vanderneersch, eds., *Sud-Ouest (Aquitaine et Charente): Livret-Guide de 1' Excursion* A4, IX Congrés U. I. S. P. R. Paris, 1976.

[22] Worrad, *Creat. Sci. Res. Q.* 197; see also letters by D. Cardona and B. Raymond in 3 *Pensée* (Winter, 1973), 48- 50; and E. L. Williams and R. J. Herdklotz "Solution and Deposition of Calcium Carbonate in a Laboratory Situation," 13 *Creat. Res. Sci. Q.* (March 1977), 192-9.

[23] Ltr. of Felix Fernando, III *Pensée* (1973), 50, citing Nat. Geog. (Oct. 1953).

in the ceiling - so writes a correspondent, U. E. Ramage, to this author - as well as in the sides and floor. In as much as these species' extinctions were quite recent, this shows that it may not take long to hollow out a cave. Furthermore, the small cave is "prettily ornamented with concrete growths."[24]   So we would appear to have a very recent catastrophic bone assemblage of animals, then or soon extincted as species, followed by a geologically instant cave-making, and prompt furbishing with stalagmites and stalactites.

     Although water may quickly hollow out caves, the role of electricity is not to be ignored. Electric fields, as Asakawa has demonstrated experimentally, enhances heat transfer in nearby gases (up to 1.5 times); liquids (up to 2.0 times) and solids (up to 1.6 times), depending upon the positioning of electrodes and the strength of the applied fields.[25] Perhaps caves are ancient hotspots, electrical calderas, where creation time is shortened by the blasting impatience of electrical arc currents.

---

[24] 8 Sept. 1967 from Ceylon; Villey Aellen And P. Strinati, *Guide des Grottes d'Europe* (Paris: Delachaux, 1975), 130.

[25] Y. Asakawa, "Promotion and retardation of heat transfer by electric fields," 261 *Nature* (20 May 1976), 220-1.

CHAPTER SIX

COSMIC AND TERRESTRIAL LIGHTNING

A powerful, highly developed and mysterious people of ancient Italy, the Etruscans, believed in the strictest set of relationships between the small Earth and the great and divine Universe.[1] They planned their cities astronomically, as did all early peoples, but, more specifically, worshiped lightning and gave "the thunderbolting god" Jupiter to the Romans. They founded a College of Lightning Arts (ars *fulminum*) at Visul. When a bolt of lightning struck, the ground became at that instant hallowed; no one might disturb it until priests made a site inspection and had concluded which of thirty types of lightning it was and what should be done about it.[2]

They dug wells to receive lightning and marked the wells with the bidental symbol of Jupiter (Zeus), a two-pronged spear. Zeus has been variously portrayed as the hurler of cosmic lightning, with a two or three-pronged spear, and even hurling a bolt whose shape was not forked lightning but like an American football, a plasmoid perhaps, a kind of lightning bomb.[3]

---

[1] Nicola Rilli, *Gli Etruschi a Sesto Fiorentino* (Firenze: Tipografia Giuntina, 1964), 92.

[2] *Ibid.*, 94-5.

[3] Ralph Juergens, "Of the Moon and Mars," 4 *Pensée* 4 (Fall 1974) 21-30; 4 *Pensée* 5 (Winter 1975), 27-39; A. de Grazia, *Chaos and Creation,* 203

All mountains were sacred to thunderbolting Jupiter. Seneca, the Roman stoic and dramatist, has him dissolving mountain ranges with his bolts.[4] The Bible says the same of Yahweh, all this and more. Psalm 97 gives us:

> "His lightnings lighten the world;
> the earth sees and trembles.
> The mountains melt like wax before the Lord,
> before the Lord of all the earth."

The Babylonians speak so of Marduk, the Indians of Shiva, the Persians of Mazda. Other gods played with lightning and fire - Hephaistos, Apollo, Hermes, etc. but Jupiter was the overwhelming lightning god. Giambattista Vico believed that lightning was less on Earth in the damp age of Saturn, before Jupiter, because of deluges. It is noteworthy that satellite maps of terrestrial lightning published this year (1981) by Orville and Vonnegut show a dearth of discharges upon oceanic surfaces.[5] Satellites have also shown that a realm of lightning bolts a thousand times more powerful than the ordinary terrestrial bolts dominates the upper atmosphere.[6]

The Etruscans said that their great city of Volsinium, by what is now Lago Bolsena, was destroyed by a thunderbolt of Mars. They believed that a portent or an inducement to the awful act came from rituals performed by their King.[7] This was about

---

for illustration. On ball lightning see A. Wittmann, 232 *Nature* (27 Aug. 1971), 625.

[4] In *Thyestes,* a drama remarkable for its catastrophic images.

[5] R. Orville and B. Vonnegut, "Patterns of Thunderbolts," 92 *New Scientist* (1981), 102.

[6] *New Scientist* (20 Oct. 1977), 150.

[7] G. P. Pliny, II *Natural History* (trans. Cambridge: Harvard U. Press, 1967) II: LIV. The translation of Rackham is questionable, if only because he has no idea that the Etruscans and early Romans, like the Hebrews and Greeks of the age, were using electrostatic machines to produce divine image and oracles.

the time that Rome was founded, likely by near descendants of fugitives from grave disasters in the Near East.[8] The famous Seven Hills of Rome themselves may be a set of extinct volcanos, according to an early French geologist. Since few scientists believe in cosmic thunderbolts, this report of Lake Bolsena has never been thoroughly investigated. The Italian anthropologist-geologist Leonardi assured me that the lake basin is a typical extinct volcano. Velikovsky accepted the lightning thesis.[9] Geographer Donald Patten calls it a meteoric crater-lake because it lacks a volcano talus, is oval shaped, 7x9 miles, and is bottomed by lava and ash.[10] Until an intensive investigation is made, Leonardi's expertness must weigh heavily in our judgement.

J. E. Strickling has guided the author to a passage in Ginzberg's *Legends of the Jews* (I, 240) where, it is said, " the day whereon God visited him (Abraham) was exceedingly hot, for He had bored a hole in hell, so that its heat might reach as far as the earth..." Was this hole dug by a meteoroid impact, a lightning stroke (downwards or upwards), or a volcanic outburst? That it would have been a sudden occurrence, and that other studies indicate probably exoterrestrial (hence volcanic) disturbances in Abraham's time and that Abraham's God was a God of lightning are bits of fact to consider with the larger mosaic being pieced together here.

Archaeologist Nicola Rilli dug in one location at Prato (near Florence) and found three distinct heavy ash layers defining three distinct periods of prehistory.[11] He found a small silo grain, intact but carbonized, a fact that he ascribed to a great fire that

---

[8] A. de Grazia, critique of *Enea Nel Lazio* (Rome: Palombi, 1981) on the Virgil Bimillennial Celebration, in *The Burning of Troy* (in press).

[9] *Worlds in Collision*, 273.

[10] Donald W. Patten, R. R. Hatch and L. C. Steinhauer, *The Long Day of Joshua and Six Other Catastrophes* (Seattle: Pacific Meridian Pub. Co., 1973), 18-9.

[11] *Op. cit.,* 88-91.

had been suffocated. Lightning fires may have played a role in the burnings.

Recent astrophysical opinion regards Jupiter as a hot hyper-active planet that exchanges bolts with its satellite Io over a distance of 50,000 miles. The bolts are frequent enough to be an arc or current. Strangely, Pliny described great thunderbolts as the "fire of the three upper planets," not to be confused with terrestrial lightning.[12] Today lightning could not discharge over the great distance between Jupiter and Earth, not unless Jupiter were to explode, a great cloud of gases that would drift between the planets and provide a conductor for the electric spark. Something akin to discharge can affect the Earth and Sun, though, when the great planet is in conjunction with Earth and Sun, as Gribben and Plageman have propounded.[13]

However, according to the theory of Solaria Binaria which we have advanced in another book, the two bodies were once nearer, there were remnants of a gaseous envelope between Sun and Jupiter, and there were sporadic efforts to push through discharges along the defunct axis of an electrical current that had once connected the bodies. Since Earth was descending upon this axis, which became the ecliptic plane it may have experienced the reported Jovian bolts. These would still be discharging from time to time, seeking to make contact with the Sun and being short-circuited by Earth and probably other intervening bodies.

It may be surmised, too, that, upon the nova and fission of Super-Saturn (Saturn-Jupiter), not only would water and debris be discharged into interplanetary space, but also gases that would temporarily afford Jupiter its chance to earn its reputation as the discharger of interplanetary thunderbolts. Not until the arc flashes had quite disappeared, the gaseous medium had been quite dissipated, and the Earth drifted out of its binary-locked,

---

[12] Patten *et al.*, 92.

[13] *The Jupiter Effect: The Planets as Triggers of Devastating Earthquakes* (New York: Vintage Books, 1974).

conjunctive orbit with Jupiter would the cosmic lightning cease to threaten the Earth with a bolt from the blue.

Replacing the binary current and magnetic gas tube were two contemporary phenomena: the solar winds and the space plasma. The solar winds are not a current, but are unfocussed particle flows and blasts. They diffuse into space rather than concentrate upon the planets. Earth receives only a very small fraction of the solar radiance.

The space plasma that surrounds the planets is composed of dissociated ionized atoms that generally do not assemble in electrical charges.[14] It protects the Earth and other planets from inducing and suffering repeated cosmic discharges. And it prevents leakage of the remaining charge of Earth, which may indeed be building up.

However, the space sheath or magnetosphere of the Earth cannot suffice as a buffer when large or fast erratic bodies approach. In the Venusian catastrophe, cosmic lightning played a heavy role. Cometary Venus, according to Velikovsky's reconstruction, encountered the Earth in the spring of 1453 B.C. and followed roughly its orbit for some days. The comet with its millions of miles of tail appeared and reappeared as the Earth continued with interruptions its rotation. On the second approach, after six days had passed, a gigantic column towered into the sky, a pillar of smoke by day and of fire by night, as *Exodus* 14: 19 describes it.

> This stage was accompanied by violent and incessant discharges between the atmosphere of the tail and the terrestrial atmosphere. When the tidal waves rose to their highest point, and the seas were torn apart, a tremendous spark flew between the earth and the globe of the comet, which instantly pushed down the miles-high billows. Meanwhile, the tail of the comet and its head, having become entangled with each other by their close contact with the earth, exchanged violent discharges of electricity. It looked like a battle between the brilliant globe and

---

[14] Juergens, "Moon and Mars," *loc cit.*, 37 *et passim*.

the dark column of smoke. In the exchange of electrical potentials, the tail and the head were attracted one to the other and repelled one from the other. From the serpent like tail extensions grew, and it lost the form of a column. It looked now like a furious animal with legs and with many heads. The discharges tore the column to pieces, a process that was accompanied by a rain of meteorites upon the earth. It appeared as though the monster were defeated by the brilliant globe and buried in the sea, or wherever the meteorites fell. The gases of the tail subsequently enveloped the earth."[15]

I would depart from the scenario mainly to suggest that the column of smoke seen everywhere was probably a mixture of the comet's tail and the "catastrophic column" (as Kelly and Dachille picture it). The main contact between Earth and Venus occurred at this point were the main discharge left Earth carrying upwards surface material and building then and there a "great chemical factory" of Venusian and Earth raw materials.[16]

Legends from around the world describe this engagement. It is the battle between Marduk and the dragon Tiamat, between Isis and Seth, between Vishnu and the serpent (or Krishna and serpent), between Ormuzd and Ahriman, between the Lord and Rahab and, the most widely known of all, between Zeus and Typhon.

Velikovsky proceeds, after citing these legends, to place the comet Typhon in the mid-second millennium B.C., at the time of the Exodus of the Jews from Egypt. Bimson has established the pharaoh of Egypt just then as the first Hyksos King by the name of Typhon.[17] Typhon is related to Typhon (South Seas), Toufan (Arabs), and is another version of the legend of Phaeton. Legends, myths sacred scriptures, and ancient historians have been mobilized to support the theory of the encounter. That

---

[15] *Worlds in Collision*, 77-8.

[16] *Target Earth*, 189ff.

[17] "Rockenbach's *'De* Cometis' and the Identity of Typhon," I *S. I. S. R.* 4 (Spring 1977), 9-10.

Venus also suffered is logical; it still faces Earth "respectfully" in "resonance", upon its near passage.[18]

Electrical phenomena akin to lightning are associated with volcanism, earthquakes, and meteoritic phenomena, including atmospheric pass-through and impact explosions. They may also be an independent "instrument of the gods," as strong or stronger than gravitation in their effects when two dense bodies approach one another closely. Further, cosmic electricity may traverse a whole star system or planetary system.

C.E.R. Bruce of the British Electrical Association for many years sought recognition of the place of electricity and lightning in the creation and destruction of whole galaxies of the universe.[19] He described lightning discharges of $6\text{x}10^{11}$ miles in width and ten times as long generating temperatures of $5\text{x}10^8$ degrees Celsius and lasting for 10^6 years or more. The discharges occur amidst accumulations of cosmic dust.

Bruce's colleague, Eric Crew, who shares his views, has given more attention to cosmic lightning within the solar system and particularly in encounters involving earth. How he handles electrical problems of large-body encounters can be exemplified in the following passages:

> If a charged body B (such as a large comet) approaches a planet A which has an atmosphere, opposite charges are induced and the atmosphere will be pulled out towards B. This increases the voltage gradient between B and the extended atmosphere very rapidly and violent discharges may take place even though the two bodies are separated by a considerable distance. The effect is intensified if both A and B have atmospheres, and even more so if they have opposite charges.

---

[18] C. G. Ransom, *The Age of Velikovsky* (Fort Worth, Texas: LAR Co., 1976), 117 interprets several studies.

[19] His basic work is *A New Approach to Astrophysics and Cosmogony*, (London: Unwin Bros., 1974); *cf* letter of Dec. 1958 in 4 *Electronics and Power*, 669-70, "Cosmic Electric Discharges."

The effect... is to cause jet of compressed material to form and for the substance to be ejected on to the negatively charged body, or the induced negative charge.

Charges induced in the solid surface of A as B approaches will cause a ground current to flow and the resistance of its path will cause the induced charge to lag behind the line joining A and B. The electrical force will produce a turning moment on A and B and the resultant motion will depend on the direction of the force in relation to the axis of rotation of A and B. The displacement may be increased if B has a crust floating on a molten interior, as the moment of inertia of this would be much smaller than that of a completely rigid sphere, even if the possible tilting of the axis is ignored.[20]

That is, in the case of the several large body encounters of the Earth, which we think may have occurred, strong lightning exchanges took place, atmospheres exchanged in varying proportions, debris flew into space, powerful ground currents of electricity followed the point of closest contact, and these currents assisted inertial forces to push crustal sections of the Earth over its plastic mantle.

Ralph Juergens' theories of cosmic electricity have been close to the historical events proposed by quantavolutionary theorists. Intimately acquainted with the experiences and ideas of Velikovsky, he worked for many years upon the basic astrophysical problems posed by the Venus-Mars-Earth scenario, specializing in the application of electrical theory.

His primary theory deals with the source of solar energy.[21] It is in one sense non-catastrophic. It is also quite new and unaccepted; yet, as he says "the modern astrophysical concept that ascribes the Sun's energy to thermonuclear reactions deep in the solar interior is contradicted by nearly every observable aspect of the Sun." Whereas the conventional theory is that the Sun derives

---

[20] 20. "Electricity in Astronomy," *S. I. S. R.* (1976-7) I: 1,2,3, II: 1.
[21] 2 *Pensée* (1972) 3 (Fall), 6-12.

its energy from a hydrogen-fusion nuclear reaction continuing over millions of years, Juergen's theory is that the Sun's surface bears a negative heavy electrical charge, which it has gathered mostly from galactic winds and from a great many bodies brighter than the Sun, and which discharges itself upon the solar system bodies. The solar radiance that strikes Earth and causes heat is as nothing compared with the galactic radiance that strikes the Sun. The Sun's bloated atmosphere is the anode; its highest levels are of the highest temperatures, which go down, rather than up, as the surface of the giant gas bag of the Sun is approached. Hence, elaborate attempts to catch neutrinos from the Sun's "solar furnaces" as they traverse the Earth must fail; if no nuclear fusion, then no neutrinos.

The Sun's radiance, varying only slightly as its total charge varies, penetrates the electrically neutral plasma of interplanetary space, passes through the positively charged outer magnetosphere, enters a neutral zone and then a negatively charged inner zone, and finally strikes the Earth's atmosphere with warming and radioactive effects. (Jupiter does not "need" the Sun's heat; it radiates several times as much energy as it receives from the Sun.)

A great proportion of all the craters and many fissures of the Moon and Mars, and, though less visible, of the Earth, are explained by Juergens as the effects of cosmic lightning, occurring during the holocene period that we are studying.

The "plasmoids" which I referred to earlier are a type of lightning conducted to Earth as "pieces of plasma." These balanced "things" of positive ions and electrons retain their identity and appear as luminous objects of missile-like proportions. They would cause impact craters or above-ground explosions that leave little trace. A second type of primeval lightning, like that known best to us, would give clear evidence of electric scarring, whether as a crater or as a jagged crack in the ground.

The jagged cracks of clefts are called rilles and are found by the thousands on the Moon. The principal candidate for the most recent creation of rilles is the planet Mars, which, following

Velikovsky's reconstruction of events, would have happened in the period 776-687 B.C. Electrons has to be torn from the lunar crust in numbers sufficient to trigger an interplanetary discharge. The Moon becomes the cathode, Mars the anode. As the charge mobilizes quickly on the Moon, it probes along lines of weakness and explodes the surface in traveling to its discharge point. It blasts a crater as it exits into space.

Again Juergen's theory is exceptional. More favored as agents are running water (now gone), erosion by dust winds, an explosion of underground gases, and the collapse of lava tubes through which liquid lava had passed. That these alternatives to the agency of eruption of a breakdown channel raise severe problem is documented by Juergen's table presented below. It may be seen that the lightning channel eruption, not entirely unknown even today on Earth, provides a better explanation of rille characteristics.

Competence of Various Theories to Explains Sinuous Rilles of the Moon

Proposed Rille Origin Theory

Rille Characteristic	Erosion by Running Water	Erosion by Ash-Gas Cloud	Formation by Gaseous Outburst	Formation by Lave-T Collapse	Eruption of Breakdown Channel
1.Width greater at higher end	C	C	O	B	A
2.Channel sinuous	A	C	O	C	A
3.Irregular crater at upper end	B	B	O	B	A
4.End of rilles at different elevations	A	A	O	A	A

5.Outwash deposits lacking at lower end	C-X	B	A	C-X	A
6."Bridges" lacking along channel	A	A	O	B-C	A
7.On-channel cratering frequent	O	O	A	O	A
8.Channel may traverse high ground	X	X	B	X	B
9.Channel may stray from dip of surface	C-X	C-X	B	C-X	B
10.Channel may follow crest of ridge	X	X	A	B	A
11.Channel may expose numerous strata	B	B	A	C-X	A-B
12.Surface strata upturned at rille margins	X	X	A	X	A
13.Clustering of rilles	C	C	B-C	B-C	A-B
14.Young rilles may cross older rilles	C-X	C-X	A-O	C-X	B
15.Secondary rilles in rille bottoms	B	C	C	C	B

Symbols:
A. Predictable on basis of theory,
B. Permissible in terms of theory,
C. Permissible, but difficult to explain,
O. Apparently irrelevant in terms of theory X.
Evidence precludes theory.

Probably the main focus of the electrical battle between Moon and its assailant is the huge crater Aristarchus. It expresses its recency by a bare uncratered floor, by giving off light and by being intensely radioactive. The greatest concentration of lunar rilles is also located at and near Aristarchus. The light bolt was estimated by Juergens at 2×10^{21} joules of energy, "a few million times as energetic as ordinary lightning."

The likely partner in catastrophe, Mars shares gases with the Moon.

> As things stand, the situation is this: Lunar finds are rich in argon, neon, other rare gases, and carbon dioxide None of these gases is known to be present in the solar wind, nor is elemental carbon a known constituent of that medium... Precisely those gases known to be present in the atmosphere of Mars -the great bulk of which has been mysteriously "stolen" away in the not-too-distant past -are also found tenaciously held in superficial crystalline layer of the Moon's outermost blanketing materials. This would be a most incredible coincidence if the interplanetary discharges described by Velikovsky never took place.[22]

We are only in the early stages of fulminology. Edward Komarek has discovered that the effects of modern lightning are extensive. When a tree is struck, surrounding trees and vegetation are affected by structural, biological, and chemical changes for a long time to come. Lightning also may fuse the Earth around. Fused sand tubes caused by lightning and called "fulgurites" are common around the world. "In one sand-dune patch of 5,000 acres at Witsands, on the southeastern border of the Kalahari Desert, Lewis estimated that there were not less than 2,000 fulgurites. Since lightning is at the present time very infrequent in this area, some of the tubes must have been formed thousands of

---

[22] Juergens, "Moon and Mars," *loc. cit.,* Winter 1974-5, 33.

years ago.[23] The fulgurites often followed bush and plant roots. Perhaps they occurred simultaneously and were one of the causes of the desert. That all deserts, whatever their origin, may be indeed new is a question worth considering.

Lightning may descend in showers. Lightning may instantly fossilize trees; a high tension wire did so too in Alberta, Canada, E.R. Milton reports. Lightning alters C14 content in trees, hence their "age" for dating purposes.[24] Recently various theories have been offered to explain the mysterious kimberlite tubes of South Africa and similar tubes in Utah. The former are like fulgurites and are found near the great diamond fields. Probably the same electrical flows that dug the kimberlites produced the diamonds. Whether this should be called "slow lightning," and discussed in the preceding chapter, or should be discussed here is perhaps immaterial at this stage of research. The Moses Rock dike of Utah is about 4 miles long at the surface, in the shape of a hook, and about 1000 feet wide. It was forced up from possibly 200 kilometers below the surface.

Komarek has come to believe that "lightning is ecologically fully as important as such better known factors as temperature, rainfall, soils etc.[25] He does not estimate past incidences. If present lightning effects must be exponentially retrojected into the past, the world would have been significantly remolded therefrom.

Juergen's theory of Moon and Mars belongs to Earth as well. The Earth must have lunar rilles in large numbers. An unknown but considerable number of craters, "river" valleys, fractures and ravines must owe their origin not to ice, water, volcanos, or meteoroids, but to cosmic lightning. In the absence of well-directed field work, not only are their indications

---

[23] "The Natural History of Lightning," *Proc. Tall Timbers Fire Ecology Conf.* (9-10 Apr. 1964), 150.

[24] L. M. Libby and H. R. Lukens, "Production of Radiocarbon in Tree Rings by Lightning Bolts," 78 *J. Geophy. Res.* 26 (10 Sept. 1973), 5902-3.

[25] Op. cit., 171.

misinterpreted, but usually their very existence remains a surmise. The present level of electrical activity on Earth does not excite research except in imaginative minds, like Ralph Juergens, Nicola Tesla and Frank Dachille.

It is well for geologists to consider meanwhile the promise of such theories. Take, for example, the consequences of the concept that the Earth's global electric potential has not been uniform throughout its history, an idea that I repeat in this book several times; consider its consequences for another insistent idea of these pages, that geological time may be grossly exaggerated.

Juergens argues that the Earth's surface potential is highly negative and low.[26] Suppose that it is lowered further. Rampant radioactivity would occur. The half-life of every radioactive atom would be drastically reduced. Radiochronometric time would be largely erased. In the opposite case, if Earth's potential became higher and less negative, polonium, for instance, which has a short life as evidenced in the geological record by the halos it inscribes upon rock, would acquire a much longer half-life and so would other radioactive isotopes.[27]

Nikola Tesla's work is acclaimed for its genius. But some of it was unfortunately cut short by a lack of funds and his growing madness. It went largely unreported and, especially because it was so astonishing, it was and is difficult to describe and appraise. Around the turn of the century, after his dramatic successes in designing and building alternating current electric motors in the East, Tesla went West to Colorado Springs and built an extraordinary electrical apparatus.[28] He set up a 200-foot-tall mast with a metal ball on top nested in a 10-foot diameter coil. At a diameter of 80 feet he provided a second surrounding coil. These were affixed to banks of condensers. A 300-volt line from a nearby power plant supplied initial impetus to the oscillator. The

[26] "Radiohalos and Earth History," III *Kronos* I (Fall, 1977), 3-17.

[27] J. J. O'Neill, Chapter 2.

[28] 4 *Pensée* 4 (Fall 1974), 30.

magnetic field created by the current in the large coil set up an
alternating current in the central coil. Over 150,000 times per
second, a charge was sent through the Earth and back up and out
into the atmosphere, discharging as bolts of lightning.

Tesla thought that such a machine oscillating through the
Earth might be tapped at a number of place through local
receivers to supply energy for local consumption. It would be a
wireless electrical power distribution system. This naive and
astounding project has not to my knowledge been seriously
considered by geophysicists and electrical engineers in these years
of energy crisis.

Nor, for that matter, has the idea of Juergens, that "once
the curtains of *thermo*-nuclear theory are drawn aside, electrical
engineers will quickly discover that the controlled-fusion reaction
they have been seeking in vain for a quarter of a century have
actually been within their grasp for at least twice that long - that a
relatively small throughput of electrical energy will release the
pent-up power of matter on a scale far beyond the most fanciful
prediction of the late 1940's."

CHAPTER SEVEN

FIRE AND ASH

"A 'universal conflagration' (if possible) would certainly not last long enough to leave any sort of recognizable stratigraphical record, whereas a few centuries or millennia of occasional heath or forest fires, during a particularly dry spell, would probably do so without requiring any special mechanism."[1]

Even to speak of a universal conflagration gives a geologist cause to blush, as Derek Ager, the author of these lines, remarks in another context. Without the "special mechanism", forest fires, started by lightning, and volcanos, started by hot spots in the deep crust or mantle, must do the full job of whatever we see as signs of burning on Earth and whatever the ancient voices are fearfully asserting. If this were all, and it certainly is not all, we would still have to ask about lightning and hot spots; neither is a simple autodynamic mechanism, as we have seen already in the case of lightning and will see in regards to hot spots.

The legendary and early historical record is replete with assertions that global burning has occurred. Writing apparently about historical experiences, Seneca, the Roman stoic philosopher, gives a common ancient view of the holocaust:

> And when the time is come when the world destroys itself to be renewed, then (Earth, seas and life) will destroy themselves

---

[1] letter, 2 *Catastrophist Geologist*, 1 (June. 1977), 13.

by their own strength. Stars will fall upon stars. And when all material things are in flames, everything which now shines according to a planned distribution will rise up into a single fire.[2]

Of course, Seneca does not declare that a stratigraphical record will be thereafter available; the Earth is "renewed," which implies that few marks would have been left upon the rocks and no bed of ashes would have formed and persisted. Where are the ashes of single or multiple events, for that matter? Sometimes they are present, sometimes not. In certain parts of the world, extensive beds of ashes of possibly local type can be found. They are thin. We can find the ashes of Troy, on several levels of destruction, but can the ashes of the countryside around be found? If not found, does that means that Troy alone was burned, or that ruined Troy alone preserves its ashes? Paleocalcinology - such a science hardly exists - will help us someday to measure the words of Ager and Seneca.

The "ordinary" fire mechanism of volcanos and forest fires sometimes incite rains, but these are hardly conspicuous. On the other hand, the legendary coupling of fire and water is so flagrant as to pass notice, except when a progressive rabbi, for example, finds it easy to explain to his children why the heavens are of fire and water; *ish-vamayin* (fire and water) make up *shamyin* (heaven) because the ancients thought of sunlight as fire, and the rains, of course, come from the sky.[3]

G.R. Carli, writing in 1780, was already asserting that "the idea of a deluge of fire and a deluge of water was present among all peoples... This idea of fire and water... seems to recall tradition of an event of which the memory has endured. It is certainly odd that the indications manifested by a sea flood should have

---

[2] "Consolatio ad Marciam"

[3] L. Ginzberg, *Legends of the Jews,* p. 7: 15, 76; *cf.* H. Tresman and B. O'Geoghan," The Primordial Light," II *S. I. S. R.* 2 (Dec. 1977), 40, fn. 102.

suggested the idea of a deluge of fire.[4] Carli cites Clement of Alexandria for the observations that Stenelas, father of the king of the Ligurians, lived at the time of the fire of Phaeton and the flood of Deucalion. So fire and flood occurred together. Reasoning from effect to cause, Carli then assigns the coal deposits of the world to burning and water acting in quick succession, a theory now coming into prominence again. He argues that only a comet could burn up the world, drop vast amounts of water, and bring great tides at the same time. Probably this line of argument will stand up: a large body encountering Earth, even if it were not dropping water or ice, would bring both conflagration and flood. Whether it crashed or not, the effects would still be similar.

Donnelly produces an abundance of legendary accounts of the world in flames: from Druid mythology, Hesiod's Greek account, the Eddas of Scandinavia, Ovid's Roman account of Phaeton, the Meso-American Toltecs' *Codex Chimalpopoca,* the Persians' *Zend-Avesta,* the Hindus' myth of Ravana and Sita, and the legends of the Tupi, Aztecs, Tacullies, Ute, Peruvian, Yurucares, Mbocobi, Botocudos, Ojibway, Wayandot, and Dog-Rib Indians, that is from one end of the Americas to the other, and across both continents. He quotes the Gothic Surt of the flaming sword, "He shall give up the universe a prey to the flames," and also the Algonquins, whose god "will stamp his foot upon the ground, and flames will burst forth to consume the habitable land."[5]

Job of the Bible hears from a retainer that "the fire of God is fallen from heaven and hath burned up the sheep [to the number of 7,000], and the servants, and consumed them, and I only am escaped alone to tell thee." (In our days, cases of a score or more animals being electrocuted by a lightning bolt are recorded.) There begins then the woes of the stubbornly patient Job against frightful divine tests. It is only one of many references

---

[4] II *Lettres Americaines* (Paris: Buisson, 1788), 309.

[5] Ragnarok: The Age of Fire and Gravel (New York: Appleton, 1883) p. 428.

to naturally caused combustion in the Bible. The story of Job may be exceedingly old; there Elohim (Heavenly One) is addressed; it happened in full Neolithic times, perhaps at the ending of the age of predominantly Saturn worship.[6]

Later in reference to fire is the "flaming sword", east of the Garden of Eden, "which turned every way, to guard the way to the tree of life." This was after the "Fall from Grace."[7] The image of a sword in the sky may refer to the Great Central Fire of early Greek Philosophy and, as we elaborate in *Solaria Binaria*, to a then intermittent arc between Jupiter and the Sun. (We treat the image in detail in *Solaria Binaria.*) The seasons begin; it must be now the period of the gods Jupiter-Jehovah, the Jovean Age I have elsewhere termed it. In a later incident, the wicked cities of Sodom and Gomorrah are destroyed by a fall of fire and brimstone and swallowed up. Then, as earlier described, the Tower of Babel succumbed to fire in part. During and after the Exodus, repeated references to the heavenly fire are encountered. It comes in all its forms; lightning, gas blasts, burning naphtha falls. These are elaborately treated by Velikovsky in *Worlds in Collision* and by the present author in *God's Fire: Moses and the Management of Exodus*. In all, von Fange quotes 37 different passages from the Bible referring to, or prophesying, destructive fire from heaven.[8]

Both Donnelly and Velikovsky claim the myth of Phaeton - the one writer for a Great Comet of an earlier age, the other for the events of the mid-second millennium, where we too have decided to place it. The Latin author, Ovid, is the principal source of Phaeton. The Babylonian cuneiform expert, Kugler has explained Ovid's as a true history of a comet.[9] Phaeton is the inexperienced son of Phoebus who demands to be let to drive the

---

[6] Martin Sieff, "Cosmology of Job" I *S. I. S. R.* 4 (Spring 1977), 17-21, 32.

[7] *Genesis* 4: 24.

[8] *Op. cit.*, 136-7.

[9] L. C. Stecchini in A. de Grazia *et al., The Velikovsky Affair,* 2nd ed. (London: Sphere, 1978), 120ff.

chariot of the Sun one day. He prevails, but loses control of the steeds and burns up sky and earth. The constellations are disturbed. The flames turn whole nations into ashes. The ground bursts asunder, the rivers dry up. Smoke billows bring darkness to the world. The ocean shrinks. Ashes cover the Earth.

Mother Earth trembles and sinks below her usual place. She pleads with Jupiter. "If the sea, if the earth, if the palace of heaven, perish, we are then jumbled into the old chaos again. Save it from the flames, if aught still survives, and preserve the universe." Jupiter responds by demolishing Phaeton and the chariot; Phaeton, his yellow hair streaming in flames, is hurled to the earth like a falling star.

The Sun, Father of Phaeton, mourns as in an eclipse. The earth was lit only by its own flames. He would not resume his daily journey until all the gods supplicated with him. The days appeared once more, and Jupiter restored order and life to the heavens and earth.

No one disputes the fact that the earth has been badly burned. Provided, of course, that the statement is properly qualified. The ocean basins are of melted rock; they are fashioned almost entirely of basaltic lava. Ocean abyssal sediments are thin and loose, and composed of organic and dust fall-out for the most part, including some products of combustion.

Of the continents, part of the surface that is exposed to view is igneous, a product of old or new melting. Another portion is metamorphic, a word meaning rock transformed mostly by heat and pressure, both old and new; this emerges from both sedimentary and igneous rocks. (It is significant that whereas observers are compelled by the sight of volcanism to say that *some* lava beds are new, they are reluctant to name any metamorphosis of rock that has taken place very recently.) Igneous rock, if not witnessed as it forms, is also invariably given old dates.

A Phoenician vase of around 1500 B.C. was found embedded in the copious lavas of the Jezreel Valley of Palestine, where volcanism had supposedly ended in prehistoric times.[10] At Nampa, Idaho, in 1889, a well-worked human image carved of pumice stone was found amidst coarse sand of an old lake bed beneath 300 feet of alluvium, lava and clay.[11] The lava had been and still is classified as late tertiary or quaternary, a million or more years before mankind is supposed to have arrived in America. The Nampa image, now lost, is disregarded; given the strong testimony concerning it, one may wonder how much of natural and human history would be erased under the same strict rules of appraisal.

Granites are the continental structure: nearly all come from an ancient cooling of molten rock. They rest below the recent igneous rock, metamorphic rock and sedimentary rock on all continents. We have direct information downwards only on a couple of miles of crust; it is considered that granites carry on down to a basalt not unlike that of the ocean bottoms. When and how the granites formed is unclear; their chemistry is distinctive.

A final part of the continents is covered by sedimentary rock. Sedimentary rock is formed from transports of materials by wind, water, and ice. Donnelly argued that much of the clay, gravel, and till that composes it descended from a cometary train recently in " the age of fire and gravel," rather than from other rock being ground up and spread around by moving ice.

From the standpoint of human primevalogy, the uppermost layers of rock and debris are highly important. These are usually termed unconsolidated, or loosely consolidated, or aggregated, or conglomerate. High energy expressions of "earth, air, fire and water" will produce large quantities of this material and their origins, dating, and relation to the biosphere are hard to discern.

Everywhere one is likely to find soil, a catch-all work for any layer from the thinnest film up to a few meters in which life

---

[10] Velikovsky, *Earth in Upheaval*, 197-8.

[11] See W. R. Corliss, ed., *Ancient Man* (Glen Arm, Md: Sourcebook Project, 1978), 457-60, from G. F. Wright, 11 *Am. Antiq.* (1889), 379-81.

forms take hold or dwell. Fossil soils often rest between layers of the several types of rock.

Besides the soil, too, exist metals, soda ash, peat, various ashes, coal, oil, natural gas, salt, and other deposits. Some of these are thermal products. Billions of tons of glassy microtektites are strewn over the globe; whatever their origin, they may have fallen in as hot rain on land and sea. Layers of ash are found over vast stretches of the oceans bottoms, perhaps everywhere, since the searches have just begun. Ash is fairly distinguishable; it is more difficult to detect whether the much more profuse sedimentary clays are not themselves in part the products of combustion, carried over and dropped upon the sea or drained off the continents onto the slopes and shelves.

On the land, too, ashes mix readily with soils and detritus to form clays. It is not impossible to detect calcination in soils and clays, but the subject has attracted few geo-chemists. Soils and young marine sediments of northeastern and offshore America reveal, under chemical analysis, evidence of a fiery origin in that they contain polycyclic aromatic hydrocarbons.[12] These are carcinogenic and mutagenic. It is possible that their incidence is world-wide. If so, it would indicate that the whole world suffered one or more fall-outs of burning or burnt material. The burning could have been caused by super-terrestrial impact explosions or gases. Or the products of atmospheric fire (burning naphthas and brimstone or sulphur, as the Bible would have it) descended. Or both might have happened. The authors of the report cited here considered the effects to have been possibly produced by giant forest fires and air transport, and unfortunately, did not consider exoterrestrial origins of the widespread combustion products, or for that matter, of the fire that consumed the biosphere. T.M. Harris,[13] in describing "Forest Fire in the Mesozoic," found much fusain in many layers at many places, including the deltas of

---

[12] Blumer and W. W. Youngblood, "Polycyclic Aromatic Hydracarbons in Soils and Recent Sediments," *Science* (4 Apr. 1975), 53.

[13] 46 *J. Ecology* 2 (1958), 447-53.

Greenland and Yorkshire. The admission that cosmic lightning and cosmic fire were prevalent at quantavolutionary points is avoided by placing layers of time between layers of ashes.

We cannot readily separate ash from human, at least not without chemical tests of a degree of sophistication hitherto undeveloped because of the theory of gradual accumulation of soils over long eons. Commenting upon Ager's search for ash, Hans Kloosterman speaks of a "black horizon" of soil "that seems to have been covered with sediments immediately after its formation," this in Derek Ager's work; and despite Ager's retreat into what Kloosterman calls "crypto-uniformitarianism," the latter defends the idea that there might be identified only" one enormous forest fire, which is moreover correlatable from Southern England to the Great Lakes of North America. Doesn't that sound somewhat like a universal conflagration?[14]

Kloosterman goes on to discuss the "dark bank" he witnessed in Brazil. Despite deliberate tropical burnings that are regular and go back hundreds of years, "no charcoal-rich layer is formed anywhere; the ash is incorporated into the human layer or washed away."

Whereupon, this author adds evidence by Wendorf, Said, and Schild that in Egypt, at claimed dates around 10,550 B.C. a burnt layer appears over a large region of the Upper Nile Valley, which the investigators guess to have been caused by brush fires, but which to de Grazia seemed to have been associated with holospheric catastrophe and world-wide conflagration and/ or incredibly heavy ash fall-out.

J. Lamar Worzel of Lamont Geological Observatory (Columbia University) published important findings in 1969, entitling them "Extensive Deep Sea Sub-Bottom Reflections Identified as White Ash."[15] The analyzed deep sea cores came from the east-central Pacific, from Mexico to Peru, an area of a million and a quarter square kilometers. The piston-corer was not

---

[14] 2 *Catas. Geol.* (1977), 14.

[15] 43 *Proc. Natl. Acad. Sci.* (15 Mar. 1959), 349-55.

long enough to probe the nature of echoes, possibly representing other ash layers, obtained from below 78 feet.

The layer of ash measured differently in the various drilled cores but ranged from 5 to 30 cm of thickness. "Since the layer is fairly near the surface and is not discolored and contains nothing but the glassy ash material, it must have been laid down fairly quickly." At depths of 1000 to 3000 fathoms, the ash was under great pressure, also the original atmospheric and hydrospheric conditions might have dissipated and disintegrated some of the initial deluge.

The fall was so heavy and quick, "that it may be difficult to ascribe it to the Andes... Perhaps sub-bottom echoes from other areas can also be correlated with this white ash layer. If so, it may be necessary to attribute the layer to a world-wide volcanism or perhaps to the fiery end of bodies of cosmic origin."

In a critique of "The Significance of the Worzel Deep Sea Ash," Maurice Ewing, Bruce C. Heezen and David Ericson, also of Lamont Geological Observatory, advanced reasons why the white ash layers might be found elsewhere: citing the sounding of the vessels *Albatross, Galathea,* and *Verna* from different part of the world, they conjectured that the same sub-bottom echoes and possible ash layers existed over much of the globe.[16] Sedimentary mixing would often subdue or annul the echoes.

> The ash deposits observed by Kuenen and Need and Bramlette and Bradley were mixed through a column of sediments several times the thickness of the original ash bed. In addition to this mechanical mixing, solution may vastly alter the sediment before permanent burial is accomplished. Devitrification and alternation, proceeding at rates dependent on the environment, may transform an ash bed into products whose origin is not readily recognized.

"Extensive ash layers are now recognized in continental areas throughout the geological record," they point out, citing C.

---

[16] 45 *Proc. Natl. Acad. Sci.,* 351-61.

S. Ross. They declare too that "ash of similar composition has been logged in boreholes in many of the dry lakes of the western United States." (These dry lakes are all very young, post-glacial.) As mentioned, Wengret and others showed extensive ashes and calcination in the Nile Valley to which they assigned fairly recent ages; one can only wonder, for similar reports simply are not registered generally, how many cuts and profiles around the world reveal such calcination and why, as has been observed, the older rock-strata show almost no calcination - except that metamorphics, granites, igneous rocks, and perhaps limestones themselves are sign of heavy thermal activity.

Until very recently, geologists, like archaeologists, have been incurious about thin beds of ashes. An alerted surveyor, Heladio Agudelo, wrote this author (Oct. 4, 1977) saying, "In my work... while helping build a new street I noticed a black line in the gravel formation." It was a "one-inch-thick black line in otherwise homogeneous alluvial(?) formation." Within several weeks it became invisible due to erosion "but it will take no bigger a tool that a hand shovel to expose it again. This is in Londonderry, N. H., no more than an hour's drive from Boston Airport." Thin beds of ashes represent enormous fire, the effects of ordinary forest and construction fires disappear quickly.

The Ewing group, quoted above, comments that "Murray and Renard identified volcanic particles in practically all of the *Challenger* surficial samples of deep sea deposits, demonstrating that volcanic detritus is an important component of modern deep sea deposits throughout the world. They suggested that the abyssal clays are largely the result of alternation of volcanic ash." Later on, the authors themselves conclude: "It is necessary to study the alterations of fine pyroclastics in the sea and to set up criteria for recognition of the alteration products formed under the full range of environmental conditions." (I proposed such procedures for heavy combustion products in many archaeological levels, exemplified in the "Burnt City" of Troy IIg).

The Worzel ash consists of colorless shards of volcanic glass without sorting by particle size. "In all important respects it is similar to material which has been classified as volcanic ash in the deep-sea deposits of the world." Analysis of the surrounding sediment in the Worzel cores indicates that the bottom waters "must have contained some oxygen" and that the sediments "probably represent no more than 100,000 years and conceivably far less." Whatever the date, mankind was very much present and concerned. Certainly years of darkness, disease, distress and terror occurred around the world with this deluge.

"The ash is entirely unlike material described as meteoritic dust. Only the wide geographic extent of this layer suggests any source other than volcanic eruptions. "To this proposition, with which Worzel might differ, given his quoted remarks, one might take exception. "Meteoritic dust" is too imprecise a term to use in argument, considering that we may have to consider lunar material and the 50 to 150 million mile tails of comets. If, as the authors grant, there is a need to examine and re-examine numerous types of sediment, there is also a need to distinguish, if at all possible, "cosmic dust" from "terrestrial dust". If world-wide volcanism can only originate from an externally interrupted motion of the Earth, or from a titanic large-body encounter, then "terrestrial dust" is also an effect of exoterrestrialism.

Heezen and Hollister write that the Krakatoa eruption of 1883 "produced an insignificant sprinkling of ash" by comparison with six great eruptions of the past "million" years that blanketed thickly the ridge and basin of the Java Trench. "Indeed how great must have been the earlier eruptions if the greatest known to man was too small to produce significant record. Powerful eruptions in the Japanese, Kurile, and Aleutian arcs have produced so much ash that these airborne volcanic products dominate the scenery of the Northwest Pacific in a belt almost 1000 km wide.[17] (We note again that human were already present during these great ash storms and presumably coining legends.) Heezen added elsewhere the Mediterranean Sea bottom as a depository of several heavy ash layers.

---

[17] B. Heezen and C. Hollister, *Face of the Deep*, 476-8.

Walter Sullivan describes " a succession of ash layers" encountered on the edge of the continental slope before striking the lava basalt of the true ocean bottom. Might this not indicate that the continental slope was laid down subaerially, collected its sedimentary and ash layers and was then inundated by the ocean? Drilling in the Atlantic "has begun to paint a picture of the awesome events that accompanied the birth of that ocean.[18]

To all of the ash layers referred to, and much more exist, one must accredit exponential ash storming that has dropped to relatively tiny amounts during historical times. Max Blumer led the discovering and detailing of paleochemicals in soils. His group found polycyclic aromatic hydrocarbon from pyrolysis in many places and wondered at the great conflagration of ancient times.

Blumer even suggested that these carcinogens and mutagens played a role in the mutation of species,[19] Beadle has explained the origins of a peculiar ancient Mexican corn as a case of thermal polyploidy, genetic gigantism brought on by subjection to environmental heat, a feat he duplicated in the laboratory.[20] Gigantism, and possibly dwarfism, and associated polyploidy in plants and animals have, then, as a possible contributing cause, heat stress.

The work of Edward Komarek Sr. on fire and lightning stresses the role of these agents in prehistoric as well as modern times. He regards many species of plants and animals as fire-prone, including mankind. They have become adapted at some time in the past to naturally caused fires and are inclined to make the best of it. Komarek has been active in instituting controlled forest fires to imitate natural fires which strengthen growth, rather than weakening it as is popularly believed; the observed quick recovery from fire is one more indication that the great conflagration can occur without citation in the geological record.

---

[18] Walter Sullivan, *Continents in Motion,* (N. Y.: McGraw-Hill, 1974), 147-8.

[19] 234 *Sci. Amer.* 3 (1976), 45.

[20] *Cf. New Scientist* (12 Nov. 1981), 433.

Fires in prehistory may have been much more extensive than they are today and their part in animal adaptations may have been considerable. He quotes Harris on "Forest Fire in the Mesozoic," where the author describes vast fusain deposits, identifies them as fossil charcoal, and says "the objection usually used against accepting fusain as charcoal produced by fire is that there is too much of it and in too many layers. It would make the past a 'nightmare.'" Fusain is intimately associated with coal beds and thus reinforces the Carli and Velikovsky thesis, seconded by Francis and Cook, that coal is what remains of a bulldozed burning biosphere, buried deeply and tamped down promptly by successive waves of other material.

Animal fossils are sometimes found amidst ashes. "Ancient Ashfall Entombed Prehistoric Animals," heads an article by M. H. Voorhies[21] where a Middle Miocene prodigious ashfall over hundreds of square miles snuffed out over 200 species at one waterhole alone. When geologist Louis Lartel was excavating Cro-Magnon man (fragments of 15 individuals) near Les Eyzies-de-Tayac (Dordogne, France) in 1868, he uncovered five archaeological layers that had been covered by ash. The Upper Paleolithic was an age of ashes too. The glacial ice, where such great sheets existed, must have been covered with ash, if today ice drilling reveals no heavy ash fallouts it must mean that the caps are exceedingly young.

Erich von Fange has come upon many a recent report of burnt sites. He mentions towns whose calcinated ruins resemble strikingly what one can read of Troy IIg, "The Burnt City," when reexamining the extensive records of its excavation. His cases come not only from the Near East but also from Western Europe and Britain, Central Africa, the Gobi Desert of Central Asia, the Mohave Desert of the American Southwest, India, Cusco (Peru), and Cete Cidades (Piaui, Brazil). The "Cities of the Plain," including Sodom and Gomorrah, flourished in an area that became a scene of utter devastation to this day, over four

---

[21] *Nat. Geogr. Mag.* (Jan. 1981), 66.

thousand years later. All that grew in this Dead Sea Rift area, all who lived there, all that was built there, were wiped out by falls of fiery debris and an upheaval of the earth; asphalt, salt and sulphur are abundantly displayed now.

The prophet Isaiah (2: 10, 2: 19) has people rushing into holes and caves when the Lord in his majesty "ariseth to shake terribly the Earth." The lowland Indians of Peru put pots on their heads and run for the hills when the earth quakes. So do Kamchatka Siberians. Against softly reasonable explanations of such behavior stand grimly reasonable ones, that in times past, earthquakes and fall-out and heavy tides came together.

Boiling seas have been observed near subterranean volcanos. That large stretches might boil is arguable. Velikovsky adduced legendary accounts around 1450 B.C. Thus, quoting the *Zend-Avesta*, "The sea boiled, all the shores of the ocean boiled, all the middle of it boiled," when heated by the star Tistrya (Venus).[22] Carl Sagan claimed a total boil-off if the Earth abruptly stopped rotating,[23] but a slowdown would bring limited surficial boiling.

Perhaps the oldest radiocarbon dates of a burnt city come from Dilmun (modern Bahrein) at the North end of the Persian Gulf.[24] There the lowest level is calcinated. It is located below a thick wall. The burning occurs over the whole area of settlement. The debris contains burnt bitumen and "black masses," producing radiocarbon dates of 19,000 to 36,000 B.P. (in my opinion, valueless). There are "strange" sand "fill" intrusions at this level that carry various artifacts and bits of copper. Below the calcination occurs a meter of sand with shards, and below that, bedrock.

The ruined mysterious city of Tiahuanaco, 18,000 feet high in the southern Andes mountains, seems to have had port

[22] In Asimov *et al.*, *Scientists Confront Velikovsky*, (Ithaca: Cornell U. Press, 1977); but see Shulamit Kogan, ltr. *Physics Today* (Sept. 1980), 97-8, repr. VI *Kronos* 3 (1981), 34-41.

[23] *Worlds in Collision*, 92

[24] G. Bibby, *Looking for Dilmun* (N. Y.: Mentor, 1969), 167-9.

installations and to have been connected with Lake Titicaca, to the north, which contains living species of oceanic type. Tiahuanaco stands on strange ground. The climate is dry, the foliage is scanty, the weather is cold, the neighboring people wretchedly poor and few in number. The top soil of the plateau is a two-foot dry deposit, now soft stone. Below it stands the lignite of charred tropical plants. Next come a layer of ash deposited amidst rainfall, and then appears an alluvial deposit. All can be considered short-term deposits of the lowlands. Combustion obviously played a large part in the happenings. In such a place, one would normally expect merely a scanty soil, windswept, on rocky ground.

Poznansky, the major investigator, detected three cultures and three natural destructions.[25] He allows Tiahuanaco a very old age, calling it the oldest known city in the world. Bellamy believed it to be a city that dwelt beneath a terrible sky, with first a satellite that closed into Earth and crashed later, and then a newly captured moon circling above.[26]

I argue elsewhere for a single event, that the Moon erupted from the Pacific Basin to occasion the destruction of Tiahuanaco; at the same time, it was elevated, but not to its present height. Another elevation might have followed in the second millennium B.C. whereupon the city was left in ashes and ruins. That is, an early Tiahuanaco might have flourished before the new-born Moon. Peruvian legend has it that before the Sun and Moon were made, Viracocha, the White One, rose from the depths of Lake Titicaca and presided over the erection of the cities on its islands and Western shores.

The conventional view classifies Tiahuanaco as pre-Inca and places it therefore in the present era. It was never an important Inca site and its resemblances to Inca culture are no

---

[25] Arthur Posnansky, *Tiahuanaco, The Cradle of American Man,* (N. Y.: Augustin, 1958).

[26] *A Life History of Our Earth* (London: Faber and Faber, 1951); *Built Before the Flood* (London: Faber and Faber, 1947), especially on Tiahuanacuo.

more than its resemblances to the earliest Ecuadorian or Mexican cultures or to the Easter Island complex for that matter. Its astronomical observations carved upon stone gates were magnificent,[27] the Incas were underdeveloped by contrast. Tiahuanaco may then be the oldest of fire-devastated ruins.

Examples of the latest possible world conflagration can be found in Greece. These would be in the -776 to 687 B.C. period, by Velikovsky's chronology, which I accept; owing to a major shift in time reckoning, most of the great destructions in these areas that has been assigned to around 1200 B.C. is now scaled down to the eighth and seventh centuries. The new great destructive sky god was Mars in many forms.[28] It was now that King Nebuchadnezzar ravaged the Near East believing himself to be the personification of the planet-god Mars-Nergal: " I am Nergal. I destroy, I burn, I demolish, leaving nothing behind me." (He was, of course, not nearly so effective as his model, and was ultimately killed).

The same age began with the downfall of the Mycenaean culture. The evidence of the destruction of Mycenaean culture by fire has been available for a long time, but put aside for lack of a cause. A. H. Frickenhaus, a German excavating long ago at Tiryns, described how he had located a burnt Mycenaean palace with a new Greek-style temple built right over it.[29] At Pylos, not far away and of the same period, fire was manifest everywhere, burnt rooms, burnt oil, fused metallic implements, scorched pots.

In his analysis of the Pylos event and others, Isaacson has substantially proven the correctness of the revised dating.[30]

---

[27] *Cf.* H. S. Bellamy and P. Allan, *The Calendar of Tiahuanaco* (1959) and *The Great Idol of Tiahuanaco,* both published by Faber and Faber, London.

[28] *Worlds in Collision,* Part II; Chaos and Creation, 235-46.

[29] August H. Frickenhaus, I. *Tiryns* (Athens, 1972).

[30] In 3 *Pensée* 2 (Spring-Summer, 1973), 26-32 and 4 *Pensée* 4 (Fall 1974), 5-20. See also my study: *The Disastrous Love Affair of Moon and Mars* (1983).

Apparently, the Mycenaean (Greek) Age changed into the archaic Greek period amidst general conflagration. But so did age upon age before, both geologic and cultural.

I have not mentioned thus far the catastrophes that ended the Old Bronze Age around 2300 B.C. According to Schaeffer: "There is not the slightest doubt that the conflagration of Troy IIg corresponds to the catastrophe that made an end to the habitations of the Old Bronze Age of Alaca Huyuk, of Alisar, of Tarsus of Tepe Hissar, and to the catastrophe that burned ancient Ugarit in Syria, the city of Byblos that flourished under the Old Kingdom of Egypt, the contemporaneous cities Palestine, and that was amongst the causes that terminated the Old Kingdom of Egypt." Egyptian Old Kingdom tombs are generally marked by signs of conflagration, Emery has discovered.[31] A great many places elsewhere must have become heaps of ashes as well.

At Anemospilia, Crete, a small north-watching hillside temple was excavated.[32] Of four skeletons unearthed, one was identified as a priest, a second as a youth of 18 who had just been sacrificed. (" The only remains of Minoans heretofore unearthed had been recovered from tombs.") He had been trussed and laid upon the altar. The sacrificial knife lay on his bones. The priest sprawled in an agonized posture nearby, intimating death by sudden collapse of the stone structure. The other persons were perhaps attendants and killed simultaneously. Earthquake was presumed. The youth, an analysis of his bones revealed, had died just before the disaster, half his body had been drained of blood, the upper-most half. Then a fire had swept the premises before the bottom half was drained.

The fire is attributed to tipped oil lamps, but, following the logic employed in my study of Trojan fires, I would suspect an external source, possibly drifting flammable gas pockets, for an ordinary fire would consume the bones, in the unlikely event it

---

[31] *Archaic Egypt* (Penguin Books, 1961), 71-3, 92, 97.

[32] Y. Sakellarakis *et al.*, "Drama of Death in a Minoan Temple," 159 *Natl. Geog.* (Feb. 1981), 208-23.

could start up in the first place. Perhaps it was not fire, but a scorching blast, that preceded or succeeded a seismic shock. Why, also, were there only three persons in the temple? Were all other people in hiding while the heroic priest and his staff went to offer the sacrifice? Why did not the people return to dig out the bodies and restore the temple? Burial was a holy obligation; an unburied priest would be a holy horror. A presumption of total desolation and death over a considerable area arises.

John Bimson, describing the recently famous Ebla excavation in Syria, finds that the proto-Syrian culture datable sometime after 2300 B.C. by Schaeffer's scheme was destroyed by seismism and fire.[33] As I stated on several occasions, this finding was predictable, for all known settlements of the time were similarly struck. It would fit among the Mercurian disasters described in *Chaos and Creation.*

The specific origins of burning are usually in doubt. Catastrophic combustion is a product of earthquake-caused fire, of fissure and cone volcanism, of lightning, of phaetonic atmospheric penetration, of typhonic impact explosion, of fall-outs of combustible materials that are ignited in flight, including gases and naphtha. Donnelly, a century ago, speculated convincingly upon the fall-outs of gas clouds from the tails of comets. His *Age of Fire and Gravel,* the culminating devastation of all human time, was pictured as a burning of great patches of the world from carbureted hydrogen.

Some kind of exoterrestrial gases are often to be suspected in great prehistoric and ancient fires. A combination of gases and lightning, if the gases are not too concentrated, will bring masses and sheets of flame, rather than explosions. I have read few convincing reports of gas and fire explosion - the Pestigo Forest Fire and the Tunguska blast, both modern, being the type

---

[33] 33. "Ebla Reconsidered," V. S. I. S. R. 2 (1980-1), 37-9; Matthias, *Ebla: An Empire Rediscovered* (London: Hodder, 1980).

of event to look for, nor have I read a report of excavation revealing an exploded city, unless some of the settlements that seem to have been wreaked simultaneously by fire and earthquake do not in fact involve earthquakes. Probably a strict investigation would discover any such explosion affecting human settlements, but the geologic causes would have to depend for evidence upon legends. A gas explosion and flash fire would leave practically no traces within a few years of occurrence.

Volcanos are more obvious sources of fire. Many a volcano has claimed its Pompeii and Herculanum. It has worked its way with mud and lava flows, ashes lofted nearby and afar, and noxious gases. It may be fissure or cone, extinct or live. One of the oldest pyramids, that of Quiquilco in America, stands almost buried in lava. It is probably as old as the oldest pyramids of Egypt.

When a great may volcanos erupt simultaneously, the effects upon settlements are more than proportionate to the effects of a single eruption. Inasmuch as layers of ashes have been discovered over millions of square kilometers of the ocean bottoms, it has to be granted that the same ashes fell upon the land and the biosphere, and upon human settlements, if such existed.

Ashes are apparent to an alerted observer when they lie in belts and heaps. But material dissolution occurs, the destruction and effacement may involve additional forces that remove the ash, incorporate it, or dissolve it. Ash may be washed away by tides, blown away by hurricanes, or subjected to these forces gradually. It may turn to clay, impervious to all but the most exacting chemical analysis and electron-scanning microscopy.

The tephra of Thera-Santorini, falling from the plinian explosion of 3000 B.P. (a less likely date is 3500 B.P.) is found in heaps, but also in microscopic form amidst debris that may or may not have been of the same occurrence, in widely separated locations. In Thera itself, one bluff is composed of pumice, the next one, higher, contains none. In Kos (Greece) at one place, 40cm of Thera ash is visible; at many other outcroppings of subsoil in Kos, no ash is visible. So in Crete, so also Anafi. Common clay is abundant on land and on sea bottoms. It contains not only the material of slow erosion and ice age drift but of sudden exponential erosion and ice cap avalanche, of volcanic ashes, and of meteoritic and other exoterrestrial fall-out.

To conclude these pages of fire and ashes, we may assert once again that the gradual processes of today were preceded very recently by quantavolutionary processes. More and greater fires burned more widely in the world than during the past two thousand years. More blankets of ashes were laid down. More settlements were ruined. That the fires and ashes may often have had ultimate exoterrestrial causes is probable. Until the basic issue of geological chronology is settled, we are not prepared to affirm that the 85% of the exposed Earth's crust which is of igneous rock is all nearly as young as the ash levels, but the possibility is real. From the standpoint of theoretical mechanics, the Earth's ash layers and all the components of soil and clay originally containing ash may have been the fall-out of global volcanism which produced the igneous rock. But we have yet not covered enough ground in our tour of the Earth's features to determine the matter. And perhaps in the end, we shall still be uncertain.

PART TWO

EXOTERRESTRIAL DROPS

Reasoning from exoterrestrial exploration and the shapes and chemistry of the Earth's crust, and with a strong assist from early human legends, the origins of some materials of the Earth's surface is assigned to fall-outs. They range from invisible gases to giant meteoroids. The geophysical column commonly displays exoterrestrial products and their effects.

We stop short of using exoterrestrial fall-out like a magician pulling everything out of a hat. If still we appear extreme, it is well to recall that the physicist Alfven theorized that the Moon was at first a much larger aggregate which broke up, showering upon the Earth the whole of our continental masses.

CHAPTER EIGHT

FALLING DUST AND STONE

When Alexander the Great asked some Celtic leaders in 325 B.C. what they most feared, expecting them to reply Alexander himself, they said it was that the skies might fall. Somewhere along the line of history, this story lost Alexander but became attached to the Celtic Gauls; the schoolbooks universally read by French children until lately began by telling them that their earliest ancestors were the Gauls whose eyes were blue, who feared nothing but that the heavens would fall on their heads, and whose huts had holes in their roofs to let out the smoke. Were the Gauls known for nothing else? The naive, simplistic image lets the children be amused. But the insistence with which this particular canard is purveyed says something about the fear of falling skies, which absurdly seems to grip even the savants in their obsession with foisting it upon their perceived ancestors and their descendants.

In the most ancient legends it is common to find references to more than comets and deluges of water. Deluges from the sky consist also of dust, loess, stones, glass, tar, oil, salt, gold, iron, ashes, carbohydrates - all of them sometimes hot and sometimes aflame. They are invariably tied to catastrophes. Donnelly collected some of the stories:

> We read in the Ute legends... that when the magical arrow of Ta-wats struck the sun god full in the face, the sun was shivered

into a thousand fragments, which fell to the earth, causing a general conflagration."[1]
[One is cautioned to read "sun" with reservations; foreigners who pass along legends are likely to make the word "sun" out of any brilliant great body in the sky. That the Sun is only one of such historically manifested bodies is the thesis of a number of studies.]

Further:

It is a belief in many races that the stone axes and celts (chisala) fell from the heavens. In Japan, the stone arrow-heads are rained from heaven by the flying spirits, who shoot them. Similar beliefs are found in Brittany, in Madagascar, Ireland, Brazil, China, the Shetlands, Scotland, Portugal etc.[2] (And the Greek Apollo is famed for discharging clouds of arrows and plagues from afar).

Also from the Aztec prayer to Tezcatlipoca:

Hast thou verily determined... that the peopled place become a wooded hill and a wilderness of stones?... Is there to be no mercy nor pity for us until the arrows of thy fury are spent? Thine arrows and stones have sorely hurt this poor people.[3]

And, of course, the Bible (Deuteronomy xxviii)

The Lord shall make the rain of thy land powder and dust; from heaven shall it come down upon thee, until thou be destroyed..."

Thus, in *Deuteronomy;* but more too in *Joshua x:*

---

[1] *Op. cit.*, 258.

[2] *Ibid.*, 258-9.

[3] *Ibid.*, 186-90.

And it came to pass, as they fled from before Israel, and were
in the going down to Beth-horon, that the Lord cast down great
stones from heaven upon them unto Azekah, and they died:
There were more which died with hailstones than they whom
the children of Israel slew with the sword.

This, it may be recalled, was the day when the Sun "stood
still", a swing-back of cometary Venus, according to Velikovsky,
52 years after Exodus, and at the least he shows that this hail was
not ice but of stone.[4]

The student of geology today is realizing that what falls
from the sky is not only nickel, iron or stone fragments. There is
a continuity of materials. P. M. Millman writes:

... physical theory, applied to the observed heights, velocities,
deceleration, and luminosities, indicates that in most cases the
mean densities of the meteoroids may be below that of water
and that they have a fragile structure with a tendency to crumble
and fragment. A small fraction, probably 1 or 2 percent,
consists of denser, compact particles corresponding more
closely to meteorites. These latter are either nickel-iron, with
densities about eight times that of water, or heavy stone, with
densities between three and four times that of water.[5]

Where does all the dust and stone rest today? It may be,
as Donnelly said it, the main constituent of the so-called glacial till
and in heaps called mistakenly glacial moraines. It may be in much
of the clay of the Earth, in red loams of many countries, in abyssal
clay of varied red and blue hue. The geologist Johan Kloosterman
tells a story from Brazil:

Early this year, Professor Doeko Goosen in Enschede,
Holland, told me that there was something odd about the iron
content of the early-Holocene coversands of the Netherlands.
These sands are thought to have been formed through a

[4] *Worlds in Collision*, 42-3, 51-3.
[5] "Meteor," 12 *Ency. Brit.* (1974), 36.

combined fluvial and aeolian activity. But in many of their soils, the amount of iron is much too high for such an origin. Moreover, the present loss of iron by seepage water, observable along many ditches, demonstrates that the original iron content must have been higher still. Weathering of minerals (loss of Silica and relative accumulation of iron) does not satisfactorily explain this anomaly. Could the iron have come from above, as a sort of ferruginous loess?

A few months later in Mato Grosso, Goosen's remark led me to look more closely at laterites profiles. I noticed an inch-thick layer of hard laterite between two layers of unconsolidated gravel; its undersurface was *smooth:* it had obviously been formed prior to the deposition of the top gravel. I traced the layer for several kilometers, and later found it in places tens and even hundreds of kilometers away, on different geomorphological levels. The only possible explanation for these observations seemed aeolic precipitation on a barren, moist surface.[6]

Doeko Goosen has gone well beyond the ordinary unsatisfying explanations of soil formations commonly employed." Not so long ago soils were considered to form in materials derived by weathering of the underlying rock. Over several decades there has been a growing recognition that much of the mantle of soil is allochtonous."[7] But where does it come from? Few are the regions where soil can be shown to have aggregated as humus from the vegetation above. The large areas of Europe and Asia covered with loess are now considered all or in part by Russian scientists as non-aeolian. This is conveyed forcefully to their minds by the presence in the loess of numbers of angular stones. Promptly we are recalled to the pages of Donnelly's old book where he insists on the exoterrestrial origin of the angular stone typical of "glacial till" and of loess.

---

[6] I *Catas. Geol.* (1976).

[7] Unpubl. mss., 1980, Soil Dept., ICT, Entschede, Netherlands.

Now Goosen advances the argument with respect to the soils that sit atop the loess. He claims that humus does not form except in waterlogged area, presently and historically unlike the Kazakhstan (U. S. S. R.) area he discusses. Furthermore, the "Chernozems," the aforesaid soil, is rich in hydrocarbons. Presumably, some of it was combusted, too. The incident of its formation was most likely a cometary encounter.

Goosen goes farther, in what approaches in fact a general theory of soils formation. Slickensides (common in cracked vertisols and related to mass movements of ash and clay), and latosols, along with much other soil with a high iron content are assigned catastrophic origins, with tides and floods in the first case, and heavy hematite exoterrestrial fall-out in the second as the mechanism.

"Dust thou art, and to dust thou shalt return." From dust to dust, goes the pathetic saying about man's fate. "To dust" we know from experience. "From dust" -what does geology say? Nothing, of course. Does mythology have something to say? Yes. One of the most popular creation legends has man being made from clay, Hebrew *Genesis*, for example. The Greek Promethean creation, for another. Moreover, the "Cree Indians believe that the flesh of those who perished in the waters of the Deluge were changed into red pipe-clay. Similar myths or echoes of myths are found in the tales of almost every nation. "So reports Bellamy[8] "We are all made from common clay," say egalitarians.

Why clay? Because, according to ordinary surmise, clay is malleable; early people would make images of clay and, projecting their desire for omnipotence onto the gods, would imagine that the gods could fashion real people from clay. Is this adequate reason? Is there additional reason to believe so?

Bellamy also asserts that the enormous and unfamiliar loess deposits, which must have formed such a striking feature of the new Earth, were regarded by the survivors as the dissolved

---

[8] H. Bellamy, *Moon, Myths and Man* (London: Faber and Faber, 1936), 241, 243.

bodies of their unfortunate brothers and sisters.[9] It is noteworthy that loam deposits do surround the remains of Peking man at Choukoutien and human tools of the Lower Paleolithic in Europe and Tadzhik (U. S. S. R.) The loess is a fine undifferentiated loam of brownish or reddish color that makes eerie standing images by its vertical pipe structure when eroded. The logical divine action, in magical theory, is to create people from the same material, especially if its origin is celestial.

To conclude our reasoning, the myth and the magical reasoning press a hypothesis upon the geologist. The origin of loess may be in an immense fall-out of dust from a comet or an explosion of Earth material into the highest atmosphere whence most of it fell back to form loess and clay covering many hills and valleys to this day. Since humans seem to recall such an event, the time might not be far off.

Donald Cyr, a California amateur and devotee of the Canopy Theory of Isaac Vail, has studied loess. He has a story to tell too.

"Loess a is mixture of silica and clay, with particle size ranging from 0.1 mm down to 0.005mm. Where loess in unoxidized, it has a greyish color, but may also be yellow, orange, or brown because of presence of ferric oxides. Deposits of loess occur in North America, Europe, Russia, Siberia, China, and also in Argentina and New Zealand...."[10]

The State of Kansas is estimated to be overlain by more than 50,000 cubic miles of loess. There is little glacial outwash in Kansas, Cyr writes, and he does not see how glaciers had the power to grind down sufficient rock within the Pleistocene age, wherein it is placed, to supply the loess. He estimates the worldwide deposits at 7,000 cubic miles per degree of longitude per hemisphere. And he suggests that the ocean "blue" mud may be part of it.

---

[9] *Ibid.*

[10] D. A. Cyr, *Annular Space Dust* (Thousand Oaks, Calif. Annular Publs., 1968).

A few more words are owing on the origins of the drift or till, before letting the abused author Donnelly stand in his solitary majesty. Many accounts of stone falls are acceptable; Corliss has compiled and introduced some of them. Velikovsky has analyzed several cases, while rejecting Donnelly as to the cometary origins of the drift. For instance, he points to 28 fields of blackened, sharp-edged and broken stones (harras) in Arabia in strewn fields of many thousands of square miles; they are not igneous; they are referred to in ancient Arabic and Hebrew literature; they originate from the sky in early historical times.[11]

Till is a stiff clay full of stones varying in size up to boulders; conventional science says it was produced by abrasion and carried along by the ice sheet as it moved over the land. So Geikie said in 1863, and the definition is still useful. Donnelly pointed out that this till, which he called drift, is not in all places where the ice was said to be and exists in other areas where no ice was supposed to have been. Till is common "over much of the most important mineral producing terrain of the northern hemisphere. Till occurs ubiquitously in Canada and Scandinavia and is present as well over significant areas of the United States, U. S. S. R. and United Kingdom."[12]

But why, argued Donnelly, was there a "driftless region" is Wisconsin, Iowa, and Minnesota.[13] And why is very little found in Siberia; there exist " the great river-deposits, with their mammalian remains, which tell of a milder climate than now obtains in those high latitudes, still lying undisturbed at the surface." So wrote James Geikie.[14] And why are "glacial" pebbles and a "terminal moraine" found on hills and in valleys of the Southern Appalachians, and where the ice was not supposed to

---

[11] *Earth in Upheaval*, 96.

[12] W. W. Shilts, "Glacial Till and Mineral Exploration," in R. F. Legget, ed., Glacial Till (Ottawa, Royal Soc. Can., 1976), 205; Also Dreimanis, *Ibid.*, II, 14-5, 42.

[13] *Op cit.*, 28-31.

[14] *ibid.*, 121-2.

have reached in Eastern Kentucky.[15] Why do glaciers today not produce true ancient-type till, that is, striated stones, drift clay, mountain-top till, and how could glaciers form sheets over 30% of the Earth's surface a million years ago, not to mention pushing boulders up thousands of feet in elevation.[16]

Crossed trains of drift occur, and are rationalized into successive advances and retreats of ice under different climate and morphological conditions. The till is not fossiliferous. Where drift and till have been found in Australia, India, and deep beds of older rock in Scotland, they were attributed to more ancient ice ages, thus scholars might conveniently dispose of all material appearing to be till. It is not difficult in historical geology to use time freely to make place for anomalies and to create events, even the greatest types of events, such as ice ages.

Using the ordinary theories of glacial geology, even though he is an exoterrestrial catastrophist, the Soviet geologist Salop has pointed out "that the Precambrian glaciations occurred under very unfavorable physical-geographical conditions. The glacial deposits are interbedded between strata indicating a hot climate, such as red-beds, dolomites, phytolite-bearing limestones (at present only found in warm, usually mineralized waters along the seashore or in tropical lagoons and hot springs), evaporites, kaolinitic sandstones and bauxite." This association of tillites with formation of warm and hot climates is typical of the Paleozoic Ice Ages too.[17]

But Salop also demonstrates that nine ice-age pre-Cambrian "intervals vary from 40 to 125 (or 180) MY and no evident periodicity can be observed." He then associates "biologic revolutions with the epochs of excessive climatic cooling usually

---

[15] John Bryson, 4 *Am. Geol.* (1889), 125-6; W. R. Jillson, 60 *Science* (1 Aug. 1924), 101-2.

[16] Chester A. Davis, 19 *New World Antiquity* (Mar-Apr. 1972), 27-43; Donnelly, *op. cit., passim.*

[17] L. J. Salop, "Glaciations, Biologic Crises, and Supernovae," 2 *Catas. Geol.* 2 (Dec. 1977), 24-5.

resulting in glaciation." Tillites are taken as the signal of an ice age; whatever the climate above and below the till, whether cold or hot, the till is supposed to designate cold.

Some association may be found among tillite beds and a) low sea-water temperatures as measured in the differing gas and mineral concentrations of stratified sea-shells, and b) "coeval strata" that "attest to the influence of a cold, almost glacial climate." All correlations are subject to variations and even to possible basic flaws in radiometric dating. The association is loose enough to permit the argument that tillites may not be associated with cold climates, hence the tillites are not deposits of ice sheets and glacier, and, further, that tillites may be exoterrestrial deposits occurring in both hot and cold climatic period, wreaking quick destruction upon the biosphere.

Cyr and Sun point out that tektites are chemically similar to loess. This would suggest a possible exoterrestrial origin for loess and a coincidence of the two substances. Tektites are jets of fused silica. They range from microscopic size to large chunks. They are strewn around the world in enormous fields. They are found in the waters and soils of Central Europe, West Africa, Australia, Indochina, Thailand, the East Indies, the Philippines, Japan, China, and the Caribbean.[18] Heezen and Hollister estimated an Indian Ocean deposit of a billion tons that they think occurred upon a reversal of the Earth's magnetic field 700,000 years ago.

Billy Glass and R.N. Baker of the University of Delaware, with D. Storzer and G.A. Wagner of the Max Planck Institute of Heidelberg, studied intensively the Caribbean-North American strewnfield.[19] They estimated the total tektite field at 10 17 grams of material, dated stratigraphically at Middle Upper Eocene. Some 6000 such glass microspherules were found in the sediment of one

[18] John A. O'Keefe, "The Terminal Eocene Event..." 285 *Nature* (1980), 309-11; Heezen and Hollister, *op. cit.*, 254; O'Keefe, ed., *Tektites*, (U. of C. Press, 1963) and Tektites and Their Origin (N. Y., 1976).
[19] B.P. Glass *et al.*, "North American Microtektites from the Caribbean Sea" 19 *Earth and Plan Sci. Ltrs* (1973), 184-92 (N. Holland).

thin core at a depth of some 250 centimeters below the Caribbean Sea Bottom. The falls apparently came either at different times, or from different phases or portions of a gigantic single incident, because there are chemical differences among the tektites coming from different strewnfields of the world.

The writers claim different times, for they hold few reservations about their dating techniques. If from different times, a Moon origin is suggested, for there could have been large meteoroid explosions upon the Moon that would have splashed debris onto the Earth. Or, since the tektites are of a material akin to the Earth's crust, they might have been a fall-back from large explosive impact encounters with Earth.

Glass and Heezen differentiated three forms of tektites found in the Far East. One was melted twice, one melted once, and a third little melted. They deduce a massive cosmic body breaking up upon atmosphere entry into two or several pieces. Of these, one would explode in the upper atmosphere, another closer to the ground and a third close to the ground.[20]

Faul says "it is an established fact that tektites fell from the sky," but show too little cosmic-ray interaction to have spent much time in the sky.[21] Although he allows a possible lunar origin for some tektites, he shows that some tektite fields are too concentrated spatially to have been flung from the Moon and that, in Germany and the Ivory Coast, a similar composition can be ventured for large astroblemes and nearby tektite fields.

No writer has considered the possibility of an origin from the fission of the Moon and Earth. If the present author's theory of lunar fission were postulated, then the composition, distribution and occurrence of the specified forms of tektites would be consonant with the event.

I think that legendary streams of cosmic arrows shot by the gods upon hapless but offensive mankind might refer to the glassier kinds of fall-out. Tektites resemble somewhat obsidian, a

---

[20] 217 *Sci. Amer.* (1967), 35-6.
[21] 152 *Science* (3 June 1966), 1341-5.

popular igneous stone for fabricating arrowheads. Tektites may fall like showers of needles, or arrows, or as arrowheads in size, weight and hardness.

The same tektites are called "Dragon Pearls" in China. Carter Sutherland in 1973 traced dragon art in China back to its apparent origins around 1500 B.C.[22] That reinforcements of the horrendous (but sometimes beneficent "Lucky Dragon") image have been supplied by various comets through the ages was documented by Dwardu Cardona (1975).[23]

Invariably the Chinese dragon is chasing a "chuh," or globe, or sphere, and "chuh" also means "pearl". "Huoh chuh" is "fiery sphere" and "fire pearl." Moreover, the Chinese also call the tektite "huoh chuh". Indians, Javanese, and Tibetans also call the tektite "fire pearl". Long before modern science became interested in tektites, the ancient Chinese (the *T'ang Annals*) knew that these 'fire pearls' originated in space." They were esteemed by priests and emperors.

The tektites fell from the sky.[24] Aerodynamic ablation experiments with tektite glass have simulated their shaping upon entry and passage through the atmosphere. They are found in recent sediments and on the surface. The tektites were not long in space, they display no cosmic-ray interaction. They are easily eroded[25] but still exist in abundance and cannot be found in fossilized beds, another sign of youth. But other tektites have received old ages, 20 to 45 M/Y, as reported by Barnes.[26] Many are around the million-year mark (Heezen Glass, Chaprian).[27] and

---

[22] C. Sutherland, "China's Dragon," 4 *Pensée* 1 (Winter 1973- 4), 47-50.

[23] "Tektites and China's Dragon," I *Kronos* 2 (Summer 1945), 35-42.

[24] 152 *Science*, loc. cit.

[25] G. Baker, "Origin of Tektites," 185 *Nature* (30 Jan. 1960), 291-4.

[26] Gill, 75 *J. Geophys. Res.* (1970), 966-1002 finds ages of only 4830 to 14600 years. B. glass *et al.* find end of Eocene deposits, 10 *Lun. Plan. S.* (1979) 434-6.

[27] B. Heezen, B. Glass, *op. cit.,* and 112 *Science News* (1977) 408 on Ivory Coast tektites.

ages of 5000 years were found by George Baker and Edmund Gill.[28] Gentner's dating by fission-track suggests a million years or less for certain groups, much longer times were assigned to others.

The tektite falls have been associated by Billy Glass and others with magnetic reversals and faunal changes.[29] A syllogism emerges: a heavy-body impact explodes tektites high into the sky; it causes reversal of the Earth's magnetic field; as the EMF hits zero point, cosmic particles, ordinarily deflected, pour down and cause mutations and extinction. Contrasting with this theory are opinions such as Lyttletons's that tektites fell from a passing comet train. However, Urey and Spencer argue that they reflect a splash from a cometary or meteoroid impact on the Earth. Moreover E. A. King: "the answer is now clear: tektites are produced from extraterrestrial rocks melted by hypervelocity impacts of large, extraterrestrial objects."[30]

Erratic bits of an exploded planet from the Mars-Jupiter interregion often fall to Earth. Some of them may also be surviving, uncaptured, terrestrial material. The tektite fields on Earth could also be fall-back from the lunar eruption. Rittmann writes: "The chondrites (of meteoric falls) correspond genetically to the terrestrial sima, and the tektites to the protosialic upper crust of the primeval earth."[31]

James Sun proposes that half a million years ago, a snowball comet laden with flammable gases approached Earth from the Northwest.[32] It shattered by gravitational force, and part crashed while part continued on. Loess was thus laid down, and in some place melted by impact into glass. Loess has a chemical composition very much like the tektites, as I have mentioned above. Aerial explosions created innumerable small glass blobs that fell to Earth.

---

[28] *Op. cit.*

[29] John Lear, *Sat. Rev.* (6 May 1967), 57.

[30] E. A. King, 65 *Amer. Sci.* (1977), 212-18.

[31] Rittmann, *Volcanoes and Their Activity* (New York: Wiley, 1962), 284-5.

[32] James M. S. Sun, 56 *Trans. Am. Geophys. U.* (1975), 389.

The investigators generally agree that tektites are earth-like and moon-like in composition. Probably, the loess and tektites arrived within the same time span after passing into the upper atmosphere following their explosion from the Earth. Either a passing large body exploded the Earth's crust to make them or a meteoroid impact did the job.

John O'Keefe links the North American strewn field of tektite and microtektite falls with the terminal Eocene (Tertiary) event, when radical climatic change can be perceived in floral abundances and radiolaria were devastated.[33] His theory calls for the tektites to assume, before final descent, a ring-like structure around the Earth. The ring might have lasted a million years and cast a blighting shadow over the biosphere.

It is apparent here, once more, that earth scientists are becoming ever more daring in their suggestions of mechanisms to satisfy the resultant state of geological facts. Just under a century ago, Issac Vail received short shrift from academicians for proposing a Saturnian ring canopy system for the globe and arguing that it was known to early civilized man and fell apart before his very eyes.[34]

Reporting systems on natural phenomena have gradually become more complete, regular, and valid. Nevertheless, the Edinburgh Philosophical Journal in 1819 issued an enchanting list of "meteoric stones, masses of iron, and showers of dust, red snow, and other substances, which have fallen from the heavens, from the earliest period down to 1819."[35] Among the exotic items were: a great fall of black dust at Constantinople on November 5-6, 472 A.D. accompanied by burning heavens; a kind of red matter like coagulated blood in the middle of the 9th century; a burning body that fell into Lake Van, Armenia, turning the waters red and cleaving the Earth in several places (1110 A.D.); gelatinous matter in India with a globe of fire; and a mixture of

---

[33] 285 *Nature* (1980), 309-11.

[34] *Selected Works* (Santa Barbara, Ca: Annular Pubns, 1972).

[35] *Edinburgh Philos. J.* (1819), 221-35.

red rain and snow whose dust contained silica, aluminum, lime, iron, carbon and loess and was coincidental with a shower of meteoritic stone over central and southern Italy in 1813. Red rains, often associated with meteors, were common. William Corliss, in his compilations, has educed much additional literature on peculiar fall-outs. Peter James,[36] Donnelly, Velikovsky and others have demonstrated the frequent occurrence of red falls in proto-history.

Much meteoritic dust falling upon the Earth is invisible and immeasurable. Meteoritic falls have been estimated at 4000 tons per year by Saukov.[37] Hughes (1976) arrives at a figure of 16,000 tons per year. Schmidt gives an average for all of geological time at $8x10^{11}$ tons per year, very much larger and based upon an exponentially leveling off of initially vast drops of material.[38] At the last rate, with a geological age of $5x10^9$ years, one would have a total of $40x10^{20}$ tons dust dropped on Earth from space. This is not far from the total mass of the Earth, $6x10^{21}$ tons. But if Pettersson is correct, the rate of accretion of cosmic dust may be about 10,000 tons per day.[39]

Micrometeorite dust has been estimated by Fred Singer[40] to fall at a median rate of 1250 tons per day or 456,250 tons per year (the rate may actually be 10 times more or less, he estimates). The calculation is from the detection of aluminum 26 abundance ratio in Pacific Ocean bottom cores. This is $4.5x10^{11}$ grams per year today, but Schmidt's estimate is only 400 tons per year today.

---

[36] I. *Catas. Geol.* (Dec. 1976), 5.

[37] In D. I. Shcherbakov, ed., *The Interaction of the Science s in the Study of the Earth* (Moscow: Progress Publ., 1968).

[38] R. A. Schmidt, "Extraterrestrial Dust as a Source of Atmospheric Argon," 151 *Science* (14 Jan. 1966), 223.

[39] "Cosmic Spherules and Meteoric Dust," 202 *Sci. Amer.* (Feb. 1960), 123-32.

[40] S. Fred Singer, "Zodiacal Dust and Deep Sea Sediments," 156 *Science* (26 May 1967), 1080-3.

If any exponentialism is part of Singer's scheme and it should be, a fairly considerable portion of the Earth's crust should be composed of gathered-in planetary dust, achieved in a fairly short time. If, for example, we had a measure showing this figure to have been 10^{20} grams per year in 500 B.C. and 10^{25} in 2500 B.C., the subsequently plotted curve would give us the mass of all of the continental crust except for the basic granite within a few thousand years. We do not have such figures, but if we consider the obsession of ancient voices with days and years of darkness and ascribe half of this to fall-out of dust, the required substantial deposits would be quickly forthcoming.

Between 1956 and 1964, W. D. Crozier collected exoterrestrial black magnetic spherules from atmospheric fall-out at two New Mexico stations, of a type noted around the world and in sedimentary rocks of great ages. These were accreting at an average annual rate of 1.04×10^{11} grams for spherules in the diameter range of 5 to 6. David Hughes considers the interplanetary dust to originate with comets and arrives at a figure of 16,000 tons per year of all sizes.

Hans Petterson, reporting upon the oceanographic expedition of the *Albatross*, disclosed a high nickel content in the Pacific Clays. Since basalt, the bottom material contains little nickel and meteoritic dust, meteoritic showers hundreds of times greater than presently observed were required to explain the abundance. The nickel abundance is also 5.5 times that in continental igneous rock; hence an exoterrestrial source is invoked.[41] Assuming the average of nickel in meteoritic dust to be 2%, he arrives at the aforesaid figure of 10,000 tons of dust per day, 3,650,000 tons per year (3.6×10^6), hence, especially if any kind of exponentialism is introduced as we go back in time, we should have the sediments of the ocean receive their quota of nickel laid down in a few thousand years.

McSween and Stolper, in their study of basaltic meteorites, which were definitely not of earthly or lunar origin,

---

[41] "Exploring the Ocean Floor," 183 *Sci. Amer.* (Aug 1950), 42-4.

abstracted a type of shergottite meteorite. This material they assign originally, not to comets, or asteroids, but to the planet Mars, which has many extinct structures and surface rocks with a known resemblance to the shergottite.[42]

The electrician, Eric Crew, has analyzed confirmed reports of ice and stone falls associated with lightning; many such were collected by Charles Fort (1874-1832) who wrote once, "we shall have a procession of data that Science has excluded... a procession of the damned."[43] Crew ascribes both pick-up and fall-out phenomena sometimes to high-speed jet occurring in and about air-to-ground fast electrical discharges.[44] Dust storms and volcanism greatly augment the fusion of particles. There may be posited that in large meteoroidal and cometary encounters, the Earth will be subject to considerable material exchanges by the electrical discharge channels occurring between Earth and the intruder.

The "White Cliffs of Dover" and other immense chalk beds elsewhere are a mixture of tiny spheres, a formless chemical mass, and organic debris, which contains some marvelously unattrited marine skeletons. How were they formed? Conventional science pleads continuing longtime deposits, but the stratification and water-current markings attesting to such are missing, nor can the preserved shapes admit to this mechanism.[45] A great updraft and precipitation is suggested, or else a dust-laden electric discharge penetrating the waters, followed by an upheaval or expansion of the bottom terrain.

---

[42] *Sci. A*41. "Exploring the Ocean Floor," 183 *Sci. Amer.* (Aug 1950), 42-4.

[43] *Book of the Damned,* repr. (London: Abacus, 1974). See also the compilations of W. Corliss (Glen Arm, Md., 21057, Sourcebook Project).

[44] "Electricity in Astronomy," I *S. I. S. R.* 1-4 (1976-7), esp. # 4.

[45] W. A. Tarr, "Is the Chalk a Chemical Deposit," 62 *Geol. Mag.* (1952), 252.

A study by L. and W. Alvarez, Asaro and Michel describes a fall-out of dust 1000 times that of Krakatoa from a meteoroid crash, which, they claim, darkened the Earth for years.[46] The crash was deduced from the presence in Italian, Danish, and New Zealand limestones of the fossil break between the Cretaceous and Tertiary periods of iridium, 30,160 and 20 times its normal background level in terrestrial rocks but characteristic of meteoroids. Spain and Holland were added by Ganapathy to the locations bearing the tell-tale chemical signals. Fish-clay analyses by Kyle and others in Denmark agreed with the limestone findings. A number of additional rare elements were also in long supply, 5 to 100 times their normal abundances.

The correlation of a fossil index set with a distinctive chemical element marks an important advance in geological investigation. A sure layer is now presumed to exist worldwide; even were it not to signal an age boundary, it would permit a tightening of identifications of relative and absolute dates of strata and species. We know that we are dealing with a uniform world-wide event, something that is only hoped for when correlating fossils and rocks. We know that the event is limited in time.

We know further that if the event is not denoted in the strata, the reason is not that the event did not occur. That is, some stratum capable of containing the iridium (or other element) must at the stipulated time have existed everywhere. Where not found, conditions for its prompt removal must have existed, or later removal must have occurred. Alternatively, the fall-out was erratic and initially directed only to certain spots by the presumably catastrophic winds and tides of the moment. Despite all this, with a dozen such exoterrestrial chemical markers, historical geology and paleontology would undergo a quantavolution.

What conclusions can be drawn from the material of this Chapter? At the least, a considerable part of the Earth's crust is exoterrestrial and has fallen as dust and stone not long ago. There

---

[46] Luis Alvarez *et al.*, "Extraterrestrial Cause for the Cretaceous-Tertiary extinction," 208 *Science* (6 June 1980), 1095-1108.

is reason to accept in general terms the multitude of legends speaking of heavy falls. Even the most bizarre material has descended during historical times and every indication points to an exponential increase in the quantity and perhaps the variety of matter with the regression of time from the present. All the seas and continents contains heavy deposit of suspected exoterrestrial origin.

Yet there is also some indication that the time of heavy falls may have been concentrated in a catastrophe or set of catastrophic climates. The "ice ages," for instance, may have been a period of combined ice and stone deluges from outer space, explaining thereby a number of inconsistencies in the terrestrial pure theory of a central focus and outspreading therefrom. The absence of fall-out stratigraphic formations in older rock formations bespeaks a primeval peace.

A question arises as to what constitutes outer space or exoterrestrialism for dust and stone falls. Under certain conditions of large meteoroid or cometary impact, and heavy multiple volcanism, exploded material can achieve extreme heights and even be lost into space. Such would be the case, for instance, were the Moon to have been exploded from the Pacific Basin. In such a case, a prolonged fall-out period of a great many years, perhaps centuries, might result. Pebbles, dust, loess, tektites and other types of matter might separately collect in orbit and shower down homogeneously, while simultaneously, volcanism would pave large stretches of the globe.

Once more, we find the gradual fall rates of the present and the more credible exponentially higher fall rates of the recent past so productive of mass and volume for the Earth's crust that a young age for the Earth or a very young age for the catastrophized Earth suggests itself. Whatever the properties of fully exoterrestrial falls to explosion and fall-back, the fall-out even will wreak havoc: darkness, lightning, winds, possible interruption of Earth motions, and biosphere destruction, plus excitation of seismism and volcanism; holospheric transactionism, that is.

CHAPTER NINE

GASES, POISONS AND FOOD

That "all things come from heaven" may be untrue, yet even in these last peaceful centuries the quantity and variety of things reported to have fallen upon Earth is astonishing. For two hundred years, scientific establishments sought to resist the flow of accounts, making out those who appeared with such claims to be culturally retarded and childish, clowns, cranks and religious fanatics. Now the door is open to claims, and some scientists are tripping over each other's footnotes in their eagerness to go to through it. Since most chemical elements and compounds can be either found beyond the Earth or conjectured to have once formed from the thermal and electrical conditions that occur exoterrestrially, scenarios of past events to explain present processes are becoming as common, prolonged, and disastrous as the "soap operas" of radio and television.

Contemporary man is motivated to come to grips with the sky by economics, politics, militarism, and the need to survive. Poisonous hydrocarbon, radiation, aerosols, carbon dioxide, acid pollution, radio microwaves, ion disturbances, acoustical turbulence, supersonic stresses in flight, and civil and military thrusts into outer space amount to a major challenge to human modes of existence. To cope with such developments, ever more scientific knowledge is required and this in turn leads to discoveries of processes occurring in outer space that influence the Earth, and thereupon present new problems and possibilities -solar energy, weather control, incursions of hitherto unrecognized chemicals and particles, and even, say some, life forms contributing to evolution and diseases. A modern pragmatic preoccupation with the skies, it would appear, is now being laid on top of the age-old preoccupation with the forces and gods believed to dominate the celestial sphere.

The gases that we discuss are mainly effective in the biosphere. We address not only their chemical qualities but their behavior in mixtures and their propulsion by winds. The poisons we discuss are cell destroying chemicals. The food consists of the rare occasion of the descent of digestible cell-building chemical compounds. Hydrocarbons are considered here as poisons; petroleum deposits are dealt with in a chapter to come. Radiation is treated as a poison, though it may be a creator at times. Electricity, as was said earlier, is everywhere and can go onstage with a number of the processes involving gaseous behavior.

Comets and meteoroids, like volcanos, can emit gases. Explorer and scientist Humboldt thought it probable that the vapor of the tails of comets mingled with our atmosphere in the years 1819 and 1823. When, on March 24, 1933 a fireball of six miles diameter sped across the American South, it trailed a tail one-mile-wide that carried a thousand cubic miles of dust. The people who were beneath its passage smelled a peculiar sulphurous odor for hours and for several days suffered from throat irritation.[1] If the intruder is admitted, one may grant the occurrence of gases. An actual impact is not necessary.

Can a gas cloud descend through the atmosphere without exploding or burning? It would have to be charged oppositely to the Earth's surface and buffered during descend by a plasma. Even under normal conditions, the positions of light and heavy gases are sometimes reversed in the disorderly atmosphere. The Great Chicago Fire, and forest fires which burned out millions of acres of land in Wisconsin, Michigan, Western America and Canada broke out on the same day in the fall of 1871. E.K. Komarek speaks of a peculiar fire weather and cites this case; Donnelly

---

[1] F. W. Lane, *The Elements Rage* (Phila: Chilton, 1965), 179.

claimed that all were due to gas drifts from the tail of Biella's Comet which had not been seen on its expected three previous visits but was glimpsed without its tail in 1872, a year later, at which time a spectacular meteoritic display occurred.[2] Donnelly offered a number of testimonials that the fires referred to leaped incessantly from different locations above the houses and forests and behaved as electricity in some ways (fusing without burning) and as a gas in others (asphyxiating people away from the blaze).

A few years later another comet neared Earth and the Earth passed through its tail. The comet broke up on September 9, 1882. Krakatoa exploded on August 26, 1883, after months of eruption. A great many people were burned, smothered in the choking gases, and nearly blinded. We should recall how the Krakatoa ash is negligible in the sea today when compared with the layers described in earlier pages. Mass asphyxiation would be a logical deduction from the conditions cited.

Just as research has shown sunspot gaps to be connected with climatic disaster, and has correlated planetary-solar conjunctions with earthquakes, it may establish that cometary passbys have occasioned violent volcanism -all of this during the uniformitarian Solarian period. All the more may have happened, then, during ancient periods of catastrophes.

Cosmic dust can be struck by particles from the Sun or stars and emit gases. David Tilles explains only 20% of the argon 36 and 38 on Earth as an effect of the solar wind upon space dust and debris. The balance he believes to be derived from an unquiet sun of long ages ago acting upon then larger dust clouds surrounding the Earth.[3] However, argon has been unexpectedly detected in the thin atmosphere of Mars, and if Mars has been recently in gaseous exchange with Earth, as Velikovsky wrote in 1950, it would have given argon to the Earth or taken it away.[4]

---

[2] Donnelly, *op. cit.*, 102-6.

[3] 3. David Tilles, "Atmospheric Noble Gases...," 148 *Science* (21 May 1965), 1085-7.

[4] L. M. Greenberg, "The Martian Atmosphere," II *Pensée* I (1976), 5-9.

Gibson and Moore, investigating subsoil samples from the Moon, found so many differences in volatile elements between North Ray Crater and other sampled locations that they concluded it to be the site of a cometary impact. They agree with Kopal that "the total amount of gas which can be acquired by the Moon in a catastrophic encounter with a comet is far from negligible."[5] The Earth is a bigger target for comets than the Moon. We would expect the Earth, then, to have also picked up many elements from foreign sources. Traces of gases and hydrocarbons were found some distance from the crater. Gases emitted by an impacting body would probably cause significant surface phenomena on Earth as well.

In the year 687 B.C., at a time when natural phenomena, attributed to Mars, were verging upon the catastrophic in many places on Earth, the great army of the Assyrian king Sennacherib was destroyed as it was preparing to assault Jerusalem. "The angel of the Lord" is credited with the deliverance from the enemy by the Bible. The angel is identified as the Archangel Gabriel. He is connected with divine fire, with the founding of Rome, with the planet Mars. It was "a consuming blast" that rabbinical sources say burnt the souls of the Assyrians but not their bodies.

An analysis is contained in Velikovsky's *Worlds in Collision* (230-41) in several fine passages. The grotesque incident was coincidental with several other documentable events around the world, and with a probable interruption in the Earth's movement. As happens when a mega-force is operating, one force incites another: the destruction might have been occasioned by gas and "celestial fire" acting together.

A charged gas would have descended, possibly lured by the concentration of metal weaponry and myriad campfires. The gas cloud would have sent an electrical leader to the camp grounds and the subsequent exchange of potentials would have killed the Assyrian host. Sennacherib the king escaped, he was probably camped high and far from the multitude of soldiers. Even in

---

[5] 179 *Science* (5 Jan. 1973), 69-71.

modern times of untroubled skies, verified reports of flocks and herds being annihilated by a lightning blast occur.

The destructive meteoroid in this case would have been a plasmoid, preserving its integrity as it passed through space and the atmosphere by the repulsion of its surroundings, but driven down to Earth's surface by decrease in the repulsion, until ultimately a "soft explosion" extinguished the oxygen available to humans and replaced it by methane, carbon dioxide, carbon monoxide or these together.

We turn next to the famous case of the mammoths, not waiting for the chapter on extinction.[6] One almost should say the "deathless" case, for it has endured the whole battle between catastrophists and uniformitarians, two hundred years - except that now it may even become the case of the "deathless" mammoth, for a late news report tells us that certain Russian experimenters are seeking to unfreeze and clone a mammoth cell with an existing elephant to give birth to a live mammoth.

Were the original mammoths gassed into extinction? Instant death, fractured limbs, destroyed sometimes in herds and sometimes alone, discovered on hills (not in river channels), some found with their skins and innards intact, several found with food in their stomachs, even their mouths, often associated with an incongruous assembly of other species, they lived and died where they were found, several still standing, one with a rooted tree buried with it. The mammoth and almost all other large animals of the same period were extincted between 5,000 and 30,000 years ago over the face of the globe. The extinctions occurred from over practically the whole arctic area and down to the southern part of the United States, Europe and Middle Asia, where their close relatives, the mastodon, now-extinct elephants, and modern

---

[6] W. R. Corliss has compiled and reprinted numerous extracts from the scientific literature, *Strange Planet* (Glen Arm, Md.; Sourcebook Project, 1975), section EBM. An important update and new material is contained in L. Ellenberger, *et al.*, "Catastrophism and the Mammoths," VII *Kronos* 4 (Summer 1982), 62-96.

elephants browsed. It is strange that no human skeletons have yet
been found, since we have their drawings of the mammoth.

Obviously if the date of each specimen were to be taken
seriously, we would have, as one writer argued, a series of local
catastrophes. All over the world, he might have added. Nor were
the frozen elephants found encased in ice, but rather in a muck of
pebbles and clay, which is the same kind of muck that is
widespread over hundreds of thousands of square kilometers in
the frozen arctic regions and contains the mangled remains of
millions of animals and plants. It is hard to dispute claims of a
sudden, widespread, simultaneous, and single catastrophe.

The assigned dates are hardly defensible. In the
preliminaries of such a catastrophe, valid carbondating would be
extincted along with the large animals. The supplemental dating is
provided by the complicated ice age series, of which more later,
but which, we can say, is something of a muck itself. With the
unreliable dating shunted aside, a global scenario can be provided,
an extravaganza, to be sure, but one is driven to it by the facts.

One may speculate that a large body passed by the Earth
perhaps 6,000 years ago. It drew up tides of water and air below
its path by hundreds of meters. It drew along and up, then, water
and atmosphere from the extreme northern and southern
latitudes. It tilted the globe at the same time. Most animals were
asphyxiated during the hours of the withdrawal of air.
Simultaneously they were deep-frozen by temperatures reaching
in directly from outer space in the range of -150°F.

The intruding body departed. The columns of air and
water collapsed, and rushed up to the north and south. The winds
and tides collected most of the dead animals, tore up the ground,
and finally deposited the remains in a muck that sometimes
reaches to 1,000 feet of depth, even to 4,000 feet in one that the
Soviets have excavated. Much of the air never returned; the supply
from the larger envelope around Earth was depleted and the
immediate atmosphere was thinned. As the legends say, it was
now the bitter, cold age of the "God of the Bright Skies", Jupiter.

The mammoths, dry frozen in a vacuum, rested in their packages of muck until the present day.

After relating so dramatic a story, it would be excessive to speak of the dinosaurs and other mass extinctions, and these shall be saved until the appropriate chapter. Other issues remain to be discussed here relating to gases and poisons.

One has to do with human experiences with atmospheric pressure, not only in moments such as asphyxiated the great mammals, but time and time again in primeval history. Sudden electrical events, not encounters alone, must have raised and lowered the air pressures under which humans lived. At times, mankind must have endured miserable headaches. Anthropologist Kennedy once referred briefly to "certain ritual practices like trepanation (which also developed obsessive proportion in Late Neolithic and Beaker time in Western Europe)." The practice extended in North Africa from the Canary Islands through the Berber lands at least as far as Egypt. It was performed in Mesoamerica as well. George Sarton writes in his history of science of prehistoric skulls that have come down to us with evidences of trepanation (trephination) performed upon them in life. The trepan is a saw for cutting holes in or removing pieces from the skull. It is a dangerous operation, hardly on a plane with piercing the nostrils to hold decorative devices. (But why are these devices so near the sinuses, too?) Extreme headaches and fury can thus be relieved. Trepanning, we surmise, was an indication that some considerable part of the population could not cope with a periodical fluctuation or definite change in atmospheric pressure.

A second issue has to do with ozone. Having discovered that aerosol devices and supersonic transports might destroy the ozone layer, several scholars have ventured to say that such events have occurred in the past. Ozone, or atomic oxygen, exists in a thin layer in the upper atmosphere, where it blocks solar and cosmic particles from penetrating to the Earth's surface, here to cause innumerable mutations and cancers. Ozone, too, is a poison in itself.

Associating ozone layer destruction with the periods of a reversal in the Earth's magnetic field and these with the extinction of a number of species, discoverable in ocean bottom drilling, Reid, Isaksen, Holzer and Cruzen theorize " that current concern about possibly anthropogenic destruction of stratospheric ozone may be well-founded since it is possible that major depletions occurring in the distant past have had profound effect on the development of life as we know it."[7] Anticipating again what is to be developed later, we can give credence to the theory, but would add that the destruction of the ozone layer will have occurred during any catastrophe involving turbulence in the stratosphere, especially with the passage of a large body.

Furthermore, the authors say, "the harmful effects accompanying polarity reversal, whatever they may be, form only one component of the total environmental stress on a given species." Beland and Russell point out that solar flares of extreme power, of a kind never observed and perhaps occurring once in 200,000 years by probability theory, would have to coincide with the reversal of GMF in order to account for a large number of species extinctions.[8]

The Sun might well have become agitated by changing movements of large bodies within its field and add a heavy dose of radiation to what might be occurring on Earth in reaction to an intruding body or bombardment of meteoroids. Ozone problems would have to take their place among many disturbing chemical and radiation changes. As Waddington pointed out in 1967, particle radiation increases inversely with magnetic shielding.[9]

Presently one speaks of background radiation, or low-level radiation, and a pressing problem of the future is how to keep radiation at the same low level at least. Sternglass finds even now indications of birth defects, infant mortality, and old-age

---

[7] G. C. Reid *et al.*, "Influence of ancient solar-proton events on the evolution of life," 259 *Nature* (22 Jan. 1976), 177-9.

[8] 263 *Nature* (16 Sept. 1976), 259.

[9] *Science* (17 Nov. 1967)

respiratory problems traceable to low level radiations.[10] Evidently both long-term increases of level and single bombardments can cause damage to most people. Latest medical reports (1983) are more ominous.

Prehistoric cases can exchange ideas with future cases. J. W. Gofman has predicted that "a nuclear-based (U. S.) economy with 99.9% perfection in plutonium containment could mean a 25% annual increase in total death rate from this source alone," amounting to over 25 million extra cases of lung cancer over 50 years.[11] One must evaluate prehistoric indications of abnormal radiation and high-energy explosions in this light.

Vera Rich, reviewing knowledge of the Tunguska (Siberia) meteor of 1908, brings forward evidence of scabrous infection of the local reindeer in that year, a great acceleration of tree ring growth beginning then, and an increase in the radioactivity of surrounding trees.[12] Another report has it that certain plants mutated as well. The event was exoterrestrial in origin and probably is of the category of "gas-bag" explosions, since scarcely a ton of exogenous particles has been recoverable from the immense scene of destruction.

Perhaps the body entered the Earth's atmosphere with great speed, electrically attracted as well as driven by inertial differences, and thoroughly ablated until it became a gas projectile without a casing, that exploded before striking. Or perhaps it was a "Sennacherib plasmoid" from its inception. Generally speaking, the radiation effect of a single meteor or cometary train passing through the atmosphere would be heavier than many hydrogen bombs (unless these latter are deliberately "dirtied" by cobalt or other chemicals) because of its great heat, its compression of the ambient air, its wide path of fall-out, and deep and large explosive cratering.

---

[10] E. J. Sternglass, *Low Level Radiation* (NY: Ballantine, 1972).

[11] Report prepared for the Committee for Nuclear Responsibility (Yachats, Oregon).

[12] 274 *Nature* (1978), 207.

During the disasters of Exodus, several documents give indications of radiation effects. The widespread "leprosy" effect may denote radiation disease, as I have explained in my study of Moses. Eating fallen quail killed many persons, reports Jewish legend. The manna, too, had to be eaten under supervision; to argue that it was" holy" and thus had to be treated ritualistically is a modern sociological notion overlooking that it might have become "holy" for several reasons, one being that priests, the savants, were called upon to distinguish the edible from the poisonous manna. The Egyptian Ipuwer papyrus conveys the impression that women became barren and that people lost their hair. The cattle herds died of scabrous diseases. The most substantial theory of Exodus times regards them as part of a much larger, a global, event, involving the close passage of a comet, so that radiation effects are logically to be expected.

Recent studies have discovered high levels of radiation in fossil flora and fauna, going back far in conventionally dated geological time. Kloosterman writes of " anomalous high radioactivity" in a fish from the same Old Red Sandstone beds in which the Pterichtyades occur, "fishes often invoked by catastrophists..." and quotes Hugh Miller (1841) on a quiet but potent agency of destruction erasing "innumerable existences of an area perhaps ten thousand square miles at once, and yet the medium in which they had lived left undisturbed in its operations."[13]   We mention the case again when discussing extinction; electric shock probably accompanied the poison, and was succeeded immediately by great tides of slurried water. In 1975, Bramlette described deep fossil beds a plankton in the sea bottom that he tied to cosmic radiation storms.[14]

Radiology is a new field of knowledge, whose development is producing a new attitude toward what can be transformed, in biology, geophysics, meteorology, and geology. Oparin some time ago began to call upon it to explain the long

---

[13] J. B. Kloosterman, 2 *Catas. Geol.* 2 (Dec. 1977), 49.

[14] 187 *Science* (17 Jan. 1975). 4172.

chain of chemo-biological events leading up to *The Origin of Life*.
He wrote of inorganic meteoric material suffering far-reaching
transformation from inter-stellar radiation before arriving upon
the Earth, of transmutations, for instance of iron and nickel into
aluminum and silieni and of these into magnesium, sodium, and
helium.

An instance of how rapidly old problems can be tendered
new solutions by seemingly remote scientific developments occurs
in the case of perhaps the most famous of fall-outs, that of manna,
ambrosia to the Greeks, soma to the Hindus, and other names to
other peoples. The insistent claim of the ancients takes on
enhanced validity in the context of operations of modern
technology.

The bits of suggestive evidence come from all quarters.
We begin with a famous 1945 experiment of S. L. Miller (in
consultation with H. Urey) and ask Bernard Newgrosh to describe
it for us:

> On the suggestion of H. C. Urey, he took a mixture of water,
> hydrogen, methane and ammonia (which were then thought to
> be the constituents of Earth's primordial atmosphere but which
> are now known to be the constituents of cometary matter),
> boiled the water and ran an electrical discharge through it
> continuously for a week. The end products were an assortment
> of organic compounds, including some sugars, cyanides and
> small quantities of amino-acids. It was the latter which evoked
> the most interest and sparked off a whole new avenue of
> research into "the creation of life on Earth." Miller had boiled
> his liquid only to prevent the growth of (and therefore
> contamination by) micro-organisms. Later experiments used far
> less energy, and it transpired that the shorter and smaller the
> amount of heat used, the greater the yield of amino-acids
> obtained since these are denatured by heat. Other workers tried
> different mixtures of gases including, in some cases, oxygen and
> hydrogen sulphide. As long as the mixture was basically

reducing in nature, the organic compounds and aminoacids were produced.[15]

M.G. Reade and Wong Kee Kuong have more recently discoursed theoretically upon methods by which carbohydrates, such as the manna which fed the ancient survivors of the Exodus disaster, could be produced with the aid of cosmic lightning.[16] Formaldehyde (a compound of carbon, hydrogen, and oxygen) is a partially combusted gas, of which "there will be no shortage. in a burning fiery cloud, almost whatever its origin." In mixtures of free oxygen, carbon, hydrogen and nitrogen, this is the only product. It has to be synthesized into sugar in an alkaline environment (already done) which is not poisonous and can be converted into starch, rolled into "coriander seed" sizes and dropped at dawn. So goes the argument of Reade, himself a confectioner and engineer. The necessary procedures and formulas are presently at the threshold of laboratory chemistry, he asserts.

Of the processes required to produce edible carbohydrates in the form described by the ancient sources, all are present in the environmental setting described by the same sources, although without making the scientific connection that present knowledge affords. The analysis of Reade is especially literal in matching edible product and the natural "chemical apparatus" within the Bible.

In a yet unpublished manuscript on the Vedas of India, Ziegler brings forward many ancient statements about dust and gases pervading the skies, including the fact that the dust was falling and carrying the dew of heavenly waters (soma) with it.

In Hindu rite, the soma-devi are celebrants of sacrifices using soma. As a libation to Agni, soma is now superseded in India

---

[15] 4 *S. I. S. Workshop* 1 (July 1981), 2-3.

[16] Wong, Kee Kuong, "The Synthesis of Manna," 3 *Pensée* (Winter 1973), 45-6; M.G. Reade, "Manna as a Confection," I *S. I. S. R.* 2 (Aug. 1977), 9-13, 25.

by *ghi*. Now the deva is a goddess practically identical with Venus, and the devi are her cohort. Venus, east and west, is worshiped at times in the form of a cow, the sacred cow of India, for instance. *Ghi* is clarified butter. The "golden calf" of the Hebrews in Exodus is the equivalent Baal-Venus image. These few (from a great many) observations are made solely to point out and complete the coincidence of a great celestial presence (a cometary body), a turbulent atmosphere full of dust and lightning, the availability of carbon dioxide, hydrogen, oxygen, methane, formaldehyde, and water in large amounts, the presence too of many enormous laboratory vessels from which would fall not one but several products, and, of course, the desperate survivors who would eat anything (regardless of its nutritional value) and reverence the imagined donor.

At the same time as the Hebrews, Hindus, Mexicans, Greeks and others were munching manna, they were vitally concerned with a certain redness in their environment. The most astonishing and fearful color had fallen out of the skies and penetrated the surface. Again we take leave to quote copiously from Newgrosh:

> Dr. Velikovsky has produced numerous citations from ancient sources to show how falls of a blood-like substance occurred when a "new" comet (later to become the planet Venus) came into catastrophic contact with the Earth: the Manuscript Quiche of the Maya, the so-called Papyrus Ipuwer from Egypt and the Book of Exodus all record the fact that the water in the rivers was turned into "blood". In addition to these examples, Dr. Velikovsky refers to the Greek myth of Zeus and Typhon, the Finnish epic Kalevala and the lore of the Altai Tartars. However, a more exhaustive survey of such legends would include the Sumerian myth of Inanna (a Venus goddess) who filled the wells of Sumer with "blood", the Egyptians story of the goddess Hathor (also Venus) whose visits to Earth were associated with the covering of the land with a blood-like "beer", and the Norse legends of the "raining of blood" associated with the Valkyries. These myths are widespread and all tell the same story. There can be little doubt that something

looking like blood fell from Venus during its close contacts with Earth.

What was its nature? Dr Velikovsky noted that it was a soluble pigment: "In sea, lake and river this pigment gave a bloody coloring to the water. These particles of ferruginous or other soluble pigment caused the world to turn red." Moreover, the accounts of Exodus 7: 24 and of Ipuwer lamentations agree that this bloody colored water was unpleasant and maybe poisonous. It is recorded of the Nile that "the river stank" (Exodus 7: 21). There was disease among the cattle which, Dr. Velikovsky claimed, was due to dust of an irritant nature.

Another writer, Peter James, asks whether legends of red falls from periods before 3,500 years ago might not refer to geological occurrences that deposited red sands or ferratites around the world.[17]

In Greek myth the Sky-god Ouranos, the first ruler of the universe, was castrated by his son Kronos and his blood fell to the Earth, impregnating it with a number of dreadful deities. To turn to Roman literature, we have a very graphic description of fall of blood in Ovid's "Metamorphoses" in his account of the fall of the Giants. "The terrible bodies of the giants lay crushed beneath their own massive structures, and the Earth was drenched and soaked with the blood of her sons." Egyptian myth tells a tale of the Sungod Re similar to the Greek myth of Ouranos - it was said that Re mutilated himself and that new deities sprang from his blood as it fell. In another Egyptian myth, Re decides to punish mankind by sending down the Goddess Hathor/ Sekhmet. She performs her task enthusiastically, gorging herself in the blood of men, but Re does not want Man utterly destroyed, and he has to devise a stratagem to stop here in her path of destruction. He mixes red ochre with beer, and pours a vast quantity over the Earth during the night, to a depth of three palms (about nine inches). The

---

[17] I *Catas. Geol.* (Dec. 1976), 5.

goddess sates herself on this "blood", and intoxicated she
returns to heaven having forgotten her task.

Newgrosh refers back to the Miller experiment, for a crucial detail
that has long gone unnoticed.

> Miller wrote: "During the run the water in the flask became
> noticeably pink after the first day, and by the end of the week
> the solution was deep red and turbid. Most of the turbidity was
> due to colloidal silica from the glass. The red color is due to
> organic compounds absorbed on the silica."

To conclude, electric discharges between the intruder and
Earth synthesized organic compounds in the cometary gases,
including an edible component and an inedible red silicate that
showered down to color the Earth and water a turbid red.
Newgrosh adds, "being organic compounds, they would be
speedily denatured, leaving no trace -except, that is, in the memory
of mankind." Also, an iron compound of partially hydrated $FeCl2$
has been reported present in heavy concentration in the clouds of
Venus today.[18] Considering that a possible source of Venus is the
"Great Red Spot" of Jupiter, together with the material already
mentioned, if this analysis remains valid, this is a significant
quantavolutionary indication, perhaps a better test than the hotly
debated question of hydrocarbon clouds.

On many occasions in the past several centuries, falls of
gelatinous material have been reported in connection with
meteors. The literature in part has been compiled by Corliss.[19]
Luminous and therefore probably electrified while falling, the
stuff is transparent and colorless, texturally a jelly, stinks when
disintegrated, and dissolves into a few grains of residue after some
hours. One may guess that the Earth's reducing hydrogen-rich top
atmosphere is carried into contact meteorically with an oxidizing
lower layer, gathering dust particles and vapor, including metallic

---

[18] G. P. Kuiper, "On the Nature of the Venus Clouds," *Planetary
Atmos.,* Intl Atmos Union, Symposium 40 (Dordrecht: Reidel, 1971).

[19] See W. R. Corliss, *Strange Phenomena,* (Glen Arm, Md., 1974), 2v.

catalysts, to form a semi-solid type of formaldehyde glob the size of a drinking cup. These are certainly poor imitations of manna, but a similar process is entailed.

To portray its relation in volume to a smallpox virus, a single crystal of salt would have to be enlarged to a five-meter cube, on a ratio of one centimeter to 1 micron (10 -4 cm) for the virus to be visible.[20] There is certainly room for viruses to ride on cosmic dust. There is not yet a definite answer to the question whether meteoroids and comets do now carry or ever have carried organic molecules and primitive life forms. Brigham, in 1881, following Hahn and Weinland, reported a collection of some six hundred specimen of fossil life obtained by analysis of meteorites.[21] Their work was discarded as imaginative to the extreme, for they were discovering corals, sponges, and crinoids. In the thirties, Lipman and Roy debated the former's findings of rods and ovoid cells in meteorites.[22] Recently, Claus, Nagy, and others have discovered inherent organic compounds, carbonaceous chrondrites, in meteoritic material.

Hoyle and Wickramasinghe have tackled the problem vigorously over the past few years and emerged with two relevant hypotheses: one that life forms originated in space and a second that plagues also descend from space. Comets carry the appropriate chemicals and can carry on the necessary varying experiments naturally, over millions of years, until "photosynthetic bacteria, able to oxidize hydrogen sulfide anaerobically," emerged.

If a cometary impact led to the start of life, the question arises: would subsequent arrivals of cometary material carry biological

---

[20] Kees Boeke, *Cosmic View: The Universe in 40 Jumps* (NY: John Day, 1957).

[21] F. Brighan, 20 *Pop. Sci.* (1881), 83-7.

[22] Work by C. B. Lipman; S. K. Roy; E. Anders *et al.* and R. L. Levy is extracted by W. R. Corliss ed., in *Strange Universe* (Glen Arm, Md.: Sourcebook Project, 1977), 2v.

or prebiological material which might affect terrestrial biology? The boldest answer must be yes; that is to say, extraterrestrial biological invasions never stopped and continue today. These invasions would take the form of new viral and bacterial infections that strike our planet at irregular intervals, drifting down onto the surface in the form of clumps of meteoritic material probably similar to those studied by Dr. Rajan and his colleagues.[23]

The authors propose a perpetual vigil and a screening of stratospheric contents for microbes. If their theory is correct, one might expect veritable plagues to have had a hand in the great extinctions of species that have marked geological history. The causes of death would not only be mechanical - flooding, wind, hailstones etc. - and radiation, but also should include "biological warfare" against the Earth. Actually there is yet another dread possibility, chemical poisons, such as cyanide.

Iridium, osmium and arsenic occur in quantities hundreds of times above the normal in strata of the cretaceous-tertiary when the dinosaurs and many other species, both terrestrial and marine, extincted. Kenneth Hsu discerns at the same time a double blow to the biosphere in the form, first, of heavy atmospheric heating owing to a cometary pass-through and explosion, which killed off large terrestrial animals, and cyanide poisoning that wiped out calcacerous marine plancton.[24] The cyanide effect would be stressed by a catastrophic rise in calcite-compensation depths in the oceans after the cyanide was detoxified.

During these disastrous events, which may have happened on several or more occasions, not one alone, the ground forces would be highly energized. Velikovsky found it impossible to determine whether, in the plagues of Exodus, "the comet Venus infested the Earth with vermin," or "the internal heat developed by the Earth and the scorching gases of the comet were

---

[23] "Does Epidemic Disease come from Space?" *New Sci.* (17 Nov. 1977), 403.
[24] 285 *Nature* (22 May 1980), 202.

in themselves sufficient to make the vermin of the Earth propagate at a very feverish rate."[25] That many forms of life are comfortably buried below ground surface is well-known. But a thermal rise, flooding, earthquake, volcanism, and electrical discharging, will bring them out in incredible numbers. Thus the frogs of Exodus, the locusts, and the vermin also. One need only retroject modern reports, and raise the scale of intensity, to imagine the succession of events. In the area of the Krakatoa explosion, the nether world of animals was stirred up even while the gases burned, choked, blinded, and smothered people.

There is normally more in the soil than the erosion of terrestrial rocks: this has become apparent. Equally, new elements are discoverable that convey surprise, mostly unpleasant. The Dow Chemical company of Midland, Michigan, has been for several years in a quarrel with local authorities and environmentalists. The latter claim that Dow has manufactured chemicals that deposit dioxins, a carcinogen, in the soils. Dow says " we now think dioxins have been with us since the advent of fire.[26] It is perhaps uninformed to discount the company's research, that is apparently discovering dioxins everywhere. Adding more dioxins to the ground, of course, makes matters worse.

A parallel can be cited from the research into "Polycyclic Aromatic Hydrocarbons is Soils and Recent Sediments," conducted by Blumer and Youngblood, on behalf of the Woods Hole Oceanographic Institution.[27] Samples were drawn from "depositional and chemical environments ranging from continental and coastal soils to marsh and subtidal marine deposits, and from high to low oxidation-reduction potentials." The PAH component is significant; PAH is carcinogenic; ancient

---

[25] *Worlds in Collision*, 192-3, 268.

[26] R. Jeffrey Smith, 202 *Science* (15 Dec. 1978), 1166-7.

[27] *Science* (4 Apr. 1975), 53; see also R. A. Hites, Laflamme and Farrington, "Sedimentary Polycyclic Aromatic Hydrocarbons...," 198 *Science* (25 Nov. 1977), 829-31.

burning may be producing some of today's cancers; it would be well to perform statistical correlations on populations, cancer incidences, and "background PAH" of soils. PAH are formed at elevated temperatures by incomplete combustion.

> Our interpretation would imply that carcinogenic and mutagenic hydrocarbons occurred on the earth's surface during geological times spans. This raises the question whether these compounds might have contributed significantly to the processes of natural selection of mutation, and to the evolution of species.

The scientists assess the possible origins of the PAH deposits. They exclude weathering, seepage and spills, they exclude biosynthesis; they doubt early diagenesis in process of formation; they settle upon pyrolysis. This burning might be thought to occur on the site, but "the consistency in the PAH distribution among our samples suggests a predominant single mode of origin;" the sites are distant from one another. The chemistry does not permit regarding the PAH as "urban air particulates." Forest fires are "possible but unproven:" low temperature burning could provide the homology among the samples and air transport of PAH carbon ash from a great central fire somewhere might preserve the similarity. The ash layers are not noticeable, however.

The authors do not consider typhonic meteoric explosions and fall-out. This could raise to great heights the combustion residue of large vegetal areas and drop it around the world. Nor do they consider a cometary pass-through with a burning hydrocarbon tail that could deliver the PAH where and how found today. The time would be recent, for the PAH are in surficial sediments.

In sum catastrophes, especially if exoterrestrially invoked, display much chemical creativity. Great typhonic explosions on Earth, probably exoterrestrially induced, will behave more modestly, but similarly. Numerous gases, poisons, and foods have fallen out in natural history, and very recently. Precarious life situations have been widely and abruptly generated. Multiple reports of gaseous and fall-out processes in space and atmosphere challenge the credibility of radioactivity rates that have been established under guidelines consistent with presently observable rates.

CHAPTER TEN

METALS, SALT AND OIL

Iron-working is "siderurgy," a word out of ancient Greece and Rome. It translates properly as the working of star-iron. The Greek word for anvil, on which iron was worked, was close to the word for a meteoric stone. The Egyptians called iron "the bones of Typhon" and "a gift from Seth," both names corresponding to bodies crashing into the Earth, devil-monster and devil-god. Meteoritic iron was known to the early dynasties. "The Jews called iron ore *nechoshet*, which literally means the 'droppings of the (cosmic) serpent,' a nonsensical term unless our interpretation of it is allowed."[1] The Jews forbade the use of iron in chiseling stones for the construction of an altar. "A similar taboo was observed in Greek and Roman cults, it was and still is widespread."[2]

But, whereas the Egyptians held an especial taboo of iron, the Assyrians did not, and M. Sieff has described how Egyptian power waned when it lacked iron and waxed, on occasion, when foreign workers and allies such as the Greeks and the miners of Zimbabwe brought in iron and worked it. The Assyrians achieved their greatest conquests at a time of grave natural disasters (the Mars-associated events between -776 and -487).[3] South and north of Egypt, iron in large quantities was found and used; in Egypt it

---

[1] Bellamy, *op. cit.*, 84.

[2] Velikovsky, *Ramses II and His Time* (N. Y.: Doubleday, 1978), 221-47.

[3] "The Road to Iron: 8th and 7th Century Metallurgy and The Decline of Egyptian Power," (In press: *Catas. and Anc. HisT.M.*)

was neither found nor used. Query: why was no distinction made between meteoritic sacred iron and mined iron? Possible answer; because all iron was known to be meteoritic. Much may have fallen in association with the activity of the great war god Mars-Ares-Nergal. Adequate metallurgy was known for thousands of years before the iron age; increased temperatures could have been devised if the will - and the material - were present.

In conventional works of human history, iron is placed as a late discovery. The "iron age" comes after the "bronze ages" which follow the "Stone-ages." These terms and divisions now only perpetuate confusion in anthropology, history, philosophy, and perhaps even in geology. Thus, a common reference, the *Columbia Encyclopedia,* thinks that meteoric iron beads existed in Egypt as early as 4000 B.C. but iron smelting not until 1900 B.C. and later.[4] Some confusion is admitted on the matter and Velikovsky's reconstruction of Egyptian chronology has added dismay to confusion.[5] Some even say that iron may have been used before bronze, since isolated iron artifacts of very early dynasties have been recovered. By the end of the second millennium, iron was in general use in Palestine and probably also to the North. A Soviet excavation has reported a metallurgical industry between 3000 and 2000 B.C. in Medzamor.[6] with steel tweezers dated at about 1000 B.C. Several experts now assert that there was no clear functional superiority of iron in the first centuries of its use; bronze was adequate even for weapons.

This all would signify a concurrent use of iron, lead, tin, copper, gold and silver by 2500 B.C. in the Mediterranean and

---

[4] R. Maddin, J. D. Muhly and T. S. Wheeler set a date between 1100 and 900 B.C. ., "How the Iron Age Began" 237 *Sci. Amer.* (Oct. 1977), 112.

[5] See fn 2, p. 5.

[6] L. Pauwels and J. Bergier, *Eternal Man* (Herts, Eng.: Mayflower, 1972), 58, 160; and also their *Morning of the Magicians* for many suggestions of prehistoric discoveries.

Middle East, also perhaps elsewhere in the world. The question arises why mankind did not use metals and invent metallurgy earlier. Could all the workable surface metals of the world have arrived from exoterrestrial sources within a brief period of late proto-history, and so vividly that the ancients even could assign separate periods for their arrival, as Hesiod and Ovid did when reporting a golden age, succeeded by a silver age, and ending in an iron age? I cannot attempt a full answer here, but would support the case for human-witnessed exoterrestrial falls.

Bellamy can again be quoted:[7]

> Gold, platinum, uranium, radium, mercury, bismuth, and other heavy metals are not detected in the surface layer of the Sun, nor of any other star. As we cannot suppose that they do not exist in those bodies they must logically be present in their cores--and hence also in the cores of the smaller cosmic bodies, planets. Therefore, the presence of heavy metals on, or near, the surface of our Earth points to strewing from without. Without such cosmic strewing no ores would probably be found on, or near, the surface of our Earth at all.

> In the south of the Belgian Congo (now Zaire) there is a zone, about 180 miles long by 25 to 30 miles wide, which contains great deposits of ores--chiefly copper, iron, tin, uranium, and cobalt. In Angola and Rhodesia, as well as in South Africa, there are smaller deposits.

> Indeed, many geologists are of the opinion nowadays that the great rich ore deposits at least must have been brought into being through strictly localized, exceptional, and briefly operative causes.

Iron, the ancients believed, was meteoritic in origin. What would they have believed if they had seen the now exposed great iron mountains of Minnesota or Venezuela? Could such mountains have fallen from the sky? Unquestionably. Asteroids

---

[7] *Op. cit.,* 197-8.

exist in the size of iron mines and contain as much iron. Would they not have exploded and dissipated into dust upon landing? Some would and some not.

A not-well-understood feature of meteoroid falls is that they can accomplish soft landings as well as hard crashes. In hard crashes, such as at Campo del Cielo (Argentina) where a number of meteoroids fell, "large masses of meteoritic iron and shale have been found in its vicinity."[8] Heide writes, "the 60-ton meteorite from the Hoba farm near Grootfontain, South West Africa, the heaviest of all known meteorites, imbedded itself in friable limestone at a depth of only 1.5 meters. The iron meteorites of Cape York in Greenland, weighing up to 30.875 tons, lay on solid gneiss rock, or were barely imbedded in moraine rubble, without any trace of an impact. Here we may guess that they fell on a thick layer of ice or snow and sank to their final location as the snow or ice melted.[9]

However, as the Mass and Velocity of the meteoroid increase, its Energy of impact increases, according to the formula $E = 1/2 \, mv^2$. The atmosphere cannot brake the body in time. Therefore, no iron masses of over 100 tons have been deemed to be of exoterrestrial origin; where such have actually fallen, and few doubt this, they have been vaporized by the impact.

In the face of this formula and the visible facts of meteoritic iron, it would appear that the large iron ore masses on Earth cannot have originated exoterrestrially. The negation, if any, depends upon variable velocity. If the falling iron mass is electrically charged, or gathers a charge, so as to render it less attractive to the Earth its velocity would diminish. Theoretically, it could waft down in a soft landing in one piece. If it crashed upon landing, it would possibly assemble itself into the form of an iron ore deposit as deluges of water and dust would fill the

---

[8] Fritz Heide, *Meteorites* (Chicago: U. Of Chicago, 1969), 44.

[9] *Ibid.*, 16.

interstices. Strange objects have been found in the midst of iron ores being mined, such as wood of recent date.[10]

Much that is meteoritic may not be discovered. On an Antarctic ice field, Japanese explorers found over 1000 meteorites, of which only one was composed of iron.[11] Were the field of stone, instead of ice, the stone meteorites would probably go undetected. Obviously we could not test all the Earth's rocks for exoterrestrial origin, especially since the tests themselves might beg the question.

Masses of iron were found lying upon a Disco Island (Greenland) shore with a great gneiss erratic boulder and associated with the talus of a basalt cliff which itself contained similar bits of iron. All the iron was termed meteoritic which led the investigators to wonder, especially since the basalt fragments were found even embedded ' inside the iron of the beach, whether the meteorite shower "occurred while the basalt was in a state of pasty eruption."[12] But, too, the range itself, though immense and tall, might have been the rim of a great impact collision and was permeated by and interacted with the exploding body.

Suppose all known meteoritic material in the world were assessed for its proportion of iron. Suppose then that one calculated the proportion of iron ore to the amount of drift, loess and homeless clay. If the two ratios were similar, the exoterrestrial thesis would be expanded to embrace the materials of both ratios. Iron in one form or another composes about 5% of the Earth's surface rocks; here is a thoroughly homogenized relationship of iron to rock. This ratio turns out to be closer to the ratio of iron to stone in meteoroids. Both ratios would be far removed, no doubt, from the ratio of iron ore to drift and loess, which would probably be one in thousands.

---

[10] Melvin A. Cook, *Prehistory and Earth Models* (London: Max Parrish, 1966).

[11] 52 *Sky and Telescope* (1976) 429 citing a report by Walter Sullivan, NYT.

[12] 2 *Sci. Am. Supp.* (1876), 510.

We can imagine, as have several scientists, that the meteoroids fallen upon Earth are those of a late planet explosion in the region of the belt of asteroids and therefore we have been sampling a planet composed as the Earth is supposedly composed, with iron and nickel core, sima mantle, and sial crust. Calculations, given this simple idea, are complex but not enough. There is too much evidence of exoterrestrial dumping upon Earth by other bodies, more of the nature of Jupiter and Saturn, to carry out this algebra of ratios with confidence.

Generally, "terrestrial" iron bodies are distinguishable in composition from meteoritic iron in that they contain either smaller amounts of nickel (about 3 per cent) or larger amounts (about 35 per cent). The meteoroids also contain some cobalt. The distinction is hardly foolproof. Generally, too, the meteoroids have encrustations attributable to their experiences in space, although this is statistically discoverable and not an absolute distinction.

Perhaps somewhere in the literature, unknown to the present writer, exists a systematic examination of the boundaries of a very large metal body demonstrating a lack of exoterrestrial experience. Nor is there a great iron body embedded in precambrian rock; nor has anyone come upon intrusive pipes of iron ore that would have conveyed metal from the core or mantle, by some combination of electrical and volcanic force.

If an alternative to an electrically-assisted soft landing were sought, one might better conceive of a welding process; gigantic lightning strokes from iron bodies in space lasting for a minute would cast molten iron ore down their path to where they now rest in heaps. Again, a study of ore body boundaries is needed. Schaeffer has written of the layers of ashes and cinder scories close in to a huge pure copper mine of Cyprus.[13] One recent theory has the same copper distilling from a hot spot of a northern fork of the great African rift. To this author, the exoterrestrial notion is as convincing.

---

[13] *Stratigraphie Comparée...* (London: Oxford, 1945). 580.

Like Bellamy, I am impressed by the fact that "there are, scattered over the Earth, a number of ore-mountains which are evidently foreign to their surroundings. At Eisenerz, in Austria, there is a huge mountain, consisting altogether of iron ore .... On the island of Elba, in Sweden, in Russia, in India, and elsewhere we find more or less considerable hills consisting of pure iron ore, mineral wonders of the world. In Orissa, India, in the jungle near the village of Sakchi, is a hill consisting of iron ore which is so rich that it yields almost 65 per cent of pure metal." Elsewhere he writes that such mountains would, upon investigation, probably prove to be 'rootless. ' He describes others.

"At Gellivara in Sweden there are enormous deposits of iron ore whose special characteristic is that they are found in floelike masses, as if they had been 'pancaked' down. At Kirunavaara and Loussavaara, in Lapland, there are similar deposits of iron ore. The 'Kursk Anomaly' in Russia consists of a mass of iron ore estimated to contain about a cubic mile of high-grade material. In the Ural area there is Gora-Blagodat, the 'Blessed Mountain, 'an iron ore mountain 520 feet high, situated in a plain. In Russia too is the Wyssokaya Gora, a deposit of rich magnetite ore, littered over a strip 40 miles long by 9 miles wide."[14]

As with iron, so with other metals: many legends have them falling from heaven. The Chinese sky dragon's "breath descends as a rain of water or of fire. Gold is the congealed breath of a White Dragon, but a Purple Dragon's spittle turns into balls of crystal; glass is regarded as solidified dragon's breath." (The tektite allusion is plain). "The dragons of mythology are often described (among the Teutons, for example) as guardians of hoards and givers of wealth." The dragons are wise in metallurgy.[15]

Donnelly says the same. He describes "Beowulf, when destroyed by the midnight monster, rejoicing to think that his people would receive a treasure, a fortune by the monster's

---

[14] *A Life History of Our Earth* (London: Faber and Faber, 1957), 196.

[15] Bellamy, *Moon...* 87, 89.

death."[16] Further, now Humboldt writes, the Scythians had a sacred gold which fell burning from heaven. "The ancients had also some strange fictions of silver which fell from heaven, and with which it had been attempted, under the Emperor Severus, to cover bronze coins." [17] An image of a rattlesnake with a tail of gold, and descended from heaven, was worshipped by the Inca as the god of riches. In the Bible (*Job* 21) it is said of the horrendous dragon Leviathan, "he shall strew gold under him like mire." And Chan reports that in ancient Mesoamerica "yellow was the color of gold, the *teocuilatl* or excrement of the gods." [18] The dragons that are the substance of most ancient myths and of children's fairy tales today tortured and enriched both the Earth and the minds of men.

Cores drilled from Antarctic sediments of Pleistocene age contained iridium and gold in anomalously high proportions. "A sizeable fraction of the noble metals is contained in vesicular, millimeter-sized poly-mineralistic grains that closely resemble ablation debris from chondritic meteorites, and there is little doubt that the noble metals resulted from the accretion of a large extra-terrestrial object."[19]

About the same time as this expedition, the largest American gold strike in a century was occurring on the Thornton-Ash ranch in Nevada. The gold was not in nuggets, but in microscopic sizes like the Antarctic find. It is extracted by crushing and leaching its host rock. Large tracts of land are being scooped out and many millions of tons of rock processed to obtain the gold. In the absence of a comparative examination of the Nevada and Antarctic discoveries, one may suspect an exoterrestrial origin of the Nevada gold as well.

---

[16] *Op. cit.,* 16.

[17] 17. *Cosmos,* I, 115. 18. R. P. Chan, *A Guide to Mexican Archaeology* (Mexico City: Minutiae Mexicanae, 1971), 75, 78.

[18] 17. *Cosmos,* I, 115. 18. R. P. Chan, *A Guide to Mexican Archaeology* (Mexico City: Minutiae Mexicanae, 1971), 75, 78.

[19] F.T Kyte *et al, Nature* (30 July 1981), 417-20.

Conventionally, studies of the origins of metals and their cultural recognition do not mention any exoterrestrial contribution to their chemistry, appearance or use. Instead, they are looked upon as components of igneous intrusions. Speaking of gold, silver, copper, lead and tin, Clair Patterson in his exceptionally important study of "Native Copper, Silver, and Gold accessible to Early Metallurgists," declares:[20]

> The primary igneous minerals of the 5 anciently used metals were generally mixed with a large number of unwanted minerals in the vein or lode. Useful igneous minerals of the 5 different metals were not generally mixed together, however. Except for close relations between lead and silver, deposits of the 5 metals were more unrelated than related in a specific region (Noble 1970). The different metals were generally successively deposited over a period of time in adjacent regions (Noble 1970). The common characteristic which bound the deposits of all 5 metals together was the fact that they were emanations derived from igneous intrusions in mountainous belts, sometimes occurring together, or nearby, or not at all.

He reports that the ratio of copper to silver to gold mined from all types of deposits in the entire world from 3800 B.C. to 1925 A.D. was 3,000 to 11 to 1, and believes the ratio not to be far removed from their natural incidence as ores. These are largely surficial, he says, even though he expects the same metals to be found in highly dispersed, fine grains throughout the crust, where their bulk would be perhaps seven million times that of the ores. "The lower the grade of ore, the more there is of it, until finally we include the entire earth's crust in our consideration."[21]

> It is likely that the greatest masses of copper, silver, gold, tin and lead ores were emplaced in the upper several kilometers of the earth's crust rather than throughout the total 35 km thickness of the continents or the thicker upper mantle.

---

[20] 36 *Amer. Antiquity* 3 (July 1971), 286-321, 288, *cf.* 294.
[21] *Ibid.*, 291.

Governing agents in this vertical distribution were abrupt
decreases in temperatures and pressures near crustal surfaces.
It is unlikely that there are any large deposits of the kind we
commonly recognize as ores at great depths in the crust,
although there are very large amounts of copper, silver, gold,
tin, and lead dispersed down there.

It seems that ores are found in a highly confused and
diversified state that does not let one assume any neat intrusion of
pure metal. Nor even is the intrusiveness manifest; the term seems
to define itself, as simply something differing from its
surroundings, not a clean belt or stratum, but as a conglomerate
chemically, physically, and morphologically.

Ore is the valued part of minerals, including metals. The
modern processes used to isolate ore are imitations of nature.
Crushing is first, where the pressures and grinding of water, wind,
and rock movements are emulated. Mineral separation follows.
Minerals of different sizes are shaken through sieves. A
hydrocyclone may be used to segregate particles by their response
to varying winds. Flotation is employed to separate the crushed
particles according to their density. The material may be conveyed
along jigging tables under running water so that high density, then
afterwards lower density material, settle. A magnetic wheel can
collect from poured minerals the magnetic ores and cast off the
less-magnetic ores. Minerals that accept water-proofing can float
in a froth while non-proofed minerals and rock sink. Once
minerals have been chemically created, by high-energy forces, the
same or a varying mix of quantavolutional forces can segregate
them.

Under these circumstances, a person of our persuasion is
likely to see exoterrestrial intruders smashed, crushed and
exhibiting metal here and there; or, secondly, rims of hardly
discernible craters containing segregated elements of the Earth's
rock mixed with exoterrestrial elements that have been subjected
to the immense heat and pressure of a crash; or, thirdly, effects of
massive electrical discharge plus fall-back of exploded earth.
(Regarding this last, and considering the unusual conductivity of

metals, have they been prepared for conductivity, like quartz semiconductors? Are we dealing with homeopathy or homology?)

The distribution of metals in the world is associated somewhat with folding and thrusting, but this may be a finder's help, not a random sample of ore distribution. More significant is the lack of correlation of these metals with volcanism or even with great faults. Why should metals congregate near circular features and basins, suggestive of astroblemes?

Flint is found that has undergone controlled heat treatment, with pressure retouching as revealed by spectroscopic experiments; this is at least Solutrean in age, 22,000 B.P. by conventional dating.[22] The skill is as complex as and less enjoyable than metalworking by heat; why then did man wait another 15,000 years to begin his work with copper, tin, lead, gold, silver, and iron? Perhaps they were not available. Or, perhaps the dating is too long and, soon after the flintworking, metalworking began, which is one logic for preserving the conventional origin of metals by casting aside the conventional chronology.

Before the ages of the metals, so-called stone age man existed. He used many different kinds of stone, bone, wood, and grasses. He designed, cut, heated, and molded them. He domesticated animals, grew cereals, performed anatomical operations with stone knives. He built cities and great monuments. He painted, danced, and sang. Coal and peat were burned. Obsidian and flint were mined; Greek myth portrays Saturn castrating his father Uranus, using a jagged-edged sickle of flint. If any amount of terrestrial iron had been present on the surface and outcroppings, why would it not have been employed? Gold, silver, copper, tin and lead were mined and used.

Mankind was ready to work and even to melt and purify iron, it seems, long before it was available. If only in order to supply the type of hypothesis that may lead usefully to historical research on the subject, I would suggest that most metals occurred around the period of the great Deluge and in the transition from

---

[22] 276 *Nature* (14 Dec. 1978), 7013-4.

Saturn to Jupiter worship, about 4200 B.C., and may be connected to a cosmic explosion that I have in *Chaos and Creation* assigned *to* a planet with the legendary traits of Apollo. It is noteworthy that the ancient metal mines of Attica had two favorite names, *Artemisiakon* and *Hermaikon*, both siblings of Apollo.[23]

John Saul drew circles corresponding to rounded features, possibly ancient exploded craters, on a topographic map of a portion of Arizona. He independently marked the location of mineral deposits on a similar map. When one overlaid the other, there appeared a significant relationship between craters and mines, with the deposits generally occurring on the rims of the circles. One circle was abundantly supplied with minerals, indicating that a certain small percentage of craters, and hence their originating body, may be heavily mineralized.[24] R.S. Diez is cited by Pauwels for arguing the origination of the immense Sudbury (Canada) nickel mines from a meteoroidal impact of pre-Cambrian times.

One can conjecture, then, about a possible ratio of large stone meteoroid impacts to large mineral meteoroid impacts corresponding to the experienced ratio of small stone to small iron-nickel meteoroid impacts. Since historical experience has been limited (explainable by the negative exponential principle), one would hardly expect historically the fall of the rarer metals such as gold, silver, and copper.

Walter Sullivan has presented in the *New York Times* of Nov. 2, 1966 a map of the world's most productive gold field below Johannesburg, which shows a large primary "bulls-eye-formation" rimmed by gold-bearing formations and a much larger 200-mile-diameter, secondary, cratered, rim-like area, also bearing gold, and asks "Did a comet create a South African gold field?"

---

[23] Advice of Prof. Merle Langdon, then of Am. School Class, Studies; Athens. "Artemisiakon" was a favorite name, the "kon" ending meaning "under the protection of," "owned by" or "discovered by."
[24] 271 *Nature* (26 Jan 1978), 347. 21. 271 *Nature* (26 Jan 1978), 347.

Unless the gold was alchemized on the spot, it might have been part of the meteoroid that crashed.

Most metals, in conclusion, may originate exoterrestrially. If an alternative must be found, it may be suggested, although hardly discussed directly here, that special thermo-electric events might produce the metals. This would constitute electrolysis on a huge scale, in a dense catastrophically formed atmospheric plasma, before or after striking.

The metal, manganese, is exceptionally terrestrial in origin. Its growth out of underseas volcanos is particularly explosive and rapid. Pure manganese is found in cones near the Mid-Atlantic Ridge. Hot water and steam percolate through lava segregating the metal and depositing it in molten pools where it cools shortly. The French-American Mid-Ocean study, "Project Famous," found manganese geysers along the Ridge in the 1970's.

Manganese is also found in nodules on the ocean floors. These, by contrast with the geyser type, are supposed to have required much time to grow. Scott and his colleagues estimated that nodules grow at rates of 1 to 10 mm/million years. They are supported by Ku, Burnett and Morgenstein, using both radiometric and nonradiometric techniques of dating. But Goldberg and Arrhenius reported finding a 50-year old naval shell with a ferromagnesium oxide coating 30 mm thick, indicating a rate of 60,000 mm/m years[25]. Heezen and Hollister point out that the rate of accumulation of manganese is a function of its concentration in water and the availability of a nucleus in the water[26]. Conventional gradualist theory cannot explain the "mystery" so well as quantavolution.

Nodules abundantly litter the deep abyssal hills. They form around a particle, tephra, a pebble, an animal tooth, a bone, or on the surfaces of volcanic or drifted rocks. The nodules should

---

[25] P. A. Smith, 265 *Nature* (1977), 582-3 reporting Scott *et al.*, I *Geophys Res. Ltrs* (1974) 355 and Golberg and Arrhenius, 13 *Geochim. Cosmochim. Acta* (1958) 153; Corliss, *op. cit.* ESS-005 doc.
[26] *Op cit.*, 424, 440.

require a very short time to form, if supplied with nucleus, warm water and a manganese rich soup emerging from fast flowing and erupting volcanos. The manganese adheres to any object and rafts to its ultimate destination far from its birth place with fast-spreading lava, which also boils out manganese accumulations as it spreads, and by swirling currents of newly forming seas around it, the same currents that hold the nuclear objects in suspension for a time. The process 'proves itself as turbulent and swift by the nuclei, which would otherwise sink in the abyssal muck if there were such and by the availability of manganese only at the hot spots of the ridge. Thus, contrary to the long-time theory of manganese formation, the very presence of the manganese nodules goes to demonstrate how rapid was the paving of the ocean basins, a topic to be treated later on.

Sodium chloride is of course a mineral compound, and not a metal. The salt domes of the world, averaging 30 cubic miles each, may carry 100,000 cubic miles of salt, about three-tenths of all the salt of the seas. Salt is not found in pre-cambrian rocks, which are said to embrace most of the time since the Earth was created. Basalt of the ocean bottoms contains no salt and salt could not have been precipitated from the melting of mantle rock.[27] Granite is also deficient in salt.

The presence of salt, like the metals, in living tissues, and therefore the need of it, does not prove its terrestrial origin. Nor should one gullibly receive the story that since salt is in our tissues, it must be part of the ancient waters that bore the first life, hence giving us proof of most ancient salt oceans. Life digests salt-free water, even ocean life. If all the water of the world were to receive all the salt deposited in domes, life as we know it might become precarious--except insofar as we constructed desalification factories to sustain it. The miracle is that salt has not killed life already, like many ancient settlements had their land sown with salt by their enemies, and thus were extinguished. Species closely resembling one another are to be found both in oceans and

---

[27] Cook, *op. cit.*, 87.

freshwater lakes and rivers. Salmon live in both oceans and rivers during their individual lifetimes. Paleontology may not be able to demonstrate the precedence of saltwater over freshwater life forms. Too, the medium of early marine life may have been brackish.

There is no apparent earthly source for salt. A Head Curator of Geology at the U. S. National Museum, George P. Merrill, long ago wrote that sodium chloride (at least the latter) must have come like meteorites from outer space and been caught up first in the atmosphere and then dumped in the oceans. By the atmosphere is implied a canopy sky. From the canopies, salt would descend with water deluges, which we shall be considering later as a quite recent event. The canopy or set of rings may have been a momentary affair or endured for centuries. The rings and body of Saturn may contain sodium chloride or its elements; the rings contain millions of small mineral objects. Legendary evidence exists on this account.

Once salt in solution strikes the ground it must run off into the basins that have water, making it salty, and also contribute with its host water to new seas. If it sinks into the ground in solution it will form a reservoir, either exposed or folded under or trapped in a cavity. In these cases, the water will boil out as steam: or it will percolate into underground and above-ground branches flowing to the sea. The salt residue will then form domes.

Cook argues that the salt domes were created in the same set of events as the deep burial of organic material of which petroleum is composed, for many salt domes act as oil traps, keeping oil from dissipation. Avalanching ice from collapsing ice caps, and sediments pushed by these, suddenly thrusted and folded salted waters that were swirling around the great movements, containing them under high heat and pressure. The trapped waters were squeezed out of insoluble sediments into their own cavity. There they evaporated quickly, leaving salt deposits. But it is unlikely that the waters of the Earth were so salty as to provide, via tides, the salt domes and still leave the run-off waters with the present heavy component of salt in solution.

Furthermore, as later chapters here will argue, the bulk of the ocean waters and ice came exoterrestrially and the salted waters mostly arrived later.

The salt may have descended both as a solid and in aqueous solution. Salt domes exist beneath the sea floor as well as below the land. Salt domes containing oil have been discovered beneath the floor of the Gulf of Mexico at 12,000 feet of depth (2000 fathoms).[28] Great salt domes have been discovered below the Mediterranean floor as well, giving rise to an idea that the Mediterranean once, 12 million years ago, became a dry basin. Why salt should not then be evaporated and laid in even layers of sediments rather than in intrusive pockets is unanswered.

In South and East Texas many cylinders of salt (with nearby anhydrite, gypsum, oil and sulfur deposits) penetrate the Earth to depths of a thousand meters and more. Kelly and Dachille ask "What could have caused these tremendous beds of practically pure rock salt?" And they write: "Our inevitable answer is the same, collision-flood. We should guess that this pan of the earth was struck by a body or bodies of sufficient size to evaporate great quantities of ocean water, both by the Kinetic energy released by the impact and by the great pool of molten lava that must have been formed in the crater. This evaporation of ocean water would have left the salt provided that it was not connected directly with the main ocean, otherwise the salt would have gone back into solution."[29]

The Gulf of Mexico does seem to have vague characteristics of a gigantic meteoroid impact. Since other salt domes have been also discovered beneath the gulf itself, one may wonder whether the meteoric body itself may not have been composed largely of salt and injected its own salt tubes into its crater basin. This would seem a more realistic scenario than the Kelly-Dachille vision of a typhoon lifting salted waters into the air, evaporating the waters, and having the salts precipitate in favored

---

[28] Oscar Wilhelm, *Geol. Soc. Am. Bull* (March 1972).

[29] Allan Kelley and F. Dachille, *Target Earth*, 211; *cf.* 205.

sequence and locale in a pure state. The fact, as they recall it, that salt is so free from contaminants (less than 0.4%) argues for the solid integrity of the salt from its initial appearance on Earth.

Legends imply my theory. Saturn was the first Lord of the Mill, a grindstone round like the revolving vault of the sky. It ground salt into the sea and was sunk in the ocean during the great maelstrom and deluge that brought the golden age of Saturn to an end. In Hindu myth, the gods churned the celestial ocean and the mill ground out salt into the sea. Norse myth has the heavenly mill churning out gold, then salt, then, sunk in the sea, sand and stones. The unhinging and failing of the Mill implies, too, a tilting of the axis of the globe, a likely accompaniment of the cataclysm.

A South American legend supplies significant detail. "The Arawak of Guyana call the Galaxy 'the Tapir's Way. ' This is confirmed in a tale of the Chirignano and some groups of the Tupi-Guarani of South America." According to Cuna tradition, "the Tapir chopped down the 'Saltwater Tree', at the roots of which is God's whirlpool, and when the tree fell, saltwater gushed out to form the oceans of the world."[30] The Cuna cosmology thus unites the idea of the tree-of life found in many places, including *Genesis*, with a Tapir-god, Saturnian-Elohim divinity, and, as the tree of life is destroyed (the old order ends), saltwaters deluge the Earth. (In *Solaria Binaria*, Earl R. Milton and the present author identify this tree of life with the legendary and philosophical axis of fire and this with the presence, until a nova of Saturn, of an electric arc-current flashing between a then-larger Saturn and the Sun, and visible to man.) In sum, various legends independently agree that the salt of the oceans came with an aquatic cataclysm in a time when mankind was an intelligent witness.

---

[30] G. den Santillana and H. von Dechend, *Hamlet's Mill* (Boston: Gambit, 1969), 247, *cf.* 146-7.

25A. *Deuteronomy* 29: 22 (Watchtower Edition); Cardona "Jupiter--God of Abraham (Part III)," VII *Kronos* (Fall 1982), 66. Fire evidence is copious in the settlements excavated at the sites.

That salt came down upon the doomed "Cities of the Plain" at a later time as well is argued by Dwardu Cardona. Yahweh threatens his people with "sulphur and salt and burning, so that its whole land will not be sown... like the overthrow 3f Sodom and Gomorrah, Admah and Zeboiim, which Yahweh overthrew in his anger and in his wrath..."

In the same work, Milton and I propose that the Noachian Deluge occurred in cyclonic form, with the salty waters hosing or jetting down at thousands of locations. If this were correct, some of the characteristics of salt deposits would be explained, such as their common cylindrical shapes and great depth below the surface of land and sea bottom. The saltwater would bore through the surface rocks under great pressure and with enough time to penetrate deeply. The water would vaporize promptly in the ambient heat and what was left of it would leak through a multitude of fractures on the margins of the deposits.

In Manchester, England, a process of making petroleum from garbage has been announced (1982). "We can do in ten minutes what nature has taken 150 million years to do," asserts a proud engineer. The oil costs half the prevailing price of natural crude oil. This price does not consider the original devastation of the biosphere that occurred with the natural production of oil. Conventional belief interprets oil resources according to an idyll, that organic rot was deoxidized, accumulated over long periods of time, roasted slowly at a deep warm level in the rocks until it turned into oil, then seeped into rock reservoirs where it was trapped to await the oil explorer of today. There is little use in our discussing this story, inasmuch as the reader will have ready access to it in many books. Here it is argued that oil is a catastrophic product and the major questions concern the catastrophic mechanisms of its formation.

The "ten-minute oil" suggests that there may be no inherent guarantee that natural oil is old. Recently discovered hydrothermal vents in the Gulf of California are producing from sediments a petroleum that is close to commercial standards. Several C 14 dates of oil offshore California and from the Gulf of

Mexico range from 5,000 to 20,000 years. Still petroleum generally is dated from two to six hundred million years; a common age given is fifty million years. One group of scientists suspects that solar ultraviolet polymerized the methane atmosphere of primeval Earth to form an oil slick of one to ten meters' depth all over the globe.[31] T. Gold believes that methane, composed of carbon and hydrogen, erupts from primeval reservoirs in the mantle; they sometimes explode from electrostatically induced sparks.[32] However, the presently continual explosions would indicate to this writer a recent origin of the methane, probably from biomass deep-buried by catastrophe. A. T. Wilson produced hydrocarbons out of electrical discharges on methane and ammonia, and claimed in 1962 that the Venus atmosphere held hydrocarbons.[33] Oro and Hart maintain a case for current hydrocarbon production on Jupiter from methane; they manufactured hydrocarbons from methane in their laboratory.[34] Libby has theorized that oil is raining down upon Jupiter today.[35]

Max Blumer, a pre-eminent paleo-geochemist, lately of the Woods Hole Oceanographic Institution, used the conventional age estimates given above in making a calculation of some social significance. Reminiscent of the Dow Chemical Company's claim about natural dioxins mentioned in the previous chapter, oil shipping interests have protested that only half the

[31] A. C. Lasaga and H. D. Holland, "Primordial Oil Slick," 174 *Science* (10 Oct. 1974), 53-5.

[32] See K. S. Lewis, 78 *New Sci.* (1978), 277 and Walter Sullivan in *NYT* (24 Dec. 1977), 1.

[33] A. T. Wilson, "Synthesis of Macromolecules." 188 *Nature* (17 Dec. 1960), 1007-8.

[34] J. Oro and J. Han, "High Temperature Synthesis of Aromatic Hydrocarbons from Methane," 153 *Science* (16 Sept. 1966), 1393-5. *Cf.* J. Oro, "Comets and the Formation of Biochemical Compounds on the Primitive Earth," 191 *Nature* (29 Apr. 1961), 389-90.

[35] C. J. Ransom, *The Age of Velikovsky* (Glassboro, N. J.: *Kronos*, 1976), 80-2.

ocean's petroleum content comes from polluting practices and the other half comes from natural leaks and seepage. In 1970, Blumer, following this logic, estimates the amount of seepage at 5×10^6 tons. Quoting then high estimates of offshore oil resources at $100,000 \times 10^6$ metric tons, he points out that all of this would have leaked out in less than 20,000 years. But, taking the average age of oil as above, 50 million years, and the claimed seepage rate, "the average offshore oil-field would have lost to the ocean 2500 times the free flowing oil or more than 1500 times the total oil existing in situ before commercial offshore oil production started."[36]

Obviously, in Blumer's view, and the publicity attendant upon the brief article indicates a wide acceptance of it, the estimate of natural seepage is ridiculously high; the polluters are responsible for the oil in the oceans. The same is true on land. Seeps are negligible because "oil reservoirs are well sealed even on the continents where uplifting and erosion should have bared oil-bearing strata more extensively than on the ocean floor." Oil leaks are frequently sealed by natural asphalt.

The quantavolutionist can address three comments to Blumer's line of argument. First, the age of oils in the sea may be grossly overestimated. Possibly the oil resources of the world are under 20,000 years old; in this case, the allegations of the seepage advocates would have to be disproven by other evidence, if at all. Second, Blumer does not deny seepage, but wishes it reduced. But he does not estimate seepage, or else, I guess, he would have to name a figure, such as one-tenth of the seepage claimed. In this case, the age of the "average oil" would drop by a factor of ten; all oil resources would be exhausted by leakage in 200,000 years. Surely he would not insist upon the fifty million years age and therefore be compelled to argue that true seepage is hundreds of times less than claimed. In other words, he is walking right into the quantavolutionary door; no significant seepage is satisfactory if conventional oil ages are to be defended. This is especially so, given that strict uniformitarian rates are not likely; no matter how

---

[36] "Submarine Seeps," 176 *Science* (16 June 1972), 1257-8.

oil is made, early seepage must have been at a faster rate than today's seepage. Even just the transfer from factory to reservoir cannot occur without large losses. Again the age of oil must drop. And of course if a quantavolutionary theory of oil formation is adopted, the exponential principle come into play: oil is made, not in ten minutes, not currently in submarine hydrothermal factories, but in very short times nevertheless.

Two quantavolutionary theories, requiring very short times, offer themselves, one best enunciated by Melvin Cook, the other by Velikovsky. Cook hints that a great deluge may have precipitated the lateral break-out of the ice caps. The vast ice avalanche bulldozed the biosphere long distances and folded it into the Earth in a heated state. Velikovsky argues for the origin of petroleum from the tail of a great comet, which he identified as an erratic Venus. Both offer short-term explanations, Cook placing the production of oil around 10,000 years ago, Velikovsky around 3450 B.P.

Cook reconstructs the oil production process as follows: around the old ice cap of the north grew a heavy biosphere. The towering ice cap, triggered by deluges, exerted fracturing radial pressures that sent great bulldozers of ice and rock in all directions to sweep up, ignite and bury deeply the vegetation and animal life. The organic matter stewed under high thermal and pressure conditions. Some became coal; some became oil and natural gas. Here is a quick "Cook's Tour" of the world's petroleum.[37]

> The most prolific oil basins of the world are those associated with the postulated major long-thrust systems described previously, namely the Mississippi valley-Gulf of Mexico system and the extensive and complicated overthrust systems comprising the great oil fields surrounding the Red, Mediterranean, Caspian and Black seas and the Persian Gulf. The southwestern USA thrust system responsible for the fragmentation in the Basin and Range province possibly contributed to the California oil basins. Another similar thrust

---

[37] *Op. cit.*, 241 ff.

system apparently generated the oil and coal provinces of Borneo, Sumatra, Java and New Guinea. These great oil and gas regions are most likely associated with sudden deep burial of marine and vegetal matter in (1) spoke-like radial thrusts from the ice sheets that began with the flood and eventually triggered continental drift, (2) continental drift itself, and (3) the Subsequent catastrophic effects of readjustment (ocean ridge and related systems). The greatest oil fields in the world, those in Iraq, Iran, Arabia and Kuwait, are apparently the result of all three of these mechanisms of sudden deep burial. The Gulf of Mexico system is postulated here to represent tremendous, sudden and deep burial thrusts contributed largely in the pre-continental drift stage, but with great contributions from both the north and the south such as to insure deep burial of sediments all along the coast and shelf of the Gulf of Mexico. The west coast of North and South America represent regions showing perhaps all of the deep burial effects: that due to welting and overthrusting in the pre-continental drift stage being strong in this region, the welting at the front of the thrust blocks during continental drift itself and the tremendous upheavals strongest here in the final readjustment stage. Perhaps the great (bathylithic) uplifts associated with the earth-circling ridge and rift system, particularly that part that cut into the continent in the western side of the Americas, contributed mostly to the deep basin structure in California, accounting for the youngest pools of the world.

Cook, then, must provide a force sufficient to initiate the break-out of an ice cap of enormous size; then a thrusting and folding of crustal rock over large distances, burying a whole biosphere of vegetal and marine life; then a cracking of the globe, sending the continents skittering from the great Atlantic and Southern ocean cleavages in a complex pattern, with a major fracture moving through most of the world along the old Tethyan sea belt. He concludes as follows:

> The physical chemistry of oil, including its formation from marine raw materials, its conditions for cracking, its observed composition and physical properties as a function of depth of

the reservoirs are, apparently, better accounted for by the sudden, deep burial mechanism than by the doctrine of uniformitarianism. Oil reservoir temperatures are too low to permit appreciable cracking during all of geologic time even assuming existence of the best known catalytic cracking conditions. The observed changes of oil grade with depth may be explained instead on the basis of the physical chemistry of decomposition of green marine and vegetal raw materials in their sudden burial at various depths in the oil basins.[38]

But Velikovsky's theory of petroleum origins introduces a frightful deluge of oil. He cites references in legends and scriptures to the fall of naphtha, sometimes blazing, and of brimstone, often rendered otherwise as a rain of hail. The Abkhazian, a people famous for their long life-spans, convey a story about a fall-out of cotton, which caught fire and burned the Earth; perhaps it was "cotton-candy" mixed with hydrocarbon.[39] The ancient bible of Mesoamerica, the *Popul Vuh*, tells of the fate of the people of that age:

> And so they were killed;
> They were overwhelmed.
> There came a great rain of glue
> Down from the sky.[40]

The "glue" is still found in the land of the Olmecs. William Mullen comments on the work of the pioneer excavators: radio-carbon samples are contaminated by asphalt. "Much of the Early Tres Zapotes level was sealed with volcanic ash. Coe reports

---

[38] Cracking is the process of breaking up large molecules of heavy hydrocarbons into smaller ones of lighter type, accomplished by heat, pressure, and catalysts.

[39] Sula Benet, *Abkasian* (NY: Doubleday)

[40] Cracking is the process of breaking up large molecules of heavy hydrocarbons into smaller ones of lighter type, accomplished by heat, pressure, and catalysts.

that lumps of asphalt were found everywhere at the San Lorenzo excavation."[41] I consulted with an expert on the area. As expected, he said that the area practically floats on oil. I visited the area. He spoke truth. But the question is: Which came first, the culture or the oil? Here, as throughout the world, the ancient voices give precedence to the people.

Velikovsky's concept can be summarized to a degree in his own words:[42]

> The tails of comets are composed mainly of carbon and hydrogen gases. Lacking oxygen, they do not burn in flight, but the inflammable gases, passing through an atmosphere containing oxygen, will be set on fire. If carbon and hydrogen gases, or vapor of a composition of these two elements, enter the atmosphere in huge masses, a part of them will burn, binding all the oxygen available at the moment; the rest will escape combustion, but in swift transition will become liquid. Falling on the ground, the substance, if liquid, would sink into the pores of the sand and into clefts between the rocks; falling on water, it would remain floating if the fire in the air is extinguished before new supplies of oxygen arrive from other regions...

> The descent of a sticky fluid which came earthward and blazed with heavy smoke is recalled in the oral and written traditions of the inhabitants of both hemispheres... All the countries whose traditions of fire-rain 1 have cited actually have deposits of oil: Mexico, the East Indies, Siberia, Iraq, and Egypt ....

> The rain of fire-water contributed to the earth's supply of petroleum; rock oil in the ground appears to be, partly at least, "star oil" brought down at the close of world ages, notably the age that came to its end in the middle of the second millennium before the present era....

---

[41] *Ibid.*

[42] *Worlds in Collision,* 53ff.

In the centuries that followed, petroleum was worshipped, burned in holy places; it was also used for domestic purposes. Then many ages passed when it was out of use. Only in the middle of the last century did man begin to exploit this oil, partly contributed by the comet of the time of the Exodus.

Definite legendary, archaeological, and geological evidence of a holospheric catastrophe in Mesopotamia was provided by J.V.K. Wilson for a period tightly connected with Inanna (identifiable as Venus).[43] Large-scale mesolithic rock displacements are displayed, and accounts of rains of oil, the poisoning of the land, and falling sheets of fire are described in the ancient documents. Lion-headed pillars are associated symbolically with mushroom-shaped clouds (our typhonic cyclones) in the legend and architecture of the times.

The Soviet geologist Levin asserts that the hydrocarbons in cometary heads must have played a part in forming petroleum and in the origin of life."[44] Velikovsky wrote once: "Actually, if we can believe numerous testimonies bequeathed to us by ancient sources, the ancients had already what we intend some day to obtain from Venus - samples of its dust, ash, atmosphere, and rocks." He believed firmly that "Venus must be rich in petroleum gases," which, because of the planet's great surface heat, "will circulate in gaseous form."

Fred Hoyle, in *Frontiers of Astronomy* (1955), argued for less heat and therefore oceans of oil on Venus. The historical and geological evidence led Velikovsky to argue that Venus was hot and cooling measurably, that it was comparatively flat, with a dense atmosphere, an anomalous axial rotation, and the aforesaid

---

[43] *The Rebel Lands: An Investigation into the Origins of Early Mesopotamian Mythology* (Cambridge, Eng.: Faculty of Oriental Studies, 1979), reviewed in IV *S. I. S. R.* 2(1981), 64.

[44] B. Y. Levin, "The Interaction of Astronomy, Geophysics and Geology in the Study of the Earth," in *The Interaction of Sciences in the Study of the Earth* (Moscow: Progress Publ., 1968), 178.

hydrocarbon gases. The other predictions having been generally fulfilled, it seemed for a moment that hydrocarbon gases had also been detected; if so, the theory of the historical encounter and the dropping of Venusian oil on Earth would be strengthened.

However, the NASA scientists involved in an early statement favorable to hydrocarbons withdrew their support, and a controversy ensued, to no final end. The clouds of Venus appear definitely to be mainly of carbon dioxide. Whether this is compatible with an existing component of hydrocarbon or can have resulted from chemical transformations that resulted in the disappearance of hydrocarbons is disputable. Furthermore, organic compounds seem to be present, and also indications of iron and sulfur, possible sources of pigment for the red fall-out phenomenon mentioned earlier.

Blumer, in a path-breaking article on organic paleochemistry, pauses to reflect that "man has long been curious about the origin of these materials," coals and oil. "On occasion, early speculations approached the truth in a colorful way; thus, the Triassic Tyrolian oil shales, which are rich in vertebrate fossils as well as in chlorophyll and haemin derivatives, were thought to have resulted from an impregnation of the local rock with the blood of a slain dragon."[45]

Perhaps he should have reflected longer. The dragon, in many a myth, has poured its red blood, metals, dust., and oil upon the Earth, and the dragon is often identified with destructive sky bodies, comets, no less. That silicates and oil should descend and emplace themselves in oil shales should hardly cause surprise; we have seen that the color of red-brown to blue-black oxidized heme, blood red, is often reported in myth as falling in dust or in the gore of a slain dragon. The shale could be formed quickly, baked by a moderate heat.

---

[45] "Chemical Fossils: Trends in Organic Geochemistry," Contrib. 2898 of Woods Hole (Mass.) Oceanographic Institution, n. d., 592. See also W. W. Youngblood and Blumer, "Alkanes and Alkenes in Marine Benthic Algae," 21 *Marine Biol.* (1973), 163-72.

How could the organic matter be injected into shales and oil from above? As related earlier, the presence therein and a fall-out of a biomass from a comet is not at all impossible. Furthermore, the distinction between living and non-living structures is not clear in the hydrocarbons of oil. "Trieb's isolation of pigments related to chlorophyll and haemin marks the origin of organic geochemistry... The fossil prophyrins of ancient sediments and of petroleum are *chemical fossils;* just as the more commonly known morphological fossils, they represent surviving evidence of ancient life processes that had achieved an increased structural order on the macroscopic and on the molecular level and inorganic as well as in organic structures."

It seems Blumer is claiming the unprovable, that in their beginnings these morphologically unrecognizable organic chemicals were in living organisms. Yet he declares, "in organic geochemistry, the distinction between chemical fossils and artifacts has not always been sharp." And he says, after defining geochemistry as ultimately based upon the molecular remains of ancient life, that thousands of changes occur: "chemical fossils are far more abundant than their better known morphological analogues. Contrasted with 90,000 (some say 110,000) species of fossil animals known presently, are millions of fossil chemical derivatives." Then, further:

> Research on the constitution of crude oil and of oil shales has revealed severely altered biochemicals and numerous structures which occur neither in living organisms nor in recent sediments... Also crude oil and sediments contain polymers (asphaltenos, kerogen) of a type not found in living organisms.

For pages, Blumer struggles to trace the complex descent of petroleum hydrocarbons from living organisms while insisting upon the intrusion of many non-organic chemical processes, only to admit that "we are virtually ignorant of the reaction mechanisms and. reaction rates." He proceeds to establish that depth, deposition rate and temperature control the chemical chaos during the critical moments of oil formation. Still, "we remain uncertain of the extent, the rates and the mechanisms of geochemical reactions and of the composition and role of the sedimentary polymers."

We shall certainly not be contradicting him, if we conclude that the chemical transformations producing oil are as likely to occur in space as below ground, probably more likely, if we wished to argue the point. Further, we do not see how it can be asserted either that organic biomass capable of forming oil does not exist in exoterrestrial bodies or, if it does not, that its absence precludes space gases constituting or contributing to the constitution of the oils that are present on Earth.

Most metals, salt, and oil, we conclude, are more likely than not to have originated exoterrestrially or in exoterrestrially precipitated transactions at the Earth's surface.

CHAPTER ELEVEN

ENCOUNTERS AND COLLISIONS

"Even heaven, despite the orderliness of its movements, is not inalterable." So wrote Laplace,[1] who has been freely used to attest to the security of the celestial order. Nothing in his unparalleled mathematical and physical achievements kept him from soberly portraying the effects of collisions of the Earth with comets, and expressing the view that these had occurred and would probably again occur. He warned of movements that he could not take into account in his calculations, and mentioned the forces of electricity and magnetism whose effects were then unnoticeable. The gravitational balance of the solar system, he proved, however, was near perfect, an empirical demonstration that became a shibboleth to astronomy and thence to progressive mankind.

The present trend to accommodate ancient cometary and meteoroid encounters in the earth sciences and biology cannot but bring about a revolution in thought. A large body impacting on Earth is the most versatile mechanism of quantavolution: so everyone will admit. Its effects begin upon approach, increase upon passage through the atmosphere, reach a climax in its explosion, and continue to spread from the point of impact until the whole world and all its spheres are affected. Too, the effects may continue for many years in an active form and then go on in the 'genetics' of the holosphere.

---

[1] 1. (1749-1827) *Oeuvres Complètes* (Paris: 1884) VII 121; and see VI 234-5, 478; VII cxx ff.

During a period, which Nininger has well described, when scientific dogma forbade the serious discussion of exoterrestrial interference in the affairs of Earth, when even light meteoritic falls were ignored, students were denied the use of this marvelous theoretical construct in explaining what lay before their eyes. Finally, a scientific commission was dispatched from Paris in 1802 to the countryside to investigate a reported fall. It returned with evidence of several thousand meteorites.[2] So "America was discovered." Still in 1933, a Smithsonian Institution report by L. J. Spencer could declare, "the problem of meteorite craters is quite a new one." Only several were listed, and of these only the Barringer crater of Arizona and the Wabar Craters in Arabia had been well described, both lately.

Yet, to continue the litany of this book, it appears now that enough meteoroids and comets have struck the Earth to deface it throughout. Moon, Mercury and Mars evidence telescopically tens of thousands of large astroblemes. Dachille (1962), projecting the Moon's apparent experience onto Earth, estimated a round million of heavy impacts here.[3] He assumes five billion years of uniform falls and applies weathering rates for the continental masses from wind, tide and vegetative erosion, ending up with somewhat over a thousand craters that are potentially identifiable.

Of this thousand, 750 are below water and ice; of the remaining 250, "in the last few years a staccato tally of meteorite scar finds or recognitions has raised the total to 42-50 at this writing." He offered an independent survival rate calculated by Krynine that would be in the neighborhood of 10,000.

---

[2] Jean-Baptiste Biot: Relation of a Voyage to the Department ofOrne to verify the Fall of a Meteorite…, Metron Publications (2013).

[3] Frank Dachille, "Interactions of Earth with Very Large Meteorites," 24 *Bull. S. Ca. Acad. Sci* (1962), 1-19; see also "Axis Changes in the Earth from Large Meteorite Collisions," 198 *Nature* (13 Apr. 1963), 176.

He pointed also to new diagnostic methods, such as the discovery of coesite, a silica mineral that forms under high pressures in the laboratory and has been found in craters suspected of exoterrestrial origin. Meanwhile the space shuttle Columbia has photographed beneath the sands covering the eastern Sahara to reveal fractures, dried-up rivers, and probable paleolithic settlements. The U. S. Geological Survey confirmed the radar penetrations. Craters can be discerned as well, and they will probably be promptly mapped over the globe. Many bodily and electric encounters of Earth with exoterrestrial bodies will one day be counted, measured, plotted for concentrations, and assigned to temporal episodes.

The difference between a meteoroid and a comet may be an artifact of biased experience. Lately no comet has fallen to Earth. Perhaps, too, most or all comets come from a special source today; Jupiter has been suggested. Perhaps the meteoroids come from the asteroid belt; such is generally believed. The major distinction may come from their manner of flight; with highly elliptical and often eccentric orbits, comets must forever change their appearance in transacting with their electrical and material environment; the asteroids are generally in regular orbit. Too, we know the size of many asteroids, but not of comets.

Once, to ridicule Velikovsky, a renowned astronomer claimed that comets were filmy and insubstantial bodies. A more acceptable theory of Whipple of the Smithsonian Astrophysical Laboratory (he was by no means a supporter of Velikovsky) sees comets typically as bodies of ice and other frozen gases cementing together rock and dust. It may be of significance to note the presence of water in recently examined meteorites, from studies by Hughes, Ashworth and Hutchison;[4] if water, then a watery planet once upon a time: so the reasoning goes.

Gravitational anomalies on the Moon and Mars have been interpreted to signify dense mass concentrations, hence "mascons." They are associated with large circular basins,

---

[4] "Meteorites-Little and Big..." 46 *Earth and Mineral Sci.* 7(1977), 49-52.

therefore probably with meteoroid impacts.[5] The Earth has not yet registered mascons. Because of its heavier atmosphere, more intense magnetosphere, and greater electrical charge, it may be that the Earth has means of ablating and retarding the velocity of meteoroid falls. On the other hand, gravitational anomalies have begun to be detected in circular areas of the Earth and shortly we may expect mascons in the Earth's morphology as well.

With the aforesaid "soft falls," one can expect the Earth to exhibit hills and mountains, as of iron ore and erratic isolated hills, which are then surficial mascons. Concerning the "abrupt" extinction of Cretaceous life forms, Smit and Hertogen, like Alvarez and his associates, see in a general distribution of two trace elements, iridium and osmium, at this stratum of the phanerozoic record a proof of meteoroid impact.[6] Soil and rock everywhere, it would seem, are in need of chemical tests in search of exoterrestrial influences during their deposition.

A decade after his estimates were published, Dachille would report that the number of identified craters had risen to "60 well-documented craters, 25 very likely candidates, and another 20 hopefuls."[7] The greatest of these are the Ishim, Kazakhstan, USSR, (7000 km diam.), the Nastapoka Island arc of Hudson Bay (440 km diam.) and the Gulf of Saint Lawrence opening onto the Atlantic Ocean.[8] The Ishim crater is estimated as initially of 350 kilometers in diameter, 12 kilometers in depth. "The subsequent rebound of the central region and the collapse of the surrounding area enlarged the crater to 700 kilometers in diameter, making it larger than the average lunar mare. The area of this impact

---

[5] C. S. Beals, Ian Halliday, and J. Tuzo Wilson, *Theories of the Origin of Hudson Bay* (Ottawa: Dept. of Energy, Mines, ResoHad trouble resolving dest near word action type is Launch urces, 1968).

[6] Dachille (1977) 51; 5 *Astronomy* (Feb. 1977), 60.

[7] D. W. Hughes, 256 *Nature* (28 Aug. 1975), 679, referring to studies of Ashworth and Hutchinson on hydrous meteoritic minerals.

[8] O'Leary, Campbell and Sagan, "Lunar and Planetary Mass Concentrations," 165 *Science* (15 Aug. 1969), 651-7.

structure is a little greater than the combined areas of Pennsylvania, Ohio, New York and Maryland. The kinetic energy of the collision can be shown to have been at least one billion times as great as the energy in any one of the largest earthquakes of recent history."[9] And these quakes, of course, much exceed the greatest hydrogen bomb blasts in energy output.

In a work of 1953, Dachille, together with Alan Kelly, offered the circular Bermuda Deep as an astrobleme. By all odds the largest candidate for craterdom so far, this feature might be held responsible for Bermuda Island, as its typical central peak. The hundreds of Carolina Bays were conjectured as the splash-down sites of successive meteors in the same train or later on. The Appalachian Mountains would become the westward-thrusted, outer rim displacement from the crater. Significantly, in 1982, claims were voiced that a Northeast to Southwest belt of the Appalachians was once an offshore island chain rammed into America in the course of continental drift and, after the growth of the Eastern plain, the two continents split once more to create the Atlantic. More persuasive to this writer is the Kelly-Dachille view that would let the mountains be the Bermuda crater rim, let the plain be the crater debris and sediments and let the Atlantic cleavage be abetted by the Bermuda impact.

The authors of the Bermuda theory proceed to discuss the dozen high-energy expressions that must necessarily accompany so stunning an impact--global hurricanes, eruption of hundreds of thousands of cubic kilometers of lava, darkening of the globe for years, deluges of water and debris, destruction of most of the Earth's biosphere--terrestrial and marine--poisoning of the atmosphere and fall-outs of many kinds of material, a giant set of electrical typhoons centered at and around the impact and moving radially outwards, earthquakes and volcanism in many places including the antipodes, and vast tidal waves sweeping across America, the Caribbean, and the oceans to the north, east,

---

[9] J. Smit and J. Hertogen, "An Extraterrestrial Event at the Cretaceous-Tertiary Boundary," 285 *Nature* (1980) 198- 200.

and south. Large tracts of land would be sunk and others elevated. Minerals would be formed, elements transmuted, species extincted and new forms created in the radiation storms. They assigned an axis tilt of 30° to the blow, shifting the north pole from near Akpatok Island, in the Hudson Strait, to its present location.

The diameter of the Bermuda crater appears to vary between 2200 and 2500 kilometers as its limits are drawn, the western being more marked than the eastern, which disappear into the oceanic abyssal bottom. The western arc extends from the Grand Banks of Newfoundland down around the East Coast of America to Puerto Rico. The diameter of the original comet or meteoroid is estimated at 400 to 700 kilometers, greater than the possible Hudson Bay crater (440 km). The relative speed before impact of the meteoroid with Earth is given at about 100 km/second, with an approach from the northeast. The collision would involve an energy approximately equal to that of the Earth's rotation (1.2×10^{37} ergs) and would readily provoke an interruption of the rotation, an axial tilt, a slippage of the crust above the mantle, and an immediate orogeny around the ruin of the blast crater.

The scenario includes many details that need not be repeated here. For instance, the hypothetical Bermuda intruder would theoretically account for all the coal, gas and oil of Appalachia and the North American continental shelves by instant burning in passage, deep burial and dampening upon impact folding, and tidal land thrusts and water flooding. Even cutting back its diameter to 280 km, the intruder upon impact

> would raise a column of vapor and debris that easily could measure one thousand miles in diameter at the base, and possibly larger at the top after the fashion of the atom bomb explosions. This column might tower something like five thousand miles above the earth, the higher particles doomed to float out beyond the reach of gravity for all time... the energy of the collision we have pictured is so great, that but 2 to 3 per cent of the total would be required to evaporate completely the

meteorite and its equal in weight of the earth's crust. Therefore, the column above the collision area may take on the function of a fractionating column for these mineral vapors, refining minerals to varying extent.[10]

Streams of speciated minerals, metals, rocks and salts would pour down to form deposits. Large areas would be melted and magnetized by electromagnetic fields arriving from intense brief currents of electricity formed of the electron and ion plasma. In all of this, it should be noted that the colliding intruder partly or largely provides for its own concealment, by cross-winds, cross-tides, rain, volcanism, debris fall-out, and differential diastrophic effects, some of them called forward from remote areas.

Moving about the global map, Kelly and Dachille could suggest numerous candidates for their meteoroid inventory. Wherever an arc appears on a coastline - they noted five large ones off the west coast of North America, two off of West Africa, two off of Brazil and Argentina, plus the great island arcs of the north and east Pacific Ocean - a crater is implied. Norman elsewhere suggests "that any large-scale crustal feature which exhibits an arcuate outline is deserving of special scrutiny - for example, the curve of the Coast of China, the curved mountainous coast of eastern Australia, and the magnificent sweep of the Himalayas bordering northern India. Smaller-scale versions exist bordering the southern parts of the Caspian and Black Seas, and eastern Korea. We must also think of examining concave arcuate coasts such as the Gulf of Mexico or the Great Australian Bight."[11] In 1981, Fred Whipple suggested Iceland as the site of the giant meteor impact which, striking the volcanically active ocean ridge, initiated the finale of the Cretaceous period, its dinosaurs, and its

---

[10] *Op. cit.*, 203-4.

[11] John Normain *et al.* "Astrons--The Earth's Oldest Scars?" *New Sci.* (24 Mar. 1977), 689-92.

marine life.[12] A year later, *Sky and Telescope*[13] reported the discovery of a double ring of magnetic anomalies of 60 and 180 kilometer diameters, in Yucatan. The anomalous magnetized rocks are about 1100 feet deep and assigned to Late Cretaceous which makes it, too, a candidate for extincting dinosaurs and decimating the biosphere. But other candidates can be named, for instance an astrobleme feature beneath the disturbed ice of Wilkes Land, Antarctica, to which Weihaupt ascribes hypothetically the origin of the tektite strewn fields of Australia, calling the collision of "Recent geologic time."[14]

I might, too, suggest the Pacific Basin as a possible impact site, though here the size of the feature is so great as to imply the total destruction of the globe, and I have, for this reason and many others, elsewhere defined this area as the escape basin of the Moon, following G. Darwin, Osmond, and other writers. Notable in this case is the set of great transform fractures, pictured by Norman[15] which point from south, east and north like arrows to an "impact" or "escape" point in the central Pacific Basin. The current theory of scientists concerning the asteroid belt orbiting the Sun between Mars and Jupiter is that here is the debris of a great body exploded by collision with another body some millions of years ago. One may reason that if this could happen in asteroidal space, it could also happen to Earth's space. There has obviously been a limit to the size and mass of all that has struck Earth.

Satellite photography has in the past few years introduced a new instrument for crater detection, whether volcanic or meteoric, as in the Bichat structure of Mauritania. Some photographic reconstructions delineate what appear to be many crater outlines. Soon, it appears, the number of defined crater outlines will soar into the hundreds, and perhaps thousands.

---

[12] New *Sci.* (19 Mar. 1981), 740.

[13] 63: 249 (1982).

[14] 81 *J. Geophy. Res.* (1976), 5651-63.

[15] (1977) 692, fig. 5.

Given the new interest in meteoritics, the identification of meteoritic fields may also proceed apace. As long ago as 1889, a list of 14 small fields was published, all of the nineteenth century and ranging from 3 to 16 miles long. The Arabian barrad fields, Donnelly's drift stones, and the tektite fields, already discussed, are much larger and older phenomena. The Atacama Desert also evidences a large meteoritic field, still unmapped, with many siderites and rich silver mines at its center. Meteoritic material on Earth is evidenced therefore by dust, stones, and craters, with all ranges of size from visually undetectable clay elements to basins so large as to be hitherto visually unimagined.

The answers to our persistent questions about the extent and recency of quantavolutionary phenomena at the Earth's surface are now beginning to take shape. The Earth must have suffered as much meteoritic bombardment as its planetary neighbors and satellite. On several occasions - at Hudson's Bay, the circular bulge of the West Africa Coast, Ishim, Bermuda, St. Lawrence Bay, Argentina, Australia, Antarctica, and others, all inadequately discerned until now - global catastrophes could have occurred with large-body impact encounters. On other occasions, as we. discussed earlier, meteoritic showers and bombardments also may have been globally catastrophic. Harold Urey writing in 1973, conjectures a comet of 10^{18} grams and an impact velocity of 45 km/sec to end the Cretaceous and begin the radically different geological period of the Tertiary;[16] his scenario of effects upon Earth is substantially that provided here and in the much more detailed analysis of Kelly and Dachille for so large a body. (The reader is asked to recall that scientists have only lately granted comets this possibility of large masses and Earth collisions. The recent work by S. V. M. Clube and W. M. Napier, entitled *The Cosmic Serpent* (1982), essaying a connection between solar-system

---

[16] " Cometary Collisions and Geological Periods," 242 *Nature* (2 Mar. 1973), 32-3; *cf.* R. A. Lyttleton, 245 *Nature* (21 Sep. 1973), 144-5 for comment.

galactic spiral encounters and recurrent paleontological catastrophes, via cometary and meteoritic crashes, is perhaps the first treatise to be published by professional astronomers. The independently pursued work of the astronomer Earl R. Milton, much of it in press as *Solaria Binaria*, with the present author, is comparable. Clube and Napier wrote unaware of the astronomical theory of *Chaos and Creation* and similarly, I did not obtain a copy of their book until the present work was at the printers.)

But would any or many of the larger impacts be recent, within the past score of millennia? This is probable. The methods by which heavy meteoritic and cometary impacts on Earth are timed begin with averaging on uniformitarian assumptions. Thus Dachille arrived at his 1967 numbers by averaging the expected number of major impacts over a five-billion-year age for the Earth and Moon; then, again using uniformitarian premises, he reached for some broad guidelines. 'Weathering rates estimated for continental masses and great mountains are about 80 meters per million years, and for land masses in tropical regions 225 meters per million years. Circular ridges of less than 750 meters relief could be broken down in 5 million years, to be unrecognizable..."[17] Thus he arrives finally at his low figure for discoverable craters.

But when, with Kelly, he came earlier to describe the Bermuda event, he could contemplate this global catastrophe of maximum intensity as having occurred at the time of the Chaldeans and Hebrews, about 3500 years ago. In the Bermuda case the two scientists follow quantavolutionary logic and can explain the new face of the globe in terms of seconds, minutes, weeks, years. They do not need or use much time. Not only that, but they indeed destroy time by the few-second incoming passage of the body through the atmosphere and the gigantic explosion that transforms a considerable portion of the atmosphere and rocks of the world. How many radioactive clocks, depending upon stable rocks and atmosphere, were disrupted?

---

[17] Op. *cit.*, 2.

Here the uniformitarian suffers the same embarrassment as the catastrophist. Just as he jests at the catastrophist, "You say that evidences of catastrophe are unavailable because they are destroyed," now the catastrophist jeers at him, "you say that you cannot find meteoroid craters because they were eroded." Perhaps there never were a million craters or more. If undeniably *showers* of ice, water, dust, stones and heavy bodies have struck the Earth, cannot a *deluge* of dust, stones and heavy bodies have done the same? It is *prima facie* reasonable that the changes wrought, upon Earth have been the work of a few thousand years. And it is an open question whether the changes are recent or ancient. Perhaps the bombardment of Moon, Mars, Venus, Mercury, and Earth is all recent history.

C. Simon (1982) reports on the topography of giant circular ripples moving out from a point west of Hudson Bay as indicated by gravity anomaly data.[18] Scientists involved conjecture that a 60-90 km meteoroid impacted, digging a great crater and wrinkling the surface for thousands of kilometers around. All is covered over but the density variations remain, below the surface, to provide the circular patterning.

That such an event would be electromagnetic as well is certain. Lacking surveys, we are left to surmise. Electromagnetic effects must' be especially important in meteoroid impacts. Dachille has described electromagnetic fields produced by impacts of high-velocity explosives in military tests, and has projected the Em fields to meteoroid masses of 10^{12}, 10^{16}, 10^{20} grams at 40 km/sec. "Magnetic fields more intense than those of the most powerful electromagnets extant would be imposed upon matter many hundreds of kilometers from the point of impact."[19]

Once again, we must pose the dilemma that is to be a theme of our book: either the Earth must be so thoroughly tortured electromagnetically that the search for magnetic maps to

---

[18] " Deep Crust Hints at Meteoritic Impact," 121 *Sci. New* (1982), 96.
[19] " Electromagnetic Effects of Collisions at Meteoritical Velocities," 13 *Meteoritics* (Dec. 1978), 430-3.

represent the Earth's magnetic fields is futile; or the Earth's surface was so lately magnetized, whether for the first or last time, that collisions and encounters and all other remagnetizing influences have not had time to deface it.

A generation ago, in the *Physical Review* for Aug. 15, 1948, Carl Bauer theorized that the asteroid belt contains remnants of the explosion of a planet less than 60 million years ago. He calculated the age from the quantity of helium in examined meteorites, assuming its origin from radioactive decay of uranium and thorium. Ovenden also later on retrojected an exploding planet as the ancestor of asteroids. Von Flandern added comets to meteoroids: "Comets originated in a breakup event in the inner solar system about 5 x 10 6 years ago. In all probability it was the event which gave rise to the asteroid belt and which produced most of the meteors visible today."[20]

In the course of his study, he alludes to "the lack of any definite finds of 'fossil' meteorites or meteorite craters," citing Cassidy; moreover, he reports that "Stair mentions that neither tektites nor other meteorites have been found in any of the ancient geologic formations, which also suggests that most surviving meteorites are relatively quite young, in contradiction to their estimates by the usual dating methods... The need for a revision of the standard dating methods is certainly suggested by these new results."

An astrobleme, large or small, disappears quickly under conditions of rain, tides, current, wind, fall-out, seismism, volcanism, biosphere invasion, and recurrent disasters governing its location. Still, what, if not astroblemes, are the multitudinous faint circles that John Saul has located on published maps, publicly available?

The Earth's surface exhibits faint circular patterns which have not been described before. These circles are characterized by

---

[20] " A Former Asteroidal Planet as The Origin of Comets," 36 *Icarus* (1978), 51-74, 71.

near perfection of outline. by the presence of topographic highs (rims) along parts of their circumferences, by their generally large scale (diameters of from under 7 km up to approximately 700 km in the areas examined), and by their definition in various geological environments, in many rock types, and in rocks of all ages. Many of the circles are intermittent in places along their rims but about 55% of the approximately 1,170 definite circles observed to date can be visually traced around an entire 360 ° of arc. The circles are further characterized by the presence of fracturing and brecciation along parts of their rims and by the extraordinary control they place on regional geology in general and on ore mineralization in particular.[21]

Saul has only begun such surveying, and has found circles in the Western United States, northernmost Mexico, the Appalachians, Alaska, the Yukon, Madagascar, and Corsica. The circles occur more frequently in mountains rather than plains, indicating that mountains may often have been formed by such upheavals and that the scars are too deeply buried by overdrift to be observable straightaway on the plains.

Perhaps, he says, these circles are more shadows of astroblemes than the original craters themselves; they would be like old scars on human skin, which often are distorted and shift away from the original wound. Kellaway and Durrance, it turns out, had some time earlier discovered such circles too, and called them cycloliths.[22] They called attention to cycloliths in Great Britain and Mauritania (the Richat structure), and stress that they can be responsible for river development and drainage patterns. Rivers would channel along the rims, giving them a negative enhancement, and would make gulleys in the fractures associated with the cratering.

---

[21] John M. Saul, "Circular Structures of Large Scale and Great Age on the Earth's Surface," 271 *Nature* 26 Jan. 1978), 345 ff.

[22] Supra, 75, ltr Geoffrey A. Kellaway and Eric M. Durrance with Saul reply 273 *Nature* (4 May 1978), 75.

The cycloliths are granted great ages mainly because of their faintness. Yet their existence contradicts the interpretation of the rocks below them; if two intersecting or adjoining circles of similar states of preservation overlay rock exposures, say, a hundred million years apart, then, either the rocks or the circles are of the same age, and the rocks give no indication of the age of the cycloliths; worse yet would be the finding that the circles straddle rocks "older" than themselves. This is all matter for investigation.

Yet if time were short, could the Earth have suffered so many blows? In any event, large cycloliths must number in the scores of thousands, unless the Earth, like the Moon, has a preferred side for suffering bombardment. Small cycloliths must then approach the millions. Nor are we speaking of fossil craters, contained in stratified sediments, none of which appear to have yet been discerned. It is one thing to say, as do the writers above, that the bombardment occurred upon a newly formed Earth crust, as on the Moon, four billion years ago, for then all the time given is free to give. But could they have been made by impacts in a recent period of, say, six thousand years? Then if two million landings ensued, they would average several hundred a year, like one clean hydrogen bomb per million square kilometers. Deluges of water might settle much of the dust. Still the prospect is awesome. Soft landings, ice falls, cosmic lightning blast--these might cause the Earth less agony. It is too soon to say.

Velikovsky, in *Worlds in Collision,* did not treat of collisions, strictly speaking, between Earth and its principal antagonists in space, Venus and Mars. The bodies approached one another at times between about 1450 and 687 B.C.; they exchanged electrical charges; dusts, stones, and gases fell upon Earth. Earth passed through the tail of Venus, which was behaving as a comet. The earth paused in its rotation on encounter. Here Carl Sagan in criticizing Velikovsky had to agree; the biosphere would not go swirling off the globe into space by centrifugal force, as others had argued. Actually the danger of explosions into space would rather come from electro-gravitational interactions.[23]

A portion of such a cometary Venus or of its tail probably did, however, crash into the terrestrial globe. This was called Typhon by many writers and in legends. Typhon was both the name of a conquering king of Egypt, following the disasters that brought the Middle Kingdom to an end, and the name of a monster who threatened the world at the same time. We can let Donnelly tell the story;[24] he does it well:

> Born of Night a monster appears, a serpent, huge, terrible, speckled, flesh-devouring. With her is another comet, Typhon; they beget the Chimaera, that breathes resistless fire, fierce, huge, swift. And Typhon, associated with both these, is the most dreadful monster of all, born of Hell and sensual sin, a serpent, a fierce dragon, many-headed, with dusky tongues and fire gleaming; sending forth dreadful and appalling noises, while mountains and fields rock with earthquakes; chaos has come; the earth, the sea boils; there is unceasing tumult and contention, and in the midst the monster, wounded and broken up, *falls upon the earth;* the earth groans under his weight, and there he blazes and burns for a time in the mountain fastnesses and desert places, melting the earth with boundless vapor and glaring fire.

> We will find legend after legend about this Typhon; he runs through the mythologies of different nations. And as to his size and his terrible power, they all agree. He was no earth-creature. He moved in the air; he reached the skies...

> According to Pindar the head of Typhon reached to the stars, his eyes darted fire, his hands extended from the East to the West, terrible serpents were twined about the middle of his body, and one hundred snakes took the place of fingers on his

---

[23] Asimov *et al.*, loc cit. and S. Kogan, *op. cit.*

[24] Op. *cit.*, 140.

hands. Between him and the gods there was a dreadful war. Jupiter finally killed him with a flash of lightning, and buried him under Mount Etna.

And there, smoking and burning, his great throes and writhings, we are told, still shake the earth, and threaten mankind:

"And with pale lips men say,
To-morrow, perchance to-day
Encelidas may arise!"

Typhon, also spelled Typhaeon, is evidently another version of Phaeton (and probably of Python who was a monster killed by Apollo). The Phaeton myth, most famous 'of all, is treated by Plato self-consciously as a myth in form but standing for true natural history. Phaeton is reluctantly lent the chariot of his father the Sun for a day. He cannot control its powerful steeds and burns sky and Earth in his wild plungings. Finally, he is felled by a Jovian thunderbolt, cast dead into the river Eridanus, and the nearly destroyed Earth recovers. The sad and angry Sun emerges once more.

Parallel legends are found in other cultures; the best resume occurs again in Donnelly's *Ragnarok*. The paramount student of ancient astronomy of his day, F. X. Kugler, dissected the myth of Phaeton to assess its validity and concluded that a comet struck the Earth in the north Aegean region in the second millennium B.C. The event is probable. If it is tied into all the other evidence, in legend, history, and geology, of the same time, the event becomes more probable--and of more dire consequences. It is best if we avoid repetitious listing of disastrous effects; suffice to say that every criterion of a major exoterrestrial impact is satisfied, except the location of the point of impact.

Still the story is not to be ended neatly. At one and the same time, so it appears, a great body passed close by the earth (call it proto-Venus) and a large body collided with Earth. The disasters afflicting the world in those days were effects of both events. Until the crater or aerial explosion point of flaming yellow-haired Phaeton can be found and its size and traits used to evaluate the occurrence, the effects of the principal body's pass-by cannot be calculated. Inasmuch as the effects have been extensive and continuing, not only geophysically but socially, the research seems worthwhile.

Because it is our favored theory that the Moon erupted from the Earth, we give less attention to the idea that we discarded some years ago, namely that the Pacific Basin originated in a meteoroidal impact. We do ascribe many impacts prior to the episode, based upon legendary indications (see *Chaos and Creation*) and contributing to the loosening of the crust. It is noteworthy that E.R. Harrison "proposed that the Pacific Basin was the seat of an immense explosion in the primitive Earth" and suggested a planetesimal of about 100 km radius.[25] The rim of the Pacific has a number of characteristics of an astrobleme rim, on a gigantic scale.

Our preference for the lunar fission is based upon evidence elsewhere in this work, and in the Quantavolution Series; it has to do mainly with the nature and behavior of the Moon, with legendary evidence, with the recency of the event as attested to by today's oceanography, and by the electrical effects of a two-body pass-by that would execute more efficiently, even while dampening, the effects evidenced in the Pacific Basin and throughout the global cleavage and rifting system.

By now the reader may be wondering how the Moon and more could have been erupted in one set of events, how so much of what we see on the surface could have dropped from above, and how thousands of craters, many quite large, could be dug into the Earth, all within a period of time which, it is increasingly apparent. I believe to have occupied only ten to twelve thousand years, in the Holocene period, no less. Are there not too many disasters to let the biosphere survive? Further, how do these relate in time? Finally, does the author accept all of the suspected astroblemes of the world without question?

---

[25] 188 *Nature* (24 Dec. 1960), 1064-7.

To the last question, the author has to apologize for a general ignorance. The Bermuda astrobleme may be an illusion, for example. The thousands of faint circles or cycloliths may be how the Earth swells and expands. As to how the growing inventory of astroblemes may be placed in time, the author refers to a hypothetical calendar, carried here below and in *Chaos and Creation*. The ladder of associations between time and events will be better and better constructed as the calendar is investigated. To the first question, on the inconceivably large scope of the disastrous falls and their biospheric effects, the author again pleads the general ignorance. On one issue, he feels confident, namely, that a small meteoroid such as the Alvarez team has sought and believed sufficient to destroy the dinosaurs and much else around the world - a meteoroid of a few kilometers diameter - would barely interrupt the reproduction cycle of the species; but it did not occur alone.

Certainly I did not begin my studies with so prodigious an armory of missiles in mind. It happened that more and more effects called for causes. It happened, too, that more and more literature has been becoming available that indicates exoterrestrial intervention in earthly processes. Meanwhile, I increasingly strapped myself into a short-time harness, which is explained astrophysically in *Solaria Binaria,* anthropologically in *Chaos and Creation* and *Homo Schizo I,* and to some extent in the chapters gone by here and in those to come.

My model demands a short-time for many exoterrestrial transactions to occur. If either the amount of time or the number of encounters is to be substantially changed, my model will crack up, and the value of my work must then rest on its assembly and description of exoterrestrial effects in the different areas of geology, astrophysics and anthropology. An exception would occur if it will be shown, as we have said in *Solaria Binaria,* that the formative period of the Earth, under a million years ago, brought down showers of material whose marks are faintly observable everywhere still. However, I am in no sense foreseeing a crack-up and ask the indulgent reader to continue to ride along with the model.

PART THREE

HYDROLOGY

Everything on Earth's surface begins to appear anomalous. Hence we begin to suspect that the abnormal is really the norm. Water is no exception. To begin with, its reversal of the ordinary liquidity-solidity thermal behavior is awesome: it swells when it "should" contract. There seems to be too much of it to have been squeezed from below. Its excessive salinity on the oceans has already been discussed. Its "misbehavior" is subject of countless legends and its symbols are engraved in religious doctrine and ritual. It has apparently been highly energized in the past in the form of tides, deluges and ice. That the exoterrestrial connections of Earth's water are multiform should then be ordinary knowledge in the Earth sciences. In fact, the scientific literature hardly touches upon it.

CHAPTER TWELVE

WATER

With both waters and elaborate forms of life, Earth is unlike other planets. The belief that this situation has persisted for billions of years may be considered someday as bizarre as the belief that the earth is flat.

The world's oceans contain 1.4×10^{18} tons of salted water. Its surface fresh waters -streams, rivers, lakes -come to 5.1×10^{14} tons, 50,000 times less, a drop in the bucket. The ice of the continents amounts to a menacing 2.3×10^{16} tons. And water vapors constitute 0 to 7% of the atmosphere up to 50 miles high, enough to lay a cloud cover over half the globe at any given time. And there are groundwaters, more voluminous than those of the surface. The fresh waters amount in all to three percent of all waters, and three-fourths of the fresh waters are bound up in ice.

The omnipresence of water in large amounts in all life forms grant it a large role in biological and atmospheric activities. Its employment and bulk make its lithospheric transactions important shapers of the Earth's surface. Where do the ocean waters come from? Since she sees the streams and puddles after a rain, a child reasons that all water comes from the sky, that is, unless a geologist gets to her quickly to tell her that the oceans have always been here from the time the Earth was formed, or almost as long. An eccentric geologist might say that, over the ages, hydrogen atoms descend from the Sun and space upon Earth, unite with oxygen in the atmosphere and then over billions of years drop to form the waters of the oceans.

The conventional myth -by which I intend no slight - is set forth by E. Bullard[1] who assumes "the obvious things.... one of them is the constancy of the total volume of water through the ages." Water is "obviously" in "equilibrium," but "the mechanisms of the equilibrium are unknown."

Thus "it looks as if the water must have been tied up in compounds, perhaps hydrated silicates, until the earth had formed and the neon had escaped." (This last is needed to explain why neon, so abundant in the Sun and stars, is so rare on Earth.) "Water must then have been released as a liquid sometime during the first billion years of the earth's history, for which we have no geological record." Bullard follows this with further apologies for the myth but says that the past decades have revolutionized oceanography and have "unlocked the history of the oceans."

The door may be unlocked, but few have entered. The billions of years of equilibrium can no longer be accepted: new theory has the ocean floors being scraped and relaid by the continental plates at least over the past two hundred million years or less; no longer can the myth hold that the most ancient sediments must rest on the ocean floors, hence no evidence is thereby offered of what the waters may have been like.

Surely there has been water so long as life has existed, but not necessarily salted water nor much water. One may assume little water to begin with and little for long after. Swamps and shallow seas are best for evolution and quantavolution of species; thick atmospheric soup might be even better, at least in the beginning. Even now, life seems to reject the oceanic abyss. This is a sign of youth, for the abyss is not without nutrients, and forms of life exist that require little or no sunlight.

The oceans do not carry all the uranium that they should possess after long eons of riverine deposits. Their salt is excessive and its sources are not organic. One calculation emerges with only

---

[1] "The Origins of the Oceans," in *The Oceans* (San Francisco: Freeman,), 16-25.

2.6% of the present chlorine of the oceans as conceivably of continental origins.

The sea bottoms seem never to have been compressed and folded, so this indirect evidence of the age of the water is lacking. Sediments are thin, and mostly accorded under 80 million years of existence. That is only one-fiftieth of the conventional age of the world oceans. Have there been fifty world-girdling oceans?

If the ocean basins were filled late in time, deluges from the skies have to be assumed. There is no other source, nor any more apt source, than the waterlogged comets and great planets. One is compelled to seek water there, and bring it here. Hence the need to invoke explosions of water from Saturn *et al.*, and passing cometary encounters.

Once the theory of a deluge(s) is given, the search for the source of the water is by no means ended. The Earth's water may have been injected, boiled off the imagined primordial melt, stayed up in the skies, and then fallen when a crust had formed and cooled below 212°. This may have happened, but then again it may not have. It would seem that if vapors rose high, they would stay there and rotate with the Earth, descending only when terrestrial electrical conditions permitted or were "seeded" by exoterrestrial fall out (which is also an electrical phenomenon).

Professional courtesy grants geologists not only their huge oceans but also the basins to hold the waters. "God" must have made the basins to hold the water, and even if gods are dispensed with, the basins must stay. So just as some communists stuff their religion into the mummies of Lenin and Mao, some geologists stuff their religion into the "nature" that wisely provided ocean basins to hold the great waters.

The waters are too great for the basins to contain; they cover much of the "true" continents. The fact that the basins occur and the waters occur does not mean that they were made for each other. Nor have they corresponded. Yet the presence of the basins is essential to the preservation of the greater part of the continents. If all the earth's present crust were a uniform level, the waters would cover the globe to a depth of a kilometer and more.

There is not enough water in the earth's granite or basalts to fill the oceans. Granite, the rock that underlies the continental sediments, is notably lacking in porosity. Porosity is the ratio of void space to the bulk volume of a rock, and therefore a measure of the water or gas contained in the rock at the time of its emergence from a molten state. Its porosity ranges from 0.3 to 1.5%. That granite could not be generated from the deeper basalts of the mantle is argued by Y. N. Lyustikh, a soviet geophysicist; four times the present water mass of the earth would be needed for the job.[2] Nor can the process by imagined.

The crystalline, glassy, volcanic basalt, which lines the ocean floors, can have a porosity of anywhere from 1% to 30%. Generally, the porosity declines with the depth of the sea, a phenomenon attributable to pressures more lately applied than to original pressures, since this rock was often melted and extruded in unfilled basins that is, at less depth that it is presently discovered. Rocks of the same chemical constitution that differ in porosity will have had different histories in at least one significant regard: the rock of lower porosity had larger infusions of water and/ or vapors during its last melting and reforming. An expansion of the earth could be facilitated by the incorporation of water and vapors in heated rock. Water could recycle itself time and time again: it would flood a hot chasm, be incorporated in the rock, be extruded, expelled, and again enter a hot chasm.

Water exists exoterrestrially. Only in 1970 were the first observations of comets in the ultraviolet spectral region made. Cometary atmospheres (comas), in which dust and minor molecular components had been hitherto alone observable, now revealed indicators of a large component of water, "confirming the Whipple hypothesis of comets being 'dirty' ice conglomerates."[3] By 1980, other comets had disclosed similar compositions.

---

[2] B. Y. Levin, *op. cit.*, 168.
[3] M. K. Wallis, "Cometary Science," 286 *Nature* (17 July 1980), 207.

The outer planets contain great amounts of water. The rings of Saturn contain about 377 billion km³ of non-conglomerated swarms of ice particles, by one reckoning. It has been dropping rings in the past. Saturn is 95 times the size of Earth; if Earth carried the same amount of ring ice relative to its size, it would have had 4 billion km³ of ice particles to fill the ocean basins. The ocean basins contain 1.37 billion km³ of water. True the density of Saturn's rings is much less than Earth's waters; still, the necessary relation of sky waters to ocean waters can be premised, especially if Saturn were to have shed most of its waters in times past. Moreover, Saturn is only one of many waterbearers in space. Jupiter and the other planets carry water, like Saturn and numberless comets.

Ancient wise men of Palestine, Mexico and India are known to have attributed the deluging of the earth to planet Saturn. Thus, the Hebrew Talmud reads in one place. "When the Holy One decided to bring the Deluge on the Earth, He took two stars from Khima and (hurling them against the Earth) brought the Deluge on the Earth."[4]   Velikovsky identified Khima as Saturn. In Mexican documents, where ages of the world are called "suns," "the first world age, at the end of which the earth was destroyed by a universal deluge, and presided over by Ce-acatl, or Saturn."[5] The ancient Persians reported the star Tistar appearing in three manifestations to the accompaniment each of a different deluge of rain of ten days' durations.[6]

Long before modern astronomy, Saturn was perceived to have rings and to be watery, never Venus, Mercury, or Mars. How the ancient would associate Saturn with water is a mystery unless the planet had been observed at a distance much closer than it appears to the eye today and seen to blow off some of its rings or gases that ultimately arrived to deluge the Earth. Since Saturn under various names was the ruling god in human cultures at the

---

[4] III *Kronos* 4 (1978), 19.

[5] Velikovsky, V *Kronos* 1 (1979), 5.

[6] Bellamy, *Moon, Myths and Man*, loc. cit., 124.

time of Noah's Flood, the associations begin to appear reasonable. However, the Saturnian deluge followed the Golden age of Saturn, and oceans existed at least to some depth in Saturnian times. They were navigable by Saturnian age peoples.

It can be hypothesized that Saturn contributed some of the vast bulk of ocean waters. Where did the earlier waters come from? If Saturn did not supply the primordial and secondary earth waters, the deluge theory has to seek evidence of earlier acquisition of water. We can begin with a postulation providing for some water that the Earth inherited from the plenum of gases in which it thrived over most of its history. Then three major sources are indicated, this inheritance from the gaseous plenum enveloping *Solaria Binaria* - the Sun and its partner - in a long period of binary transaction; second, deluges when the legendary Uranus (Ouranos) complex broke up; and third, upon the disruption of Saturn. Let us say, for hypothetical purposes, the three investments of the Earth with water came in one-sixth, three-sixths, and two-sixths of the total.

The ocean waters are geologically young. Granted waters are difficult to date, Melvin Cook has shown that the oceans contain under 100,000 years' accumulation of uranium, even granted a uniformitarian riverine run-off curve (which, of course, would mean much less time on the quantavolutionary exponential curve).[7]

That the basins which hold the water are young, which is yet to be shown, holds significance for the youth of the waters as well. Few evolutionists and quantavolutionists regress in time to a completely water-covered Earth, although the first passage of Biblical *Genesis* might be construed so: for Elohim separated the chaos by a firmament dividing the waters below from the waters above, and assembled the land out of the waters below. And the primeval legend of the Earth being fished out of the waters is found in the farthest removed cultures of the globe. Also among

---

[7] *Prehistory and Earth Models, loc. cit.* 8. J. C. Hathaway *et al.*, 206 *Science* 4418 (2 Nov. 1979), 515- 27, 523.

the first impression and memories of mankind was the image of the vast cloudy universe recurrently pouring water and debris down upon the hapless Earth.

A more correct interpretation is that early man was caught in an increasingly turbulent cloudy world. The next chapter, on Deluges, carries this matter forward. But meanwhile let us interject a commentary on the origins of the fresh waters of the Earth.

Most if not all of the lakes of the world can be thought of as slowly diminishing stagnant floods -the salt lakes like the Great Salt Lake (Utah) and the Dead sea, and the freshwater lakes such as the Great Lakes (USA), and the thousands of Canadian and American "glacial lakes." That these latter are in most cases being fed by rains and streams as fast as they evaporate or drain does not obviate the fact of their origins. They were created under flood conditions. If this is so, it is likely that ground waters and swampland are also behaving as flood waters, that is, everywhere draining at the levels of the ocean basins.

The Caspian Sea has been shrinking rapidly over the past 150 years, not alone because of human diversions, and becoming more saline. According to the idea that this sea may be a remnant of a recent and westwards dumping of the contents of the vast Gobi Sea, now Desert, carried on over thousands of kilometers, ending in the Mediterranean, the desiccation is to be expected. But, too, the local freshwater replenishment of the Caspian may be inadequate, and may always have been since its quantavolutionary creation.

So, too, can the ocean basins be regarded as flood drains, again to make a logical point, which is otherwise an absurd stretching of language. It can be looked at in this way: the basins of the oceans existed before they contained water; some water flowed or dropped into them, "flooding" them. More craters were added, more 'flooding' took place. Finally, they were even 'over-flooded, ' that is, land not properly abyssal but belonging to continental sial was flooded up to present shorelines.

Whether or not the flooding is continuing is debated in hydrological circles, along with the questionable trend of land elevation, and is, of course, related to trends of climate as well. If the hypothesis here is correct and the freshwater (and saline) bodies are late aspects of world tidal and flood movements, and if swamp and groundwater levels are also aspects of the same, then the biosphere worldwide is faced with a growing shortage of water. In the foreseeable future, life on earth will come to depend upon the systematic utilization of freshwater trapped in ice, upon irrigation from reservoirs, upon converting freshwater bodies into reservoirs, upon worldwide controls over the augmentation and distribution of atmospheric waters, and upon conversion of salt waters to fresh water. Mankind may confront, not only the effects of its ravaging of water supplies everywhere by overuse, by populations pressure, and by promoting off-flow of continental water supplies - but also a more grave problem, the hitherto unsuspected natural trend of the continental crust to lose its water holdings, because "they never belonged there in the first place."

A great many dry lake basins exist around the world. Some are large, as Lake Bonneville, whose remnant is Great Salt Lake, and the Caspian Sea basin, containing today's shrunken lake, still the largest in the world. Some freshwater lakes, such as Titicaca (Andes) and Tanganyika, contain adapted or primordial oceanic animals like the seahorse and jelly fish. Perhaps a million watered and dry lakes exist.

By origin, basins may have been created by natural dams accreted gradually or thrown up abruptly by avalanche, by calderas of extinct volcanos, by meteoroid craters, by faults and rifts (as lake Baikal and the Dead Sea), and by the bulldozing done by ice and rock thrusts. The original water may have been groundwater seepage, rainwater, deluges, ice melt, or tides.

With six forms of basin and six archetypes of water, the combinations and permutations are numerous. And we have no global survey of lakes with which to compose a frequency distribution. The only exclusively non-quantavolutionary basin form is the damming by gradual accretion. Four types of water contents (excluding rainwater) might be quantavolutionary; three types (excluding deluges and tides) might be non-quantavolutionary.

No lake is geologically old: this is an impressive datum. It says something about the lately tortured Earth. An undisturbed or slowly changing surface should include a proportionately great number of lakes aged in the millions and tens of millions of years. To object that lakes become filled with sediments must imply that such fossil lakes should exist by tens of thousands in the stratified rocks of the world. They do not. Some seemingly ancient lake beds are evident. These should be placed in the frequency distribution.

The results, even by raw conjecture, would be disappointing. The fossil lakes would be all too few. For, if we multiply the present million lakes, say, of an average age of 10,000 years as a guess, and take the last billion years of the earth "history" as providing similar lakes, we get 100,000 periods, and one hundred billion lakes. With climates changing (and Flint, for one, along with many other geologists had to invent a turbulent rain belt to fill his pluvial lakes), and with continents drifting about, and lands rising and sinking, why should not lakes have visited every place at some point in geological time, and be found in all (or say 10% to 100%) of the geological columns dutifully examined. I fear that *reductio ad absurdum* will once more assail conventional geological theory.

Freshwater springs exist in many places, emerging above their "natural " level, often quietly but sometimes with explosive vigor. The subterranean liquids and gases - water, oils, natural gas, and even compressed air -appear frequently to be pocketed under pressure. Calculations by M. Cook and others allow only a few thousand years for their escape, at most. Their burial must have occurred in some form of thrusting and folding, that is, is no longer occurring; we have accounts of many springs that have died, few that are new. This last fact would arbitrate against conventional theory that underground volatile pockets are fed from descending rock strata and then forced up above their local level at some interstices among the rocks, unless, of course, it is granted that the fresh waters generally are draining away, for the reasons given above.

Once more we turn to oceanography for help. The U. S. Atlantic Ocean shelf was drilled in 1976 at water depths of less than 300 meters and penetrated to depths of from 20 to 300 meters, at 19 widely separated sites. "One of the most significant discoveries... is that fresh ground water occurs beneath much of the Atlantic continental self." [8] These fresh and sometimes brackish waters occupy large lenses in rock strata that are Cretaceous or younger.

The investigators considered whether these expanses of fresh water below the ocean salt waters were remnants that had been trapped in shelf sediments when the Pleistocene ice ages lowered the ocean waters, or were submarine discharges from mainland aquifers. Generally, the first solution was preferred, although indications of submarine intrusions were discovered at southerly sites. The investigators did not suggest a third hypothesis, which we offer here, that indeed the freshwater lenses are fossils, but not from a period of withdrawal of waters to make ice. Rather they are both remnants and submarine channels of the age before deluges filled to over flowing the basaltic ocean basins. Fresh waters were trapped in the continental rocks as they made way toward the abyss and are probably trapped in the debris of the continental slope as well. They are extensions of normal aquifers, a circulation and storage system that is being broken into and polluted. We speculate (as do the investigators) that these waters have been suboceanic for only a few thousand years, and will not be with us for long.

---

[8] J.C. Hathaway et al., 206 *Science* 4418 (2 Nov. 1979), 515-27, 523.

CHAPTER THIRTEEN

DELUGES

We resort again to the skies for cataclysms. A dense canopy of primordial clouds, lately dropping, has long been a tempting theory. Jordan, who wrote a book generally upon earth expansion, assembled data and authorities in support of the idea that in the Devonian and Carboniferous age there was "a world-wide uniformity of climatic conditions from the furthest south to the furthest north."[1] A cloud cover of a thickness of perhaps ten kilometers was deemed possible, leading to the warmth and precipitation that grew rapidly the huge forests of the carboniferous period where, he pointed out, the trees carried no seasonal rings. R. Potonie is cited on the evidence for low light intensity in those times.

Jordan favored Dirac's hypothesis of a declining gravitational constant. This would permit a larger solar constant in earlier times, which would have brought on the vapor cloud canopy. At some point the gravitational grip relaxed and the rings and clouds descended. Jordan was not concerned with the speed of drop or the basins required to collect the waters or with the recency of the translation from sky to Earth. However, the sky-drops may not have been so long ago. Rich and specific traditions of great celestial waters and deluging of the whole earth convey a strong presumption of truth. Prehistoric floods are believed in by many peoples who have suffered in historical times floods of only trivial consequences. Not even psychoanalytic theory, which is the

---

[1] Pascual Jordan, *The Expanding Earth* (1971) (orig. German ed. 1966).

most penetrating critic of delusions, can locate a psychic source of the flood complex; the waters of the sac in which we all swam in embryo are believed to have been a soothing, not devastating, medium.

Scholars have repeatedly analyzed much of the surface of deposits of the Earth and reported them to be the result of universal deluges; just as often they have been rebutted by scientists who see in their studies the hand of religious authority. The greater the controversy, the less immediate the conviction that my few paragraphs here can convey. Nevertheless, I will state that an unbiased scientist must today admit that the action of heavy, large-scale floods produced by vertical and lateral rushes of water can, in a holistic context, account for numerous deposits and land forms around the world. A presumptive and perhaps invalid stretching of time can only stagger the events so as to deny them simultaneity and hence grand scope. Or, in keeping with legends, the events can be concentrated, but the intervals of quiescence then may be stretched out greatly. Or, finally, both the events and the interims may be condensed in time, a view preferred here.

The sources of huge flood waters are limited. They may occur from the sudden collapse of an ice cap such as that of the Pleistocene, which covered, it is said, 30% of the Earth's surface. They can be exoterrestrial -from a comet or exploding body of the planetary system. They can descend from a onetime far-flung vaporous canopy. They can be mobilized as tides from an interruption of the Earth's motion, a tilt of the Earth's axis, or a drag induced by a giant passing body. They can, also as tides, be generated from a heavy meteoroid impact on the ocean, directly and also indirectly as in all cases above, from the winds, rock shifts and seismism accompanying them.

Deluges and tides both cause flooding. Some distinctions are necessary, though, for the next chapter continues this one with the story of great tides that swept the Earth. "Deluges fell." We should preserve the strict meaning of deluge, as a cataclysm, a "down fall." That is, a deluge is defined as an immense rain or fall of matter from the sky. A flood tide is a body of water in motion. A flood is a raising of water levels from rain or tide or both. In this chapter, only the vertical flood, the true cataclysmic deluge, is considered; in the following chapters, lateral floods and tides are treated.

Diderot's *Encyclopedia* (1751-1765) carried an article on "The Deluge" written by a young French engineer and soldier, Nicholas-Antoine Boulanger. Going beyond Newton's disciple, Whiston, who had explained the Deluge by a comet, he then wrote the first scientific work uniting the four factors; comet, flood, terror, and the origin of religion. G.R. Carli followed in a few years with additional world-wide legends and geological evidence of catastrophe. The ancient reports of universal catastrophe, both men reasoned, bore the stamp of truth.

In the century that followed, the natural and psychological sciences separated themselves from history and legend. The Biblical Deluge, for example, was steadily diminished and even dismissed as a fairy tale. It became a local flood along the Euphrates River, an account which the Hebrews picked up and patched into their holy scriptures. The influential geologist Seuss opined that "the traditions of other peoples do not in the least justify the assertion that the flood extended beyond the lower course of the Euphrates. More recently, the great floods that moved over the Indus River centers of India in the second millennium B.C. have been explained by Raikes as the effects of the bursting of natural mud dams. Such floods, goes the conventional belief, typified in the work of D. Vitaliano, occurred elsewhere from time to time and were exaggerated out of local pride.

Anyone who has experienced heavy rain and flood is keenly aware of the damage and the fright that come with the prolonged precipitation combined with the rising and swirling waters. Individuals and towns do not forget them easily. But no culture makes of any such weather event a centerpiece of their history as human beings. No matter how disastrous (as for example, was the Yangtse flood that killed an estimated million people in 1887), unless a flood practically obliterates a culture, or is accompanied by compelling foreign "divine" phenomena, it does not mark indelibly the social memory.

Donald Patten lists sixty-eight deluge traditions on six continents. He might have named many more. For instance, twenty-five of them come from the Americas; but Marie and Richard Andress, folklorist and geographer respectively, found forty-six in the New World, almost twice as many accounts. But Bellamy estimated 500 deluge myths coming from 250 peoples or tribes. The probability is high that every culture can recite the story of a universal flood which practically nobody survived.[2] The deluge is frequently pictured, too, in ancient and modern art. A. Durer and Leonardo da Vinci painted their images of it, both making it a kind of typhoon. And indeed, in the ancient Chaldean story of the flood of Xisuthros the node of the Deluge is spoken of as a waterspout that "swelled up to heaven "and struck fear into the gods; the god Ea pleaded that any and all disaster be visited upon men, but nothing so terrible as "the waterspout of the Deluge."[3]

In every ancient legend of great waters descending from the sky, a few survivors live to tell the tale. At any rate, so it seemed to the survivors. But given any tiny sum of survivors in various parts of the world, one has the basis for survival of the human race. Even a single couple procreating successfully can set off a population explosion within a few generations. The mathematics of reproduction are such that some eight billions might theoretically come forth in a thousand years. That is over twice the present population of our crowded world of today.

While catastrophic forces work on exponential curves, so do populations of all living forms. Indeed, unwilling as they may be to accept such a defense, one of the best arguments for

[2] The Biblical Flood and the Ice Epoch, *op. cit.,* 164-6, 52. Bellamy: Moon, Myths and Man, *op. cit.,* 120.

[3] Kelly and Dachille, *op. cit.,* 241.

Darwinian adaptation is the capacity of all living things to increase from a pair to billions in a numbers of years. There would be no need for exponential population growth under uniformitarian conditions. But population explosions themselves are an indirect proof of catastrophes.

Since the time of Boulanger, quantavolutionary thought has arrived at a number of additional conclusions about the "Deluge." These are at odds with conventional science, yet have been using more and more the findings of conventional science.

Boulanger and others have talked of "the" Deluge as if there were only one, whether unique in occurrence or unique in size. Most of the ancients spoke of periodic flood catastrophes. The Greeks spoke of three great floods, Deucalion, Ogyges, and Dardanus. The first two have been tied to great floods of Exodus times, the mid-second millennium B.C.[4] According to Philochorus (3rd c. B.C.), "deluge-swept Attica remained without a king for 189 [or 190] years " in the wake of the Ogygian Flood.[5] Sextus Julius Africanus said that "all the former population of Attica was killed in the Ogygian deluge and the country remained uninhabited for 270 years."[6]

The Flood of Dardanus was probably of the 8th century B.C. The story of Atlantis may be contemporary with the Saturnian flood. We note that the Atlantic Ocean was called the Sea of Kronos. Atlantis would then have sunk in the flooding of the continental shelves by the Noachian Deluge. In a prescient line, Bellamy thinks: "*Genesis I* is a dragon myth without a dragon, a deluge myth without a deluge."[7] This would be the initial deluges of the first, Uranian period of Chaos. The Greek myths of Ouranos and Okeanos were concerned with universal deluges of

---

[4] By Velikovsky in *Worlds in Collision*, 148-52.

[5] H. S. Bellamy, *The Atlantis Myth* (London: Fabar and Fabar, 1948), 145.

[6] *Ibid.*

[7] *Moon, Myths and Man, op. cit.*, 178.

the earliest catastrophes, involving the breakup of the Super-Uranus partner of the Sun.

Diluvians are of several minds. My view is that the deluges were numerous, with two great peaks. This view has at least the advantage of including all known and suspected deluges in human memory. As pointed out earlier, various high energy expressions such as typhoons and volcanic explosions invariably pick up and drop huge amounts of water and are at least localized deluges.

The first peak, the Uranian, consisted of a series of drops of sky-held waters, occurring from the beginning of the holocene period when set at 14,000 B.C. and continued for several thousand years through the lunar fission. Deluges of stone and dust (or mud) occurred simultaneously. The second peak may be placed at the end of the age of Saturn and can be identified as the flood of Noah (sometimes calculated at 4000 B.C.). Dense material fall-outs of catastrophic extent occurred at the time of the heavy-body encounters with Venus and Mars, in the second and first millennia B.C. These were exoterrestrial. In these cases, described in *Chaos and Creation,* as well as on a number of other occasions, universal and local conflagrations and explosions caused damaging fall-outs of material that was raised from the Earth. The gravest such occurrences would have been the fall-back of some of the material that was erupting to form the Moon, around 11,500 B.C. Huge falls of insects, fish, frogs, etc. would have certainly constituted terrifying spectacles over less extensive areas, and were sometimes the cause of plagues.

Issac Vail, an American naturalist, in 1874 proposed that the Deluge of Noah occurred "as a philosophical necessity, arising from a world-condition that no longer obtains .... A vast cloud-canopy of primitive earth-vapors, such as now envelop the planets Jupiter and Saturn, lingered as a revolving deluge-source, in the skies of antediluvian man - a source of primeval rains, snow and hail, competent to produce all the floods, and all the Glacial Epochs the earth ever saw, and that this last fall of those primordial waters deepened the oceans many fathoms."[8] Vail was a polymath whose analyses of myth were superb. Unfortunately, a fire consumed his principal manuscripts and he was compelled to rewrite them from memory, and then only in part, omitting many citations of sources.

Vail calculated the fatal flaw of the conventional theory of the ice ages; the incapacity of the Earth internally to generate enough heat to lift the waters and convey them to where they would form ice. And, had a mechanism to lift such masses been employed by exoterrestrial sources (although noone considered this possibility), then the poles as well as the Equator would be consumed by heat. The only alternative, Vail thought, was a pre-existing high set of Saturnian rings which descended into Jovian cloud bands and then fell upon the Earth as snow and ice in the polar regions, to which they were deflected by the Earth's magnetosphere.

Vail thought that the vast changes recorded in ocean and terrestrial life proved that a canopy had existed and had from time to time dropped part of its contents upon the earth. He pointed to pre-existing tropical conditions uncovered throughout the globe as proof of a "greenhouse" climate in which the clouds diffused the sun's heat and maintained even temperatures everywhere.

Vail did not introduce heavy-body encounters into his model of heaven and earth. Yet there is yet another possible source of a deluge, terrible beyond all others. If a passing body were attractive enough to disrupt, dislodge, and explosively pull into the sky portions of the earth's surface, it would also extract water and ice directly from the earth. The portion of the water that did not follow the intruding body into far space beyond the earth's grasp would fall back upon the world as a deluge or circle the earth with the moon and ultimately, if disturbed, fall.

---

[8] "The Misread Record," p. 1. Most of the specific allusions in these next paragraphs are form Vail's *Selected Works, loc. cit.*

Vail was not specific as to why the canopies would ever fall. He appealed to a "natural" and "divine" order or process happening over long ages, without external intervention. If the rings had moved with the Earth like the Moon does, they would hold their orbits similarly. Their fall would be at best exceedingly slow and the climatic ages that they would produce on earth exceedingly long, too long for any catastrophic theory. However, a collapse would be rapid under certain conditions. The globe or canopy might change its motions and/or electrical charges. Both would occur with large-body encounters and dense-material fall-outs and radionic storms. A great meteoritic explosion, a phaetonic atmospheric pass-through, and a bombardment of particles would singly or in combination, and in proportion to their volume, precipitate deluges upon Earth.

Now we see a complex of possible events: that "heavenly waters" (canopies) might have existed, that they might have fallen, and that explosions might have produced them and/ or brought them down along with exploded waters. The mechanisms are described more precisely in *Solaria Binaria*.

We continue Vail's account: the most ancient of East Indian gods was Varuna, whose name means the "surrounder" or "concealer." He is the regent for the Sun. The root syllable "var" means water, hence "he who covers the heavens with his water canopy." Ouranos is the Greek equivalent: this Heaven-god, ancient Hesiod's *Theogony* tells us, came from far away to embrace "Mother Earth," Gaea, and "lay close about her on all sides around." The most archaic deity of the Latins was Coelus, ruler of heaven (Coelum), who like all the other heaven-gods, was ultimately banished. The Kojiki, holy scriptures of Japan, maintains that the gods, in the earliest days, brought the heavens and earth very close together. Two light-gods then ruled the world from their "floating bridge of heaven." Later, heaven "began to retire and eventually passed utterly away."

In the Hebrew Genesis, the Elohim (the Most High) created the Heavens and the Earth. The Heavens were a "firmament" placed "in the midst of the waters." The "there-waters" (Shimayim or Heaven) existed with lights but not with the sun and moon, for they are not mentioned in the opening passage of Genesis. The Assyrians said also that the sun, moon and stars came into view only when the monster foes of order were dislodged. When the Scandinavian heaven, Asgard, died with the gods, during Ragnarok, "the Sun and his legions came riding through the gap in shining array."

The name "Yahweh" came later when the skies were opened, just as names of the leading gods changed in all cultures, with the coming of a new age. In Greek terms, Kronos (Saturn) became Zeus (Jupiter). When Kronos was removed by Zeus, Zeus removed also his own younger brother Poseidon from Heaven and sent him to rule the terrestrial waters. But note that Okeanos (the *Ocean)* had, as a rebellious Titan, already been expelled from Heaven before Poseidon left it.

So the Great Deep of the earliest religions was a watery sky. The final waters of the Great Deep were broken up at the time of the Noachian (or Poseidon) Flood. But there was "a long, long time when floods were the order of the day."

If I may refashion the theory of Vail, in the light of what I have written elsewhere, I should suggest that (a) self-conscious myth-making mankind was born beneath a high canopy of rings and clouds, without a visible Sun; (b) deluges began and a visible Uranian Sun and the present Sun appeared; (c) the Uranian Sun went nova, the Earth bore forth the Moon and cleaved, while undergoing further deluges that partially filled the newly formed ocean basins; (d) the heavenly clouds remained to some extent thereafter (during the Golden Age of Saturn when the world lived tropically); and then (e) the second great Deluge came, which was the Noachian deluge.

Jewish legends of the earliest period of man go beyond the Bible in defining a cosmic catastrophe prior to Noah's Deluge. It may be called the Enosh Catastrophe, for it happened during the time of Adam's grandson, Enosh. Since I have designated the full self-awareness of modern man (in *Homo Schizo I and II)* as part of the early catastrophic scenario of a binary nova of Super-

Uranus, and suggested that this was accompanied by great flooding, and that the Moon eruption and Earth cleavage (*Chaos and Creation*) also brought down to Earth great deluges to fill the ocean basins, perhaps Enosh belonged to one of these eras. The second is preferred if only because in legend and scripture Adam (mankind) was self-aware and active, and had been evicted into a hard world from the Garden of Eden, which represents a catastrophe of a universal globe-tilting kind.

The legends say that mankind's attention was riveted upon celestial events; idolatry (implying deviant sky-body worship) and gods (the same, but lawful) were active and importuned. The terrestrial effects were said to be threefold: the sea transgressed its bounds and a third of the Earth was flooded; "There arose mountains, valleys, and rocky ground, whereas prior to that everything had been smooth and even...; man's stature was shortened."[9] Ignoring the last, which is for another book, we are left to conjecture original or successive (Uranian) deluges possibly in conjunction with the eruption of the Moon and the cleavages of the globe, at which time great orogeny occurred and much of the land was thrusted and folded. O'Gheoghan points out that two deluges were attributed by Phoenician sources to the planet El (Saturn, possibly our lunar Super-Uranian and Super-Saturn novas).[10]

The Greeks had a god who was a son of Ouranos. His name was Okeanos and his behavior was consonant with our theory. Okeanos, writes Giorgio Santillana and Hertha von Dechend, dwelt originally in heaven.[11] He was the rivers of heaven who flowed down from the sky to earth. He was the "beloved end of the earth, ruler of the pale" and his name, too, is derived

---

[9] *Ginzberg, Legends of the Jews* (Philadelphia: 1909), V. 152, note 55. Quoted by B. O'Gheoghan, "Notes on a Possible Pre-Deluge Catastrophe," III *S. I. S. Rev.* 2(Aut. 1978), 36.

[10] Op. *cit.*, and see H. Tresman and B. O'Gheoghan, "The primordial Light?" II S. I. S. Rev. (1977), 35ff.

[11] Op. *cit.*, 190-1.

etymologically from "heaven." Jane Harrison also found that "Okeanos is much more than Ocean and of other birth."[12] He was the "daimon of the upper air," of the stratosphere, of the binary system's atmospheric plenum in our interpretation. According to Homer, the universe took the form of an egg that was girded about by Okeanos, the Generator. And Socrates in *Theathetus* says, "When Homer sings of the wonder of 'Ocean whence sprang the Gods and Mother Tethys' does not mean that all things are the offspring of flux and motion."[13] "Mother Tethys" is the ancient sea that in my opinion preceded the earthly oceans, and was the central body of water of Pangea, as the wholly land-covered Earth may be called. A whole subsequent paragraph of Santillana and von Dechend bears quotation:

> The authority of Berger can reconstruct the image. The attributes of Okeanos in the literature are "deep-flowing," "flowing-back-on-itself," "untiring," "placidly flowing," "without billows." These images, remarks Berger, suggest silence, regularity, depth, stillness, rotation--what belongs really to the starry heaven. Later the name was transferred to another more earthbound concept: the actual sea which was supposed to surround the land on all sides. But the explicit distinction, often repeated, from the "main" shows that this was never the original idea. If Okeanos is a "silver-swirling" river with many branches which obviously never were on sea or land, then the main is not the sea either, *pontos* or thalassa, it has to be the Waters Above. The Okeanos of myth preserves these imposing characters of remoteness and silence. He was the one who could remain by himself when Zeus commanded attendance in Olympus by all the gods. It was he who sent his daughters to lament over the chained outcast Prometheus, and offered his powerful mediation on his behalf. He is the Father of Rivers; he dimly appears in tradition, indeed, as the original god of heaven in the past. He stands in an Orphic hymn as "beloved

---

[12] *Ibid.*, 189.

[13] *Ibid.*

end of the earth, ruler of the pole," and in that famous ancient lexicon, the *Etymologicum magnum,* his name is seen to derive from "heaven."

Boreal means "northern." It also means "bore," a "hole". Both of these prehistoric meanings refer to the first human sense of direction. As the clouds that surrounded man 's early cultures began to break up and descend as deluges, the first openings of the sky were in the north (to those living above the Equator). Uranus, in the late Roman Empire, was still pictured as a god cloaked in clouds.

The Hyperboreans were people who lived farthest north. Their legends said that the great light (commonly, but mistakenly translated as Helios) arose and also set but once a year. So time-cycles were possible in the brilliant peak of illumination.

Most legendary clues seem compatible with the model being tested here--of an early cloud-covered greenhouse world, now broken through and deluged by water, fire, and rocks; of clouds lowering upon a smothering Mother Earth; of the beginnings of reliable changing lights and planetary figures in the Boreal hole; of a rapid development of thought and culture; of the retreat of Ouranos (Uranus) and the appearance of Kronos (Saturn).

But then also the land of Pangea was being flooded and the ice was piling up in the polar regions. Life forms retreated steadily southwards. Then came a Lunarian catastrophe, the worst, followed by the full mild, misty "golden age" of Saturn (Kronos). Again, disaster, with the Noachian Deluge and the coming of electrical Yahweh (Zeus-Jupiter) to the force.[14]

Afterwards, sunshine, dryness, lightning, thunder and the present rain-making cycle governed the atmosphere. Vail put it one way: "All through the Ouranian and Kronian ages, the thunderer [Jove] was silent." I would say that these former ages were fully catastrophic in their beginnings and end, and cosmic

---

[14] The author's *Moses* examines the electrical associations of Yahweh.

lightning and pandemonium were present, but that a fairly clear and dry world was the scene for the working out of Jupiter's divine character.

The first fall-out of sky-waters must have been limited - one sixth of today's total, we guess - because, as we argue later on, they descended upon a world largely without basins to receive them. The world would have drowned without the basins. Nor did the second fall come at one time but over a period of centuries prior to and after the forming of the basins. Even then, if the waters had not fallen partly as ice upon the caps, where it did not melt, then too the world would have been swamped.

The deluges would not amount to much rain if they were spread out over thousands of years. This, of course, was not the case, but is worth calculating. We assume that the original Tethyan Sea, shallow but globe-girdling, held one-sixth of the 1,347 million cubic kilometers of water contained in the present oceans. Further, we assumed that two-sixths of the present oceans came down in subsequent deluges of Noah and thereafter. Ice caps (now 1/200 of the total waters) are ignored, so, too, possible expansion of the Earth during the period, and also the rain cycle that would be occurring all the while. We allow ourselves 6,000 years to bring down new waters equal to half the oceans today, that is, 673.5 million km^3. The annual average quota becomes 112,250 km^3/y, which turns out to be only 22 $cm^3/cm^2/y$, when it is averaged over the Earth's surface. This is much less than the average rainfall around the world today, which can rise well above 200 cm^3 in a number of localities such as the State of Washington or Hong Kong. Evaporation and precipitation would add to the figure. Further, most important, most of the deluging might occur in years, not millennia, and then we should have to resort to a dynamism unlike ordinary rain, and resembling more ropes, hoses, and cyclones of water at many locations.

The ancient Scandinavians called snow the "pus of the gods." Something is to be said about snow and ice deluges soon. In many places, however, the waters of the deluges and floods or tides were heated. Rains came down in gobs the size of a man's head and were at times boiling hot, according to the Zend-Avesta of Persia. Josippon bin-Gorion repeats a Jewish myth: "The fountains of the deep broke up first. Then came the flood from above. Then fire fell also, and rain, boiling hot."[15] Bellamy writes that "quite a number of peoples report not only a Great Flood, but specifically a flood of *hot* water." American Indians of the West claimed that the waters of the Great Flood were warm. The Voguls of Finland said a great fire raged over the world first and was followed by a deluge of hot water. Then the hot waters raged across the land. Fire mixed with the water-- even their rafts caught fire, they said. Amerindians of Brazil said that the Sun was a cauldron of boiling waters that tipped over.

Saturn was the chief sun in ancient legend, it should be borne in mind; several recent studies have established this identification (see *Chaos and Creation*). Saturn, successor to Uranus, was both an early sun, a bright binary partner of the Sun, and flared magnificently when it went nova just before its deluge waters struck the Earth. Moreover, while lightning would unquestionably have played about the deluge scene, the fires and heat connected with the deluge and flood waters would be associated with the debris of the nova and the heavy volcanism which, as one Jewish commentator wrote, sprang up on all sides.[16]

The Feast of Lights (Hannukah) and the Christmas Light festivals, as well as the Hindu, Roman and other Saturnalia derive from the brilliant seven-day display of Saturn in nova, before the deluge struck. Frazer give us a Jewish folktale to conclude our instances of sky-associations for the Flood of Noah:

> Now the Deluge was caused by the male waters from the sky meeting the female waters which issued forth from the ground. The holes in the sky by which the upper waters escaped were made by God when he removed stars out of the constellation of the Pleiades; and in order to stop this torrent of rain, God

---

[15] Bellamy, *M. M. M.*, op cit., 124-5.

[16] Velikovsky in V *Kronos* 1 (1979), 9.

had afterwards to bung up the two holes with a couple of stars borrowed from the constellation of the Bear. That is why the Bear runs after the Pleiades to this day; she wants her children back, but she will never get them till after the Last Day.[17]

In *Solaria Binaria,* which is the heavily astronomical work of the Quantavolution series. Milton and I formulate the dynamics of the deluges. 1 mentioned earlier that the form which the deluges of Uranian and Saturnian times took was probably cyclonic, with the waters jetting down, as fountains or as liquid meteoritic fails. This would be a necessary assumption for biosphere survival and for disposing of the huge quantities of water involved. At the same time, we must speculate upon the lithospheric effects of the thousands of jets or spouts. Where are the visible effects today?

Perhaps the myriad rings faintly visible on satellite photographs of the Earth's surface (as reported in earlier pages) represent cyclonic craters formed by the jets and soon filled by aquatic tides and earth flows. When I first began to study the incidence of meteoroid impacts, I was pleased at each new discovery. But as the number of indicated craters grew larger and larger, 1 began to wonder how the Earth could have been so completely bombarded yet its biosphere could have survived. Cosmic lightning bolts and plasmoid lightning balls supply part of the answer. A liquid bombardment might also be an answer. We shall have to await a more extensive survey of the surface halos of the Earth.

---

[17] Folklore *in the Old Testament* (1981), I, 143-4.

CHAPTER FOURTEEN

FLOODS AND TIDES

Paleontology is based largely upon the classification and ordering in sequence of marine fossils. Cuvier, one of its founders, claimed as the best evidence of universal floods, that land animals were always found in association with marine fossils. Terrestrial strata were laid upon marine strata which were superimposed upon terrestrial strata. In 1796, he named three ages and three catastrophes, evidenced by three quite different 'aggregates of species. Man appeared following the last of these, he believed. Today, many fossil deposits consisting solely of land animals can be pointed out, but the presence of marine fossils in all regions of the world and at all altitudes provides an unending source of doubt. The Earth has had to be made mobile, with sliding land masses and sinkings and rising, to explain this fact, and with great stretches of time to accomplish what several very general tides, directed by exoterrestrial bodies, might in theory accomplish in short order.

Strictly speaking, floods are waters 'seeking their own level. ' 'Gravity flow' is implied, whether a high cresting river is over-flowing a town's streets or waters from all Sides are rushing down into a huge basin from which the Moon has been wrenched to form an ocean. Phenomena often called 'floods' might be more carefully denominated deluges, tides, and tsunamis. Remaining as floods would be barrier-bursting avalanching floods, the aforesaid floods from the rising and sinking of land (elsewhere treated), the varieties of rain-fed waterdownslides, the rising of waters below the ground from higher waters of distant sources and. more obviously, the melting of ice.

Tides, on the other hand, are moving waters led by other moving forces. We are not concerned here with ordinary lunar tides, of whose perplexities I. Michelson writes, "We are to this day unable to decide whether high tides occur when the Moon is in the meridian or whether the exact opposite, low tide, is more nearly correct."[1] The implications in this state of affairs, that electrical fields are operative, etc., are not germane here.

The palaetiology of flooding is no less complex than the lunar tides. Possessed of records of the Nile, Thames, Mississippi and other river flows, one can make predictions of some value concerning their behavior in the near future. Given a case where long-term records are not available, it is easy to make errors both about past and future behavior. For instance, the Pecos River in Texas flooded severely in 1954. older techniques of paleohydrology had assigned a frequency of recurrence probability in the millions of years; newer techniques reduced the recurrence interval to about 2000 years.[2] Such cases should be borne in mind when considering the probable dates of prehistoric floods: are we viewing a 10 million-year effect or a 2000 year one? Are we dealing with a rapid series or very gradual pulses?

More important to geomorphology are the tides of the great tsunamis and the tides of an Earth that is losing its balance by some external intervention. On several occasions, the Earth has had not only its waters diverted up and around, but also its very crust, this too constituting a tidal movement of land.

A comet with a nucleus as large as the Earth would from 50,000 miles' distance pull up ocean waters to a height of several miles at its focus. An exact calculation requires many assumptions; approximations of such encounters have been figured by persons as eminent as the mathematician Laplace. Hoerbiger and Bellamy more recently have calculated the tides engendered by a capture

[1] *Pensée* (1974), 71.

[2] 2. 215 *Science* (Jan. 22), 4531.

of moon-sized satellites. If one is pondering the escape of a Moon-sized mass from the Pacific Basin, a larger body, closer approach, greater mass, and favorable electrical conditions (greater attraction) must be conjectured. Atmosphere, water, the crustal rocks, and the upper mantle must participate in the tidal action-- indeed the tidal force would extend through the whole globe, and the concept of tide becomes as strained as the globe itself under the postulated circumstances.

Should such an event have occurred, and it does seem the most plausible method of providing the Earth with its satellite, the tidal pull would have dragged the surface waters everywhere towards the node of escape. Thereupon, as the intruding body moved on, the tidal force would relax and the tidal waters would rush back in great rings around the globe, reverberating for large but diminishing distances until they should accommodate to the new complex of Earth motions and the tortured terrain.

However, our model here and in *Chaos and Creation* calls for a small portion of the Earth's present waters having been available for the tides caused by lunar evacuation. Less waters would yet have been available for the tides that would otherwise reach miles into the sky. Nor, for that matter, were the mountains elevated to their present heights, but rather were only then forming under catastrophic diastrophism.

The Saturnian or Noachian Flood some thousands of years later than the postulated lunar tide also would have had major traits of a tidal disaster. Patten estimates aquatic tides of 5,000 to 10,000 feet above sea level and extensive tides of magma beneath the crust. This "breaking up of the fountains of the deep," he says, might account for 99.9% of the flood waters of the Great Flood of Noah, leaving only 0.1% as deluge waters from the skies. His schedule of events follows Davidson, Stibbs and Kevan and is useful.[3] During forty days the rains fell. For another 110 days, flood (tidal) waters continued to rise. Next, 74 days were occupied in the "going and decreasing." Not until another 40 days passed

---

[3] *Op. cit.*, 65, 61.

did Noah send out a raven. Then 21 days were taken to send out three successive doves. A further 86 days occurred before the total experience ended. Thus 371 days passed.

If the Bible is historically accurate, even only generally so, a tidal catastrophe is depicted in which rains played a minor role. Even granting that all the overrunning of the land and climbing of mountains was accomplished by tides, there remains in mind a question respecting the origin of the oceanic waters. The continental slopes and shelves were permanently inundated at some point in time, and this seems the most reasonable time for the job. The quantity of water required and mode of deluging are difficult to conceive. E.R. Milton and I finally settled upon introducing waters sufficient to cover the slopes and shelves at this time, despite the enormous bulk required to raise ocean levels by thousands of feet. We reasoned that, if all of this water were not introduced here, we could not find legendary substantiation for it elsewhere.

Having the waters descend was more difficult. As Kofahl has clearly shown, so heavy a deluge in the short period of forty days might practically wipe out the surface of the Earth.[4] So, as already indicated, we relied upon a few bits of evidence to consider and adopt the typhoon mechanism, having the waters streaming down in thick columns dispersed around much of the globe. This would have the advantage of letting much of the Earth go relatively unscathed. An average of one typhoon for every 100 square miles on the globe's surface would provide all the new water needed to cover the continental slopes and shelves. Preceding and successive deluges would make less severe the requirement. So would, of course, an increase in the 40 days and nights of rain that the Bible allows for the Deluge. A reason for acknowledging the many days of rising and falling tides is that, subsequent to exploding its waters upon the Earth, a major

---

[4] R. E. Kofahl, "Could the Flood Waters Have Come from a Canopy or Extraterrestrial Source?" 13 *Creation Res. Soc. Q.* (March, 1977), 202-6.

portion of the fissioned Super-Saturn may have pursued a path paralleling the Earth's for some time before overtaking and passing the Earth. This or another major portion finally receded into a position beyond Jupiter, and probably even retained its identity as the retired god, god of the underworld, the god placed in bonds by the new king of the gods, Zeus-Jupiter-Marduk-Yahweh.

Early students of Siberian geography, working without an ice-age theory, observed from geomorphology and fossil conglomerates that in the far north a gigantic tidal wave had recently been propagated. North-south tides of this size strongly suggest an axial imbalance of the Earth. Water in the bottom of a rowboat splashes towards someone climbing up from the side, and splashes then back and forth, as he gets on or drops off. The enormous fossil aggregations that, with a sand admixture, compose whole offshore islands, testify also to tidal action proceeding northwards and then withdrawing.[5]

A change in the speed of rotation of the globe, for which an exoterrestrial large-body encounter must be presumed, necessarily entails large tides. Some writers, including ourselves, have surmised a shift from 360 to 365 days a year around the eighth century B.C. Putting aside the more plausible cause of orbital recession, and laying the burden of such a shift upon a speed-up of rotation, with shorter and more days, the sea level would be theoretically raised by 118 m at the Equator and dropped by 227 m at the poles. So calculates V. J. Slabinski, assuming a water-covered Earth and implying instant time.[6] The "historical belt" around the world in the Mediterranean, Near East, India, China, and Mesoamerica would have noted "moderate" drops or rises of 35 m or less.

If an axial tilt occurred at the same time, counterrailing and aggravating motions would have occurred. Presumably, too, the "solid crust" would soon warp and flow to erase much of the

---

[5] Velikovsky, *Earth in Upheaval,* 7-9, 38-9.
[6] C. L. Ellenberger, ltr., VIII *Kronos* (1982), 94-5.

change. Some orbital change, as stated above, probably would alter the calculations, too. The several factors at work highlight the problems of conceptualizing and calculating the effects of encounters, but heavy tidal movements must be assumed.

The legends of tides number in the hundreds, but they are usually hard to allocate to periods of time, particularly in this incipient phase of the science of quantavolutions. When the Biblical Book of Exodus says, "The waters were a wall unto them on their right hand, and on their left," tidal behavior is suggested at the critical point of the Venusian comet, about 1450 B.C. by Biblical-derived dating. And the Psalms are chanting of the same event when "He made the waters to stand as a heap..." And the Midrashim comment likewise, "The waters were piled up to a height of sixteen hundred miles, and they could be seen by all the nations of the earth." (Though here we are bothered by the height and wonder whether, with the tides, there was a cyclonic tube reaching into the far heavens, the famous column of smoke by day and fire by night, that guided the Hebrews in Exodus). Also, in China, if the time of Emperor Yahou belongs anywhere, it belongs around the time of Exodus; and there the waters "over-topped the great heights, threatening the heavens with their floods."[7]

But when the Lapps recite how the angry god Jubmel raged against the wicked, and, "foaming, dashing, rising sky-high came the sea-wall, crushing all things," we are not sure that this is the time of Exodus or earlier or a combination of later and earlier events. So it goes around the world. The tides are there: immense, overpowering everything, wrecking the surface, launched by the gods, accompanied by fire and wind; still each legend has to be examined carefully before assigning it to a given catastrophe.

The Jubmel legend ends up as sophisticated language, as good a poetry as ever written perhaps, but it is not the language of the time of the event. Even the Biblical language is not the Exodus language. All the accounts are much later than the events. So the quality of the language does not date the legend. I think

---

[7] Velikovsky, *Worlds in Collision*, 70-6.

one may accept, however, that the tides were overwhelming at Exodus-time.

They were also present at other catastrophic intervals, and particularly in the Lunarian Age. The Noachian-Saturnian Flood was a deluge and tidal flood. The *Popul Vuh* of the Meso-Americans speaks of the god Hurracan as the driver of disastrous winds and tides, but sounds as if it were reminiscing about events of the early primordial period, our Lunarian episode.

The peculiar image of the walls of water parting gives pause, too. It is not only Biblical but, for example, Inca; near Yucatan, twelve roads of escape were opened through the sea to let pass certain peoples from the East. Can tides behave to create passages? The answer must be "yes." Not only is there a typical shore withdrawal before a tsunami; the tsunami can occur in a series. Further, the immense expressions of energy in tides, as in winds and earthquake, sometimes act to spare the most incongruous as well as precious things. Cows have been picked up by cyclones and set down miles away without injury.

When Krakatoa exploded, the people of Batavia a few miles away braced for a gigantic tidal wave that never came. Yet the wave wiped out other villages not far away and raced across the oceans to frighten Indians and Africans. There are parts of the Aegean islands that were scarcely mounted by the towering wall of water that set out with hurricane speed from Thera-Santorini around 1000 B.C. Tides rip, cross, translate, and in other ways convey their force. During the flood of Manu (Saturnian flood, probably about 4000 B.C.) hurricanes and turbulence surrounded the boat of the Indian Noah.

The skies are full of motion and the mover's body is itself moving. The atmospheric is raging with currents of wind and electricity. The Earth itself is moving. The celestial actors in the scene are imposing or withdrawing forces. Hence, exoterrestrially induced tides will not behave so simply as tides operate with the regular passage of the Moon or of a single earthquake. They will draw startling geometric figures. No one would have been more amazed than the Jews themselves, to have survived the double-walled water passage into Sinai. They lost, according to legend, the vast majority of their people to the waves that swallowed the Pharaoh's warriors. It is logical that few might reach the "Promised Land."

The "great spark" that Velikovsky says struck the walls of water and caused them to collapse upon the hapless pursued and pursuers is attributed by him to a discharge of cosmic lightning between Earth and Cometary Venus, releasing the attraction between the two bodies. It is well to note in this connection that an American Pima Indian myth paints a similar scene.[8]

> There were three warnings from an eagle of great flood.

> Suddenly a terrible roar paralyzed men with fear. A green water-mountain rose over the plain. For a very short time it seemed to stand upright like a wall -then it was split by a vivid flash of lightning, and plunged forward like a ravenous beast. Only one man escaped, keeping afloat by clinging to a large lump of rubber or pitch.

The flood of Noah is an example of both deluge and tide. If it were purely a deluge, how would the Ark end up on a tall mountain of Anatolia? (How would the boat of Manu, the Hindu Noah, end up in the high Himalayas, for that matter?) Even the heaviest deluge could not over-fill the ocean basins and cause the waters to ascend the highest mountains. The waters would run off, carrying any barges downstream, or else the world would be permanently drowned.

Alternatively, the mountains would have appeared in the course of the deluge (because the continents were on the move) and afford anchorage and survival. Or else the deluge was accompanied by tidal rises of the waters of the Earth owing to the electro-gravitational attraction of close-in celestial bodies. Or else all three events happened more or less simultaneously: the deluge fell; the lands moved and rose; and a tidal force (the same that was

---

[8] Bellamy, *M. M. M., op. cit.,* 257.

causing the deluge to fall and the lands to move and rise) drew the waters up to the heights of whatever mountains pre-existed or were appearing.

The Bible contains many specifics, almost as if it were, as Patten says, an eye-witness account. His is probably the best all-around analysis relating to the Flood. He establishes it securely as a tidal flood, "a universal, global Flood, and that it was caused by the interacting gravities of two astronomical bodies of planetary dimensions - the Earth and the astral visitor. Since the Earth possesses two fields, one gravitational and the other magnetic, there were two kinds of celestial conflicts with the intruder."[9]

The question of "how few" were the survivors need not detain whether scores or thousands -but they certainly were widely scattered about the world. The following quotation from the ancient Nicolaus of Damascus seems reasonable:[10]

> There is a great mountain in Armenia, called Baris, upon which it is reported that many who fled at the time of the Deluge were saved; and that one who was carried in an ark came on shore upon the top of it; and that the remains of the timber were for a great while observed: this might be the man about whom Moses the legislator of the Jews wrote...

The steady increasing and decreasing of waters is a tidal as well as a deluge phenomenon. The ten-month duration assigned the flood seems more to indicate a long-range tidal attraction of a celestial body; a flood, even if universal, would not take so long to recede as the 74 explicit and 90 additional implicit days before the full grounding of the Ark.

The archaeological history of the deluge has been controversial. It has been reviewed by M.E.L. Mallowan and H.J. Lenzen, among others, and Robert Raikes has supplied a critique

---

[9] *Op. cit.*

[10] Book 96 (lost) quoted by Josephus, *Antiquities of the Jews*, by Whiston, and by Patten, *op. cit.,* 61.

of the theories.[11] What is generally discoverable in the Middle East is a seeming succession of water-destroyed levels in many excavations dated in the period 2600 to 3500 B.C. Raikes accepts these datings. I cannot, for I am compelled by many other considerations in this book and others to assign the Biblical Flood to a time 500 to 1400 years earlier. That humans were civilized before the Flood is undoubted. Whether there exist excavations from this period among the Middle East excavations has to be determined by examining one site after another.

Judging by the way the tide advanced and retreated, there would not have been a total dredging and destruction of already buried antediluvian sites but probably a complete extirpation of diluvian settlements. There should therefore be a rupture and hiatus between ante-diluvian and post-diluvian cultures. Probably the distinction ordinarily made between Paleolithic and Neolithic ages directs itself unwittingly at this catastrophic break.

Hence the Great Archaeological Debate over the Deluge of Noah has probably not been treating of the Deluge at all, but has been trying to force lesser floods of later eras upon the legendary accounts of the great Saturnian floodtime. Nor was Velikovsky of a precise opinion in these matters. It is in the hiatuses between Paleolithic and Neolithic that one must search for evidence of the Noachian-Saturnian-Gilgamesh-Manu world flood.

Tides may be aquatic, but readily transport denser bodies. The velocity of water is as significant as its volume for carriage. Moving currents carry to the sixth power of their velocity. If a stream of volume "X" were to move at 2 km/h it would carry 64 times the load it could carry if it moved at 1 km/h. Tidal transport is scarcely less powerful.

---

[11] R. L. Raikes, Unpubl. paper, "Ecological Role of Extreme but Predictable Climate Events on Prehistory with some examples, for comparison, of Unpredictable Events and Their Consequences;" "The Physical Evidence of Noah's Flood," 28 I *Rag* part I, 52-63.

Tides can stretch for great lengths and in all directions. Those who like to imagine that the Exodus tide was limited ignore the evidence that the Red Sea was in motion. Moreover, they overlook the fact that unidimensional tides are practically restricted to hurricanes. A splash, a large-body pass-by, an explosion or a deluge summons a 360-degree tidal effect.

The speed of tides is swift unless remote bodies are their cause, as with the daily tides of the Moon. The appearance of the tidal effect during the Exodus, long after the first plague signaled the approaching comet, indicates a remote and approaching body. The Navajo say that on the occasion of the world flood (which cannot be precisely named) the animals had been running from east to west for days before they saw a semi-circle of water moving, like a mountain range, towards them from the east. By the next day the waters were upon them and only those who had reached the nearby mountain-tops survived. The tidal flood was preceded by a bright light in the east, an indication that an incandescent body was in the sky. Again the speed was relatively slow compared with the tidal waves from hurricanes, explosions, earthquakes, and falling bodies.

The amplitude of tidal waves will vary greatly. Historical explosions have raised waves of 85 meters, as in the Krakatoan case. Earthquakes, as in Alaska, have done as much too. The Thera volcanic tsunami of circa 1000 B.C., is thought to have raised higher tidal waves than Krakatoa. As we have said, an exoterrestrial body may raise tides kilometers high. Adding to the rain-flood from a deluge would be the flash-flood, the destruction wrought by fast-draining rain waters. Ancient times witnessed flash floods of great scope and intensity under deluge conditions. Heavy deluge waters filled the rivers and ocean canyons of the world; they poured off the mountains in the Deluge of Noah, and legendary heroes from Columbia, China and elsewhere earned their glory from engineering the escape of the floods.

A non-tidal moving flood is caused by the bursting of barriers: a natural dam blocks and collects water and then collapses. Some of the behavior and landscaping to be expected of great tides and floods are exemplified in the Channeled Scablands (Wash., U. S. A). They are 15,000 square miles of effects of a barrier burst flood; they were not made by a tide, not directly at least. Some 100,000 miles of this section of North America are thickly covered with lava, in places more than 10,000 feet thick, which can be ascribed to the immense volcanism incurred when the American continent traveled westwards over the global fracture of the East Pacific area. This might have been around 11,500 years ago, not the 10 to 30 million years conventionally given to the set of events. The whole area was then covered with silt and loess.

The Scablands are a water sculpture of this lava surface. Expert opinion asserts that a barrier of ice corked a mountain pass and caused a Glacial Lake Missoula to form. The Lake was half the volume of present-day Lake Michigan, but pitched high above sea level. The lake, it is thought, was of short duration and finally overflowed. The water cut through the ice cork. (The immediate cause may have been Earth movements.) "Within a very short time - perhaps no more than a day or two - the ice dam was destroyed and the contents of the lake were released."[12] So reads a tourist bulletin on the area. A maximum speed of 45 miles per hour has been assigned to the resulting flood, and a maximum rate of flow ten times the combined flow of all the rivers of the world today. A luxuriant biosphere was wiped out, including large mammals, camels, bison, antelope, and, to my thinking, humans. I add "humans" partly because a doll was found in clay below 150 feet of lava, not far east of the same lava field, at Nampa, Idaho.

The flood plucked and transported huge blocks of basalt. It flayed the basalt of its skin of loess. It dug channels in the basalt more than 200 feet in depth, and one of 8 miles in width. It made instant falls and plunge pools and eroded them backwards quickly. When the waters slowed they began to dump debris, some 500

---

[12] The *Channelled Scablands of Eastern Washington* (U. S. Govt. Printing Office, Wash. D. C., 1974).

square miles of it, to a depth of over 125 feet. The flood crest
lasted a day or so, the main flood 2 to 3 weeks. Today, a satellite
photo taken from 569 miles up shows the ramified and interlacing
channelways of the flood cutting through the loess into the basalt,
and then generally the unvegetated region around them.

The barrier-burst flood theory originated with Professor
J. H. Bretz of the University of Chicago and was not accepted for
many years because it was catastrophic.[13] In fact, the theory can
be pressed further in the direction of radical catastrophism.

First there are the reaffirmations of certain catastrophic
doctrines. Energy kills time. Buttes, ravines, and river channels can
be carved from dense rock in days. A biosphere can be destroyed
down to bedrock in a single rush. Broad river channels are
sculpted immediately through deep soil and loose rock. Giant
gravel ripples are laid down; hills are fashioned; long steep slopes
are fashioned *à la minute*. Heavy stones are sown far and wide, the
famous "eccentrics." Basalt is stripped to form monumental
columns.

A catastrophist still may not rest content with the analysis.
Why, he can ask, is the volcanic base of the region timed so long
ago and why is the volcanism supposed to have required intervals
of thousands or millions of years to be laid down deeply? What
water did in a month could be equaled and surpassed by lava in a
few years.

It is thought that glacial Lake Missoula formed 18 to 20
thousand years ago. Also it is said that several smaller lakes had
formed in the same way and been discharged in the same manner.
That is, the glacial ice lobe plugged the escape gap and pulled the
plug several times. The previous logic holds here too: ice can form
slowly or fast; climates change slowly or fast; plugs must be pulled
in tempo with these fluctuations.

---

[13] J. H. Bretz, "The Lake Missoula Floods and the Channelled
Scabland," 77 J. Geol.. (1969), 503-43. The original work was published
in 1923

Moreover, plugs can be placed or pulled tectonically, perhaps without the use of ice; the Earth shakes and gaps are blocked; another shake and the blockage bursts. More generally, suppose that the lava-paving occurred in the first phase of "Lunaria" (11,500 to 10,500 B.P.), after the Moon explosion, global fracture and the mountain-building thrusts and folds from the north. The high canopies are still descending and drenching the northern areas. The waters drain down the old raised glacial valleys and new ravines. The tectonic scenario of Lake Missoula goes into effect.

The area through which the flood raged is tipped to the southwest and the waters of the flood drained that way. The land is supposed to have tilted after the lava beds were laid. The tilting actually might have been responsible for the uncorking of Lake Missoula. Such extraordinary seismism would have been heavily felt in the Lake area.

Nor may the heavy loess coverings of the basalt give more than brief pause. Credited to wind-blown erosion material, it is not clear where such heavy dust would have originated or what climate brought such strong winds to transport it. Wherever it came from should contain the "mother lode"; where is it? This deep frosting was laid down by exoterrestrial sources, a cometary train, some would say. Others may claim that the loess or silt is a deposit from the unutterably greater thrust and fold phase of the ice cap avalanche and crustal movement, with contributions of ashes from biospheric and volcanic fire. By the time the scablands were etched upon the surface, the fires had been banked and the Earth was settling down.

The Scablands, we recall, are supposed to have registered several floods in succession from the same general source, glacial waters. I collapsed these somewhat and placed the Uranian-Lunarian deluge-avalanche-uplift period earlier. The Saturnian deluge and tidal flood would have come later, and contributed to a huge rise of waters drawn by a passing comet, which moved from place to place, drawn upwards and penetrating barriers and then withdrew as the attractive force was withdrawn. I have not attempted to say whether the Venusian episodes drowned and scoured the Scablands; when one thinks of the shrinking times allotted to ice ages, Lake Lakontan, Niagara Falls, and a great many "post-glacial" lakes, one should not be surprised if the Scablands Flood was a much later event and that my guess is too old.

Across the world from the Scablands are Mesopotamia and India, whose peoples claim great floods as part of their historical experiences. These floods -were they originally from deluges or tides? Comparisons with the Scablands may be useful. In all cases, the tradition claims several great floods. Just as the Greeks had at least three floods, the Indians seem to have had their flood of Manu and the flood of the Gariga region, both described in the Puranas.

Both were disastrous, and we need not doubt that, as with the Scablands, other floods occurred from time to time. A similar series seems to have happened in Mesopotamia, where for centuries controversy over the number and extent of floods has raged.

However, a hydraulic engineer and scholar, Robert Raikes, has given close attention to the literature of archaeology and to the topography of the reported events; Raikes favors a non-catastrophic approach which, to his annoyance, has been deemed by many others to be a catastrophic approach. So he is in somewhat the same seesawing position as Bretz of Scablands fame.

Let us take up the Indian case first. Here, on the one side, are the true catastrophists, religious or scientific, who say that the Indus civilization was wrecked by the mid-second millennium Venusian events -mostly earth movements and tidal floods. In full opposition, the uniformitarian extremist would be one scholar (Fairservis) who deems the Indus culture to have declined because of economic extravagance and poor ecological practices, until finally the Aryans of the northern plateau could swoop down upon the remains.[14]

The area under discussion is of great size. The influence and interconnections of the Indus and probably pre-Indus culture were most extensive -at least from today's Iran on the north to China on the east, to Arabia and Africa in the west and south to the islands of the South Seas.

Raikes finds in the Indus River Valley evidence of repeated flooding and of attempts to build against the flooding, until finally about 1500 the Valley was abandoned. He finds reprehensible "a general tendency to ascribe the abandonment of prehistoric sites to climate changes" without quantification of the degree of change beyond normal variations; also quite wrong is "the over-simplification which is to ascribe abandonments of sites to regional, or even world-wide periods of tectonic catastrophes."[15]

"Many archaeologists believe that at Mohenjodaro an extreme flood event or a series of them account for the great depth of silt/clay which has buried 11 or 12 meters depth of occupation levels under the present flood plain." Raikes traces the cause of flooding to "a combination of tectonically caused damming of a part of the Indus south of Mohenjodaro coupled with the division of Indus flows between the Nara channel and that of the Indus proper." Behind the tectonism may have been a rising seacoast, together with "extensive mud extrusions (including mud volcanoes) still active..."

"Both the flood deposits and the evidence of rebuilding occur at a great many different levels." Thought Raikes, perhaps the people built, were flooded, rebuilt, and so on, always keeping just above the new water levels. But why did the act not go on indefinitely, so that when the river finally settled itself the people

---

[14] See Gil. Possehl, "The Mohenjo-daro Flood," 69 Am. Anthrop. I (1967), 32-40, opposing views such as Raikes, 66 Amer. Anthrop. (1964), 284-9 and see below, fn 16.

[15] Op. cit., fn 10 (unpubl. paper).

might be still around and flourishing? They either abandoned the culture or they were destroyed. One can imagine that silt (loess, clay) can be laid down by comet trains. Also from far off multiple volcanism and cyclones. Or the tectonism, that Raikes tries to contain, was far more extensive. The seacoast and mountains were rising rapidly. Dams were tectonically built and burst as at the Scablands. The elapsed time from damming to filling to flood "would have been very short," in Raikes own words.

Raikes suggests similar events at Chanhu-daro. He refers to "other uplift episodes," in the same article. And in another to "a general, if less marked," raising of the Indus flood-plain to the south, at Sehwan. He believes that "there has been no climatic trend toward either wetter or dryer conditions since Harappan times," so again turns to a stress upon tectonism.

Many sites, particularly in the Baluchistan region, north of Mohenjodaro, show signs of a destruction by burning. Harappen centers were not flooded. Abandonment was sudden in these and other places after which they stood empty for centuries. Yet "one fails to see any evidence of the hill raiders who supposedly brought Harappa to its knees."

B.B. Lal turned his attention to the phenomenon of a wide scattering of copper pieces and Ocher Color Ware in the present Delhi area of India. They are found over a huge area of 60,000 km 2 .[16] At Bahabrabad, for example, the pottery and copper objects had been strewn in a level six meters below ground, and had been covered by sand, pebbles and earth. The hypothesis was a veritable "deluge." Tectonism is blamed, with or without a deluge, possibly through the mud dam mechanism or river diversion.

The Indian flood area, whether once devastated or several times over, includes the famous fossil beds of the Siwalik hills. These are foothills of the Himalayas, north of Delhi. They are crammed with hordes of specimens of a great many species. Many

---

[16] " The Mohenjo-Daro Floods," 39 *Antiquity* (1965), 196-203, 203.

of them appear for the first time in these beds and are extinguished in them, so far as paleontologists know.

In the *Geology of India,* D.N. Wadia writes,[17] "This sudden bursting on the stage of such a varied population of herbivores, carnivores, rodents and of primates, the highest order of mammals, must be regarded as a most remarkable instance of rapid evolution of species." Tortoises of over six meters, two dozen species of elephant, pigs, oxen, and apes are scattered about. There are signs of earthquake, folding of the land, perhaps folding and deep burial of animals.

Similar deposits are found 1300 miles away in Burma, cut away to view in the valley of the Irrawaddy River. Two great zones of fossils are separated by 4000 feet of sand. Petrified trees pervade the fossils in the thousands. Writes Velikovsky: "Animals met death and extinction by the elementary forces of nature, which also uprooted forests and from Kashmir to Indo-China threw sand over species and genera in mountains thousands of feet high."[18]

Other instances may be added to extend the area involved in disaster much further, probably to the limits of proto-Indian civilization, and indeed throughout the world. The dates are hinging upon 1500 B.C. in many instances. Therefore, it would seem reasonable to place Raikes' work on the revolutionary shelf; try as he may to limit it, his evidence and own conjectures press in the direction of general catastrophe.

What emerges from Raikes' complex analysis is that in the Old and Middle Bronze Age -and particularly at the age-break between Middle and Late Bronze -there is proof of various terrific floods to which all known settlements succumbed. Raikes inclines, after considering six possibilities, towards a land subsidence on a large scale complementing a land rise to the east.

He does not mention the backup of river waters that would occur from Thira-type tsunamis driving north through the

---

[17] " A Deluge? Which Deluge?" 70 *Amer. Anthrop.* 5(1968), 857-63.
[18] Velikovsky, *Earth in Upheaval,* 79.

Persian Gulf, although the evidence allows it. Such tides could come from a Typhonic impact explosion, a poseidonian earthquake, or a large-body encounter producing an axial tilt or interrupted rotation of the globe. (One notes the level of ashes and char beneath the flood level of Shurrupak. It does not appear to have been an incendiary blaze.)

He does not consider canopy water-drops, but insists upon retrojecting uniformly precipitation rates from modern times. Although the evidence of the period which he is examining is disordered and prejudiced already, yet the evidence that he must confront shows a flooding that is utterly devastating, and unexampled in recent times. But still, he draws back from catastrophic conclusions, loath to abandon the dogma that catastrophe could not have happened, and certainly not an exoterrestrial one.

Since large upthrusts of the Himalayan mountains are now being dated to post-glacial times,[19] since even mountains much higher than the Siwalik foothills contain "old" marine fossil beds, since the Siwalik-type beds are so young even when conventionally dated, since evidences occur of huge waves of translation moving from south to north in India and leaving great moraines (including the Siwalik-type hills), since Neolithic stones are found in the loess of the Himalayas and since great human cultures were flooded over and probably deluged as well, one is entitled to the quantavolutionary hypothesis: a series of abrupt, intensive, wide-scale changes overwhelmed the Indian subcontinent.

Frantic proliferation and extinction of species occurred, while India broke from Africa and crashed into Asia, while tides moved over the land, ramming, ripping, rising, and drowning, while the land raised up in a great arc into Asia, while hominids, then humans, entered and built cultures that were then destroyed and recreated. It may be that from this part of the world will come the easiest and fullest proofs of revolutionary

---

[19] *Ibid.*, 21.

primevalogy, of a succession of geological and cultural ages coinciding with the successive disruptions of what had been *Solaria Binaria*.

Dwarfing the Scablands and Indus barrier floods was the Gobi Sea flood, which may have been connected with the complex Noachian Flood. Thomas Huxley wrote the first scenario of the event. Bellamy refurbished the story in this century.[20] The Gobi Desert, which the Chinese call "the Sea of Sand," was once a great body of water. Numerous settlements lined its shores. Then suddenly it was emptied in a huge barrier-type flood. Its cultures disappeared along with a great many other settlements along the line of the flood. The western barrier of the Gobi Sea broke between Tian Shan and Altai mountains, and rushed through where today remain the waters of Telli-nor, Ebi-nor, Ala-kul, Sasyk-kul and Lake Balkhash, much of it now saline and disappearing. The great flood spread out into a "Sea of Turkestan" and then drained down into the depression of the Aral and Caspian Seas.

It then poured out between the Ural Mountains and the mountains of northern Iran, descended west through the Manych Depression into the Valley of the Don, the Sea of Azov, and the Black Sea. The areas of today's Romania and Bulgaria were temporarily part of a greater Black Sea. Soon it overflowed at the straits of the Bosphorus and pushed through the Dardanelles into the Mediterranean region. The Aegean and Eastern Mediterranean lands were flooded.

Next the Adriatic River, possibly the legendary River of Eridanus, and nowadays the truncated Po River, was turned into an Adriatic Sea. The Ionian Sea overflowed and the land bridges between Italy and Africa were covered with water. The shelves of the region of Tyrrhenia were submerged, the survivors driven to the high places of the Italian peninsula and islands, and contact was ultimately made with Gibraltar.

---

[20] Bellamy, *M. M. M., op. cit.,* 308-16.

The Sahara basin may have been filled with water upon this occasion, to have become the ancient sea of Triton. It was this Tritonian Sea that figured in the mythical birth of Goddess Pallas Athena (the planet Venus) and I think that it was around 3500 B.P., therefore, when the Tritonian Sea broke out and threw itself into the Atlantic Basin. Ancient Saharan ruins and the art of the Ahaggar mountain caves amply testify to the ancient cultures there between 4000-1500 B.C.

The elapsed time for the 4000-mile journey from China may have been months or years. The drainage of the several temporary basins established *en route* from East Asia to the Atlantic Ocean occupied centuries.

Barrier-burst floods and tides must have been numerous, we conclude, because of the mountain-building, severe faulting, deluging, and other movements and outbursts that were occurring. Both actions would have been quite unexpected and erratic. They would have devastated the biosphere. Evidence of both effects comes sometimes from jumbled deposits of animal bones and wood. These locations consist of different species, that were killed suddenly (not by men), by the hundreds or thousands, and were transported to the location, by tides of water but in some cases also by hurricane and cyclonic action. In the Yukon Valley of Alaska, bulldozers scraping for gold have removed bones by the ton and drills have picked up bones hundreds of meters below ground. Such evidence exists around the world, and much more will be said on the subject in Chapter 26.

The number of fossil deposits will probably be extended to many hundreds of cases in the future. Still, most deposits would have been destroyed at the moment of catastrophe. Fires would have burned others. Impenetrable ice covers many bone piles. A succession of revolutionary actions would have blown to bits, dissipated, ground up, converted to fuels, washed into the sea, and deeply buried many others. The scenes at bone deposits are impressive: they are worldwide; they are found at low and high altitudes. Strange bedmates are discovered: ostriches and foxes; mammoths and lions; peacocks and horses; elephants and sharks.

Anthropologist Frank Hibben surveyed the bone mucks of Alaska and heard of similar deposits in nearby Siberia. The Arctic Ocean is in fact rimmed by the bones of many millions of animals. Hibben weighed the possibilities: hunters' overkill, ice flows, natural death, volcanic ash burials (ashes are abundant in the muck), volcanic gases? The mystery seemed to him unsolvable. He wrote of it in 1947; he revised his work in 1967.[21] There is no indication that he had heard meanwhile about Velikovsky, Hapgood, Patten, or Cook who were offering solutions to the mystery in terms of Cuvier's century-old expression -"revolutions of the globe."

Derek Ager, with a mind and eye for the catastrophic occurrence, remarks that "tsunami, ' or 'tidal waves' as they were for long misnamed, have an immense effect on shorelines, both in erosion and in the shifting of great quantities of sediment."[22] But what parcel of land in the world has never experienced a tsunami?

"It is generally accepted that tsunamis are usually triggered by earthquakes or violent volcanic explosions. It is also possible that they can be produced by the slumping of large masses of sediment in water..." Or by meteoroid splashes, we might add, or hurricanes and cyclones. "Though infrequent, there are certainly enough of them for geological purposes. From historical records it can be deduced that there have been more than two hundred notable tsunamis in the last two thousand years; this would allow us more than 100,000 in a million years."

Then move the continents a little here and there, raise and lower shorelines, change climates a few times, and add ten, fifty, a hundred million years. We have millions of great tsunamis to work with. Obviously the whole surface of the Earth will have been worked over a number of times by ordinary, uniformitarian waves. Thereupon add all the other high-energy forms: deluges,

---

[21] The *Lost Americans* (NY: Crowell, 1968).

[22] Op. *cit.*, 45.

exoterrestrial impacts, volcanism, and so on: it is a wonder that the crust of the Earth is not a homogenous finely ground mixture of all past life and surfacing rocks. Now add great catastrophes elaborated in this book and the homogenous mixture should be guaranteed.

That is, stratigraphy is hardly understandable by following uniformitarian principles, if we acknowledge what scientists have all along been discovering, but more recently have become acutely aware of. Even if, as Ager writes, "the changes do not take place gradually but as sporadic bursts, as a series of minor catastrophes," the strata of the Earth do not make sense.

Those who believe in major catastrophes interrupting huge serene tracts of time may be wrong, because they must add to the effects of the great disasters the effects of a multitude of minor ones called for during great stretches of "peaceful" time. The result would be a homogenized crust. The effects of the forces that have operated are such as to suggest for the Earth a short and recently catastrophic history. The Earth's surface still retains its forms and fossils because its tortures have been clustered and have occurred following a short total Earth history.

CHAPTER FIFTEEN

ICE FIELDS OF THE EARTH

The earliest humans had to contend with growing ice caps and glacial fields, or at least some force that created their effects. Did the Great Ice Ages really happen? For a century the confident answer of science has been "yes." The idea is fetching; so much ice surrounds the north and south poles now that it seems reasonable that once there was even more, and probably once there was less, or none at all. At peak time, an estimated 30% of the Earth's land surface was covered by ice, three times the area occupied by ice today; this was as late as 11,000 years ago, or so it is believed.

When Emiliani discovered evidence that the Gulf of Mexico was for a time freshwater, he posited a rapid end to the Ice Ages and a flooding which may have drowned the mythical Atlantis culture, since the time (ca. 11,600 B.P.) conforms to Plato's date of the disaster. The surmise engendered sharp criticisms, allowing even historians to get into the act.[1]

It seems that everyone believes that the ice cameth and each has an individual scenario, which is not complete unless it contains quotas of confusion and contradictions. If one wishes to spend a lifetime solving a puzzle while wrapped in an enigma, a career in paleoglaciology is recommended. One can scarcely blame an amateur from enjoying and even tolerating Donnelly's old idea that the ice ages never existed. Next best, one can call down the

---

[1] 189 *Science* (1975) 1083; 193 *Science* (1976), 1268-71.

ice (or most of it) from outer space, as we do here. And so does Patten. Third best would be the Milankovich theory which depends upon cosmic perturbations in Earth-Sun transactions, but lets Earth manufacture the ice. John and K. P. Imbrie have updated and defended the theory, which, highly complicated in itself, is also confounded by the uncertainties of paleoclimatic studies.[2]

Hard evidence that a set of ice ages occurred falls into several categories, as follows:

Certain northern lands near the present ice are rising, as if a large load had been lifted from them. They seem to form arcs with Baffin Bay as an old geographical pole and center of an ice cap. (The western rising arc is separated from the eastern arc, as if they had been pulled apart.) An issue occurs if one asserts that the rising would ensue from a shifting of the Earth's axis and North Pole, regardless of the presence of ice.

Far to the South of the present Arctic ice, and far to the north of the present Antarctic ice, the rocks and soils show peculiar qualities. Huge areas of rock are scoured and scratched as if some gigantic force has scraped over them, now advancing and then again retreating. Immense fields of stones (or drift) have been pushed and shoved into place, as if by moving ice. An issue occurs if one asserts that tides and exoterrestrial stone fall-outs had produced the fields.

Glaciers, formed on mountaintops around the world, take their origin usually in a U-shaped nook of a mountain. Their ice forms and slowly slides downwards through valleys, carrying drift and ending in melting waters. They abrade and pluck the drift as they go along. They are broad, and they terminate in broad curls, from which streams form and run off. Many "extinct" glacier

---

[2] Ice Ages (Short Hills, N. J.: Enslow, 1979); *cf.* Ian Cornwall, *Ice Ages* (NY: Humanities Press, 1970); Björn Kurtén, *The Ice Age* (NY: Putnam, 1972); Clifford Embleton, *Glacial Geomorphology* (NY: Wiley, 1975); Salop, *op. cit.* introduces cosmic disturbance as causes of glaciation, too, as does Pattern, *op. cit.*

forms exist, indicating that once there may have been much more cold and ice. That is, unless these "fossil" glaciers were pointed towards the sun in a global Earth tilt, and melted, or once were a part of a large crustal lateral avalanche that thrust whole areas away from the polar regions. Or unless exoterrestrial ice were dumped upon higher places and melted away from lower places.

Heat is required in large amounts to raise water for the snow falls over glaciers as well as polar regions. Some say the heat required would be too great for the biosphere to tolerate unless the snow gathered by very slow increments; there is evidence that "glacial ages" came and went rapidly. Further flora and fauna of the glacial age seas are arctic types; then where were the sufficient warm seas whose waters would evaporate and stream polewards as clouds? If cold water and snow fell from high cloud canopies, it could persist at higher altitudes and latitudes and accumulate and flow.

In many settings, such as Cape Cod, Massachusetts, large plains end on the downslope with a number of ponds and layer upon layer of sands, gravel, and clay. In it are scratched stones and finely ground glacial flour. It seems that an ice sheet had once moved downwards on all sides from a northerly direction, acting like a glacier on a grand scale.

Humps, low ridges, occasional erratics (rocks foreign to where they are found), and kettle pools (some dry) are scattered along the hypothetical front of the glacial sheet and might well have been produced by the forward march and retreat of the flood of ice.

Furthermore, an ice sheet that moved down into North America all across the continent blocked all northward flowing rivers; it created many lakes, some extinct like Lake Agassiz, others extant like the Great Lakes. The ice sheet forced a southward fanning out of many rivers, away from the ice front, to carry the melt waters. Once again, much, if not all, of the work assigned to ice could have been performed by winds, tides, exoterrestrial fall-outs of pebbles, dust and ice, extreme precipitation, and axial tilts of the globe.

I have not mentioned climatic changes: a very cold climate, as evidenced by the kinds of fossil flora and fauna discovered in old beds, indicates that a great deal of ice might have been nearby. Nor have I ventured to say when the ice ages happened and how many of them there might have been.

Full justice cannot be done here to the case for the ice ages. The conventional literature does so. But because some of the ice age reasoning falls victim readily to catastrophic claims, it may be time to advance the cause of quantavolution. Here three different positions are held: one is that the Ice Ages did not occur. The second says that they did exist but were sudden events, beginning and ending in disaster. A third admits their slow development but claims that they ended in catastrophe.

Ignatius Donnelly is the best older critic of the very idea of ice ages. (Douglas Cox has recently presented strong persistent objections to the reality of the ice ages.)[3] In *Ragnarok: The Age of Fire and Gravel,* Donnelly asserts first of all that there is no evidence of the ice ages in the cold Siberian wastelands and parts of Alaska that stretch up to the present Arctic ice. This is true enough. But, most catastrophists believe that a sudden tilt of the Earth occurred in the last ice age and hence these areas had *not* been so cold before then.

However, Donnelly proceeds. He argues that the debris of the called ice age -the pebble fields, erratic stones, and vast clay and till deposits -are not caused by the movements of ice at all. Rather they are the stuff of which the long tail of a comet is in part composed and it was a comet that devastated the earth in the early memory of mankind.

Little was known of comets and comet tails in his days. Until the past few years, scientists generally doubted that such substantial material was being transported around the heavens. Indeed, Velikovsky came in for much ridicule when he wrote in the nineteen fifties, in much greater detail and with stronger

---

[3] 13 *Creation Res. Soc. Q.* (June 1976), 25-34.

evidence, of the substantiality of comets. (He did not adopt Donnelly's anti-theory of the ice ages, however.)

Today the immense material potentiality of comets is scarcely doubted. Ice and gases, and otherwise terrestrial minerals found in meteoroids, are now accorded comets. Yet Donnelly's theory has not been seriously criticized; we forget that geology once got along without the ice ages, and that the inventor of the ice age theory, Louis Agassiz, was a catastrophist. The immense drift and till deposits could have come from exoterrestrial sources.

Although the analogies between glacial behavior and ice sheet behavior are numerous and strong, it is possible that the ice did not exist and that the dead glacial moraines are merely evidences of a cold climatic episode or episodes, not direct proof that they were related to a larger ice age sheet that blanketed millions of square miles to a depth of a kilometer and more.

Moreover, since the poles are flattened a bit from the spin of the Earth, would not the old polar areas of a perhaps faster spinning Earth be still relaxing into a spherical form? This would give a false impression of heavy ice caps having been removed. Further the weight of the Wisconsin ice cap would have been 3.10^{23} grams and 10^{33} ergs of heat would have been required to melt it. The melting would have taken at least 30,000 years, yet there is near to a consensus even among uniformitarian geologists that the ice cap disappeared rapidly, catastrophically. And the arctic land rising, mentioned earlier, appears to have begun only 10,000 years ago.

Why I do not accept Donnelly's theory despite its brilliance has to do with the correlative evidence going far off the straightforward discussion of ice ages. Some of the reasoning emerges when the theory of Melvin Cook is explained. Cook writing in the nineteen sixties, accepts the evidence for huge ice caps at both poles. Further he seeks no exoterrestrial power. His theory is nonetheless the most perfect of catastrophic models yet advanced. Ignoring the beginnings of the ice ages, but pursuing their end, his story commences with the great ice caps.

These, he says, by their enormous and accumulating weight, bore down upon the crust so heavily as finally to cause a rupture of the rim of the crater. The ice caps avalanched. They scraped the earth as they moved. They acted as gigantic bulldozers that caused mountain ranges to be thrust forward and buckled and folded upwards. Giant floods from the rapid melts swept the earth.

The globe fractured and caused the continents to spread apart rapidly. The Atlantic Ocean and the Arctic Ocean were opened up. In the end the surface of the earth was greatly changed. A great many land and life forms, together with cultural centers, were destroyed in the process. As the huge ice blocks descended, they turned over the biosphere and folded it to create coal and oil deposits in a geological "instant." Waters that were buried deeply are still rising under pressure. Yet the end came quickly, occupying a few years, not millions of years.

The legends are definite but seemingly too rich. The northern peoples talk of terrible ice falls and winters, far beyond historical experience, and perhaps long before history as we gauge it. In Old Norse, the language of the Edda epics, snow is called *eitr-ornir,* "white pus of the dragon." Martin Sieff writes: "Saturn is the solar system's 'treasury of snow'... The Greeks associated the planet Saturn (Kronos) with snow and hail, which were thought to be the planet-god's weapons; Nonnos told of the "shining victory of Zeus at war and the hailstorm-snowstorm conflict of Kronos..."[4]

Could the ice have fallen from the skies? Examination of glaciers shows that there is a gradation of consistency, from fresh fallen snow to dense ice, the dense ice being older. No question but that, if snows fell heavily they would promptly turn into ice. Further, the greater the falls, the swifter the glaciers would move and the longer and greater their moraines.

Moreover, why should the ice ages occur in extremely distant as well as recent ages; how do they come and go in stages,

---

4. *S. I. S. Workshop* (Mar, 1978), 4.

and concentrate most recently in a million years of the recent Pleistocene epoch (which is the typical allotted time)? The Sun is invoked. Whereas, on the one hand, the Sun is credited with great stability, on the other hand it is presumed to have stoked its furnaces from time to time, causing the ice to form. But back again. If the Sun cools, the equator cools; if the equator cools, waters evaporate more slowly; there is less to be carried north and to drop in the form of snow.

Continental drift has been argued as the cause of ice ages: "The ultimate cause of glaciation is thus seen to be movement of continents into appropriate latitudes... And much of the fossil evidence upon which the time-honored concept of Tertiary 'cooling' has been founded could be nothing more than a reflection of drifting of what are now the northern-hemisphere land masses and ocean floors toward the pole and hence into cooler climes."[5]

Another theory holds that a huge number of tropical volcanos erupted at once, which threw vast amounts of water into the air, which, because the upper atmosphere was darkened, caused less sunlight to bombard and warm the Earth, which finally caused the vapors to fall at the poles in the form of snow and ice.[6] Also, Hibbin attests to many burials of Pleistocene animals in ashes that fell *after* the ice ages.[7] It should be borne in mind, however, that extensive simultaneous volcanism, as well as the ice ages, points to exoterrestrial forces impinging on Earth.

The solution must be catastrophic, it appears, but must take a special form, which elsewhere we have called *Solaria Binaria*. If it is consolation to the reader, explanations of "the ice ages" have generally been bizarre and fantastic. Nothing less may be expected of our theory here, unless, of course, the reader is

---

[5] C. B. Beaty, "The Causes of Glaciation," 66 *Amer. Sci.* 4 (July 1978), 452-9, 458.

[6] J. R. Bray, "Volcanism and Glaciation During the Past 40 Millennia," 252 *Nature* (20 Dec. 1974), 679-80.

[7] *The Lost Americans*, 163.

conversant ahead of time with our work. It is not unreasonable, we argue, to postulate a primordial age, as recent as 14,000 years ago, when no ice caps existed. The Earth would have been generally comfortable. It would be also enveloped in the gaseous atmosphere of the binary magnetic tube. This Uranian heaven blocked direct sunlight, but afforded an equable climate to the Earth.

The binary tube atmosphere would itself have been maintained by the same electrical and inertial forces that kept the Earth in rotation and orbit. Then the solar system as a whole was disturbed by the failure of one of its parts. The part that failed was the counter-solar or Super-Uranian node of the binary solar system. When the electrical current between the Sun and Super-Uranus diminished, the magnetic field around the current diminished. All the bodies that circled around the current ceased orbiting around the axis between Super-Uranus and the Sun and descended radially to the plane of the ecliptic. They began to find new individual orbital paths around the Sun. They moved out towards larger orbits.

The atmosphere, a remnant, specially attached to the Earth, of the old plenum atmosphere, drew more closely about the Earth. "Heaven came down to embrace Earth," to paraphrase the Greek myth. The clouds were pierced by material erupted from the disintegrating Super-Uranus and blown down the magnetic tube between the binary partners. Some of it precipitated upon the surface of the Earth. The Earth could not melt much of the ice, most of which fell at the electrically least-guarded poles. The now direct sunlight helped the friction of the fall to vaporize and precipitate some of the ice as rain. Flooding began at the edges of the forming ice caps. The time postulated for these events began about 14,000 years ago.

Within a few centuries the threat to life on Earth became extreme. Great ice blocks covered the extremities and local regions of the globe and threatened ultimately to make contact, erasing practically all life. At the same time flooding spread throughout the world.

If one-third of the globe was covered by ice at the time of maximum advance, according to conventional theory, ice was piled three miles deep at the poles; there was twelve million cubic miles of ice. For a hundred years catastrophists and disbelievers in the ice ages have pointed out that an incredible power (heat and winds) was required to evaporate equatorial water, lift it, and transport it to the polar areas. The world would have burned up at the equator while freezing deeply at the poles. The idea supplies its own contradiction; yet it is the accepted theory, that molecule by molecule the water evaporated, drop by drop it condensed in vapor clouds, ton by ton it fell - all off and on for a million years and more. Then the mechanism was turned off, rather suddenly; much of the ice melted and the oceans rose by several hundred feet several thousand years ago.

Direct exoterrestrial deposition of snow to form the caps follows from the heat requirements to evaporate, lift, transport and condense as snow the contents of the ice caps. The surface heat requirements might have stressed the biosphere life tolerances. Further, in order to raise the required mass of water, the clouds transporting water from tropical to arctic regions would become so dense that heat from the Sun of today would cease to penetrate to the surface with sufficient energy to continue the lifting task. The latent heat of aqueous vapor at the tropics is 1000° F. A pound of water vaporized at the Equator has absorbed 1000 times the quantity of heat that would raise a pound of water in temperature by one degree Fahrenheit.

An exoterrestrial catastrophic solution is called for, from beginning to end. The time to erupt the Moon arrived with a passing great fragment of Super-Uranus. The Earth's crust burst. Lava had to flow in endless streams. Great volumes of sky-borne ice must have fallen and participated in the bursting mechanics. Cook has figured the needed forces, but we should add an initial impetus from the eruption and blow-off of the Pacific crust. A fracture shot to the old north pole and down the Atlantic, thence around the world. The ice avalanched. It fed the boiling sea bottoms to help them settle and expand. Much was then evaporated and precipitated again by the conventional method, but under catastrophic conditions. Finally, the new world surface shaped up and stabilized. The precipitous curve of disaster dropped exponentially to the slight level of activity where it could be mistaken for a linear uniformitarianism.

It was thus that the worst and best accident happened. The earth cleaved, lost most of its continental crust, and the ocean basins began to form. This greatest of all catastrophes removed the ice and permitted life to survive; it became the greatest of all blessings. A date of 11,500 B.P. may be ascribed to the event.

The ice caps, as Cook has so well calculated the scene, collapsed and avalanched upon all sides, moving into the great chasms of boiling lava directly or through floods that rushed over the land and plunged down into the new oceanic chasms, carrying debris to form slopes. Hundreds of deep canyons were grooved into the land and slopes around the world, where they remain today, "fossils" from the time of ice age collapse and of the filling of the ocean basins. The ice caves were formed -solid ice from the ice ages sandwiched in between layers of once boiling lava flows, still intact, though hollowed out somewhat, now refrigerating food and supplying age-old spring waters.[8]

Geologists have counted and recounted the number of ice ages and of interstadials, the periods between stages. John Gribben, in a recent work on *Forecasts, Famines and Freezes,* counts ten ice ages, of which one lasted only for a century or less. Paying no attention in their "petrofabric analyses" to our impression that fossil "glacial and stream deposits" could just as well come from comet-tail or meteoritic splashing, geologists saw breaks of climate in the interruptions of moraines, where now a swelling and then a shrinking may appear. In the soil found squeezed between strata of glacial debris, there is also the suggestion of successive ice ages. Even the Arctic Ocean is said to have been free of ice in Pliocene and Pleistocene times, on the basis of calcareous nanno-

---

[8] Patten, *op. cit.,* 120-4.

fossil deposits below the present ice.[9] And another study, of the Labrador shelf area, based on fossilized sediment cores, argues for an ice-free sea extending back 21,000 years from the present.[10]

Over a mile deep in the Greenland ice field around Dye 3 radar station, Greenland, ice cores are being drilled, extracted and analyzed.[11] From its rock base upwards, the ice is expected to afford 100,000 years of Earth history and the beginning of at least the local ice age (cf other estimates of 1 to 3 million years and our own of 14,000 years). Oxygen ratios in sampled slices of the drilled ice are calculated to determine climatic trends and time scales. The units are "annual" ice varves. As depths increase, the distinctions blur. Dust ratios are used as indicators of heavy volcanic events in the world.

The stratification challenges any quantavolutionary attempt, as here, to explain the ice accumulation as a brief episode. Obviously the ice under examination did not fall as blocks, at least not most of it, or, if it did so fall in the region, the blocks splattered and connected up or flowed afterwards under weight, internal pressure and heat, picking up atmospheric exposure and dust.

Heavy snowfalls, whirled about by heavy winds, would, however, establish the great depth in short order, in dozens or several thousands of years, with present snowfall adding steadily to the basic conserved precipitation. It is noted that at an estimated 10,000 years, the "ice age" deposits of tiny crystals end and the large ice crystals of the present era begin. For those who are disturbed by only 100,000 years for the Greenland ice cap, because of ice age theories of a million years, there is the consolation that the ice beneath relentlessly squeezes out to form icebergs that search out more southern climes.

Interpretations that seek a long drawn-out succession of uniform deposits may be an illusion of sorts. The evidence rather may indicate the erratic character of the ice falls, both in intensity

---

[9] T. R. Worsley and Yvonne Herman, 210 *Science* (17 Oct. 1980), 323-5.
[10] G. Vilks and Peta J. Mudie, 202 *Science* (15 Dec. 1978), 1181-3.
[11] See Walter Sullivan, *N Y Times,* Aug. 9, 1981, 1, 24.

and distribution over the Earth's surface. It may also indicate a wobbling of the axis of the Earth as its electrical fields changed and its motions within the solar system altered. Whether the globe changed geographical axis once, with such gradualness that it scarcely wobbled, or whether it changed once quickly and wobbled several times before settling down in its new position, or whether the geographical axis changed several times in several hundreds or thousands of years, an illusion of several ice ages and subdivisions thereof might be fostered.

Faced with the problem of explaining the chalk cliffs of Etretat (France) across from Dover, which are laminated, French geologists have tried to establish a correlation between the laminations and the oscillations of the axis of the Earth. The oscillations occur some 23,000 to 41,000 years apart, the sedimentary layers are individually accorded 20,000 to 40,000 years. *Voilà*, as the Earth rocks, the sea level and the biological activity of the sea rise and fall, as evidenced in the layers. "But how explain that such feeble orbital variations should be capable of engendering such important changes? The problem," wrote a group of French editors led by Serge Berg, "is far from being clarified." Surely so; however, not only chalk sediments but also ice layers could be deposited in a short time if the wobblings of the axis were greater and more frequent, as is demanded in quantavolutionary theory. Strata of all kinds can be laid down quickly, including strata that reflect and measure falling snow and ice.

Furthermore, as we have pointed out, unfossilized till deposits, possibly themselves exoterrestrial, are used to denote recent and ancient ice ages. ".... The Huronian super-group in the south of the Canadian Shield presents this evidence most unambiguously. Three tillite levels are reported from that region .... corresponding to three glacial periods separated by epochs of warm or even hot climatic regimes which lasted some tens of millions of years."[12]

---

[12] Salop, *op. cit.*, 23 ff.

So, too, around the world, on every continent (whence geologists have deduced shifting sidereal poles); thus "two principal tillites are dated isotopically at 870-820 MY and at about 680 MY." These statements, by a pronounced quantavolutionist, L. J. Salop, evidence the overall grip of conventional scientific theory on the scientific mind, for it would be only consistent of Salop to query the origin of the tillites and then the conventional view of many ancient and modern ice ages. The correlation of tillites with ice ages is deceptive of time and causation. Why not repeated switches of a comet tail?

A late report, in the newsletter of *Science and Technology* (54: 2, 1982), describes an area of the Huqf Desert of Oman where tillites on striated bedrock - taken as glaciation - seem to be associated with oil reservoirs, and the complex is pronounced Early Permian (-158 my), when coal is supposed to have formed as well from tropical vegetation. We see no contradiction in ice striking hot tropics, provided the ice comes from the skies, and provided that along with the ice one brings down stony till to gust along, scratching the rocks. Here, however, one may dispense with the glaciation, which is predicated upon the till; ice may or may not have fallen. One may also tie in the oil deposits with the exoterrestrial source of the till, a comet. One may, moreover, hold in abeyance the dates assigned to the events; the time may have been only thousands of years ago.

The ice ages, then, may be a product not of a million or more years, but of several thousand years, from 14,000 B.P. to about 9000 B.P. At this latter time, there began a settled and milder age, with a subdued binary, an equable climate under still cloudy skies and two suns, the Sun and Saturn. This would be the renowned Golden Age of Saturn, of which so many legends speak, an age following the revolt that dethroned the god Uranus, the age before another great catastrophe, when the gods warred again and Jupiter removed his father, Saturn. There occurred huge inundations, brighter skies, and the present ice caps developed, shaped around the present geographical poles. The Antarctic cap is largely contained on a land mass with an ice flow over its boundaries and into the sea. The Arctic Sea was almost entirely a swamped continent, despite the rifts through it, and received its ice directly upon this land in the transition from Saturn to Jupiter. The extreme conditions of Earth fracture and ice avalanching encountered in the critical period beginning at 11,500 B.P. would have destroyed all ice. Evidences of mild climate and an abundant biosphere are present in both polar areas, some of this presumably from the Saturnian interlude, most from pre-lunar times.

Thus far, no human vestiges have been discovered where once the Uranian ice cap lay. The turbulent moving ice would have erased all such evidence down to a considerable depth of rock, even in the absence of land thrusts, flood, wind and fire. Certainly humans retreated to warmer climates in the face of the icy tempests. Still, primates, proto-humans and *homo sapiens* lived among the animals whose remains have been found under ice and permafrost. Whether a long-term date (like two million years) or a short-term date such as I suggest here is adopted, these species existed before the ice and they may one day provide new fossil discoveries.

There is an old map, called the Piri Reis map, that shows perhaps the coastline of the Antarctic continent as it would have appeared in the interim between the settling down from the great ice cap collapse and crustal shifts of Lunaria and the new ice caps of Jovea that remain today. That would be during the "Golden Age" of Saturn. The Piri Reis map is the subject of a book by Charles Hapgood, who also provided a singular theory of ice cap avalanche with a mechanism different than Cook's. (Einstein thought Hapgood's idea that the ice cap would have shoved the continental crust on a wedge principle to be mechanically acceptable.[13])

I incline to the view that the map, which was drawn up from various old sources a few years after Columbus anchored off Santo Domingo, plots the shores of Antarctica well because, during the Saturnian period, a mild cloudy climate prevailed, the

[13] The *Path of The Pole* (Philadelphia: Chilton, 1970).

southern oceans and shores were free of ice, and navigation was well developed. Probably human settlements then existed in Antarctica as they did in many places in the far north that are now encased in ice or permafrosted.

We speculate that the "ice ages" did happen, first in Uranian, then in Jovian times. Much of the Earth was frozen. The ice was mainly exoterrestrial. Vsekhsviatskii writes of Saturn that "observation of its rings over the past 300 years have shown that during this time the middle line has moved 0.17 of its original distance closer to the surface of the planet. Therefore, one may suppose that in a matter of some 1800 years a large part of the material in Ring B will fall onto the surface."[14] We are here back to visualizing Vail's canopy drops, from primeval sources, or as a way-station. But a Saturnian explosion, not a falling of Saturn's rings, deluged the Earth with ice. Saturn's rings today may be fall-back debris of the same incident, still falling back.

The ice came down with falls of gravel and tillites. The great Ice Age extended from about 14,000 to 11,500 B.P. During this time the Earth was wobbling, the atmosphere turbulent and the deposits of ice were eccentric. Life would have been exterminated by the spread of ice and flooding if the greatest of all catastrophes had not cleaved the Earth and formed the ocean basins. Then ice and waters avalanched or fell into the basins as these grew in size, filling them ultimately over their brims.

The present ice age began in proto-historical times. Saturn's explosion drenched Earth with water and ice and the terrestrial axis tilted as a result of the explosive force. The age of "Jupiter of the Bright Skies," as the Greeks significantly denominated him, began; the skies were clearer and the climate colder because of the tilt; the high canopy was almost quite gone leaving merely the present upper atmospheric levels and magnetosphere of Earth. Ice began to gather around the northern

---

[14] "Physical Characteristics of Comets," (Moscow, 1958), NASA-TTF-80.

and southern poles, drifting over the cultures of the age of Saturn.[15]

[15] Flavio Barbiero, *Una Civiltá sotto Ghiaccio* (Milan: Nord, 1974).
16. 178 *Science* (13 oct. 1972), 190-1.

PART FOUR

CRUSTAL TURBULENCE

"Crustal turbulence" connotes rock movements, to be sure, but much of this book concerns the same, and we may only plead that there is especially more of these phenomena in this section concerning earthquakes, volcanism, rising or descending and swelling or shrinking earth masses. "The continent of Atlantis" is still mostly a name for geographical and legendary indications of crustal turbulence; it finally sank in one furious climactic day, so said Plato, after exhibiting the above-mentioned four forms of turbulence, plus continental cleavage, floods, pandemonium and extinction. We shall be criss-crossing as usual among the high energy expressions of nature, watching all the while for exoterrestrial interventions.

CHAPTER SIXTEEN

EARTHQUAKES

The ancients may have been more familiar with earthquakes than modern man:

> ... The earth shook and trembled... the foundations also of the hills moved and were shaken... Then the channels of the waters were seen... and the foundations of the world were discovered.,. The mountains skipped like rams... [The divine power] removes the mountains... overturns them... shakes the earth out of her place.

In these and many more lines, the Hebrew Psalmists commemorated times of catastrophe. World myth contains thousands of such songs and stories. Some can be located in time; most cannot. But, little by little, the science of myth will move to help the science of the Earth; and geology will move to interpret mythology. Then the ages of quantavolution will assume a clearer shape.

The deep valleys, rifts, and canyons of the globe will soon here be assigned to the greatest of movements. The Earth cleaved; the continents broke up and were rafted into place. At the same time and on later occasions, many places on Earth sank into the depths. These might all be called earthquakes, although they are global events.

A great but conventional earthquake would be described as in the following testimony of a resident about the New Madrid, Mo., earthquakes:

The first shock came at 2. a. m., December 16, 1811, and was so severe that big houses and chimneys were shaken down, and at half-hour intervals light shocks were felt until 7 a. m., when a rumbling like distant thunder was heard, and in about an instant the earth began to totter and shake so that persons could neither stand nor walk. The earth was observed to roll in waves a few feet high, with visible depressions between. By and by these swells burst, throwing up large volumes of water, sand, and coal. Some was partly coated with what seemed to be sulphur. When the swells burst, fissures were left running in a northern and southern direction, and parallel for miles. Some were 5 miles long, 4 1/2 feet deep, and 10 feet wide. The rumbling appeared to come from the west and travel east. Similar shocks were heard at intervals until January 7, 1812, when another shock came as severe as the first. Then all except two families left, leaving behind them all their property, which proved to be a total loss, as adventurers came and carried off their goods in flat boats to Natchez and New Orleans, as well as their stock which they could not slaughter. On February 17, there occurred another severe shock, having the same effect as the others, and forming fissures and lakes. As the fissures varied in size, the water, coal, and sand were thrown out to different heights of from 5 to 10 feet. Besides long and narrow fissures, there were others of an oval or circular form, making long and deep basins some 100 yards wide, and deep enough to retain water in dry seasons. The damaged and uptorn country embraced an area of 150 miles in circumference.[1]

Earthquakes are most simply thought of as movements of large bodies of rock, whether of a few tons or of the whole Earth. The rocks flow, flex or fracture. There may be two sets of rocks that split and separated in times past, or which do so now: one moves up and another down; or one slips alongside the other. Or one or both sets move apart or one or both press together. Or one crawls over the other. Earthquakes may combine these movements, so that one, or two, or all may happen at once.

---

[1] The account of one Godfrey Le Sieur, in E. M. Shepperd, 13 *J. Geol.* (Feb. 1905), 46-7.

The duration of the movement may be of seconds, or minutes. There may be a single shaking or a series going on for days, and again repeated months later. (The ancients cried to heaven over interminable tremblings, as when the Egyptians suffered them during the days of the Hebrew Exodus.)

Electrical fields gather and play about the scene, beforehand, during, and afterwards. The world may seem to be glowing with fire in the distance. The ground sends up thunder and groans. It screams. It makes rattles like volleys of gunfire. Winds spring up and blow hard. Waters are agitated; tidal waves sweep over the land; well waters sink; rivers stop flowing or change their channels.

Animals often sense an earthquake in advance and show distress. Birds fly far, mammals run off, lizards crawl out and away. People are of course terrified by the trembling, they pray, they condemn their sins and those of other, they swear to reform, and curse their government; they help each other or stand stupefied or behave like zombies.[2] When the rocks move, man's world shakes and shatters.

Any force that disturbs the rocks causes the earth to quake. Pumping radioactive wastes deep below ground caused earthquake tremors in Colorado a few years ago. A dynamite explosion or a small meteoroid impact will cause one. Frequently earthquakes are associated with volcanos. A map of the earthquake belts of the Earth is practically a map of the areas of volcanism. The same forces must cause both. The primary force could be an old one, unsettled, that is still working upon the rocks. Or it could be a new force. But perhaps the old and the new force are identical: the new occurs now for the first time; the old is what occurred some time ago. Is not the earth very old? Should it not have settled down? Should not the rocks be stable? If so, then force from nowhere is impossible.

---

[2] U. S. Government Printing office, *The Great Alaskan Earthquake of 1964* (1970).

Most of what is known empirically of the globe comes from earthquakes - earthquake shock waves to be more precise. Hence, it is difficult to talk about how the interior of the globe causes earthquakes, if indeed it does. Seismic waves can be made to register their occurrence and intensity on seismographs set up to record and calibrate them. Many thousands of earthquakes, mostly non-damaging, are thus registered around the world each year. They shake the housing of a heavy pendulum which, itself unmoved, marks the shaking on a graph; a reading of the graph indicates the magnitude of intensity on the Richter scale.

The patterns of seismism around the world in recent history are easily described now. One simply follows the Tethyan world belt, the world-girdling fracture (noting a greater intensity where it passes beneath the land), and the island arcs off of East Asia.

Cases such as the New Madrid phenomenon mentioned above are less effected, although a Mississippi Valley "earthquake region" has recently been described. Applying the quantavolutionary ideas, one may point to recent "Ice Age" shifts of the Ohio and Mississippi Rivers, which certainly denote earthquakes, and to the great load of detritus that the lower Mississippi basin must be bearing: "Atlas Shrugs." For, in a brief period, a large part of the North American continent surface rushed toward the Gulf of Mexico in a slurry of ice, water, stone, vegetation, and soil. If enough freshwater entered the gulf to freshen it, as Emiliani found, enough debris would accompany the flood to burden the region and deform and fracture its rocks.

In a second indicative, a severe earthquake struck north of the Adriatic Sea in the Friuli region of Italy. Shocks were felt simultaneously in the Upper Rhine Valley just northwest of the Alps. We conjecture that a branch of the African rift crosses the Mediterranean, runs up the Adriatic Sea, and emerges from beneath the Alps (which have overrun it) as the Rhine River Valley, emptying its waters into the North Sea. All of this is quite recent. The Rhine canyon cuts far out into the bottom of the North Sea, revealing its very late sub-aerial existence. Dutch

geologist Doeko Goosen claims that the Netherlands suffered
earthquakes more frequently in earlier time.[3] The Fourteenth
Century saw the erasure of many areas and villages. The Alps, of
course, make up a heavy load upon the underlying rifted area of
the crust.

The greatest known earthquake was registered variously
between 8.25 and 8.9, in Chile on May 22, 1960. On the Richter
scale, each higher unit stands for a ten-fold increase in wave
amplitude, and this represents a .32-fold leap in radiated seismic
energy. The numbers move arithmetically from 0 to 8.9 but the
magnitude increases exponentially; for example, an earthquake of
8.0 is 10,000 times greater than an earthquake of 4.0 and the
energy release much greater.

The 1906 San Francisco earthquake might have reached
8.3 on the Richter seismograph scale, which registers the intensity
of vibrations alone. Its equivalent in the more descriptive Mercalli
scale would be 11 (out of a possible 12). The present top of the
Mercalli scale reads: "Damage total. Waves seen on ground
surfaces. Lines of sight and level distorted. Objects thrown
upward in the air."

In the Assam earthquakes of 1950 "rivers were dammed;
major floods drowned the countryside; mountains and hills split
open and square miles of their surface covering were stripped off;
rain came down as mud owing to the dust-choked air; and the
geography of the region was permanently changed."[4] It was
recorded at 8.6; it is obvious that the measuring scale is a crude
indicator of real events.

There may very well have been in recent times
earthquakes of great force that do not register beyond the
recorded limits of the seismographs, as conjectured by Chinnery
and North.[5] Actually what is today meant by earthquakes is an

---

[3] "A New Model for Level Areas," Vitgeverij Waltman: Delft, 1974.

[4] Lane, *op. cit.*, 211.

[5] "The *frequency of Very Large Earthquakes*," 190 *Science* (19 Dec. 1975),
1197-8.

earth movement defined by modern experience and measured by instruments calibrated to this experience. Because of the rareness with which earthquakes of magnitude over 8.0 on the Richter scale have occurred in the brief 75-year record of various measurements, "many investigators have concluded from this result that earthquakes... greater than 8.6 or so do not occur..." However, as it is likely that earthquakes of this intensity occur on the average once a decade, it is also probable that ones of greater intensity (with a seismic moment of 10 31 dyne-cm or more as compared with the Chile 1960 earthquake of 2.5 x 10^{30} dyne-cm) can occur and may even be expected over a fifty or hundred-year period. If larger earthquakes occur they might cause destruction far greater than hitherto experienced and "may cause a considerable excitation of the Chandler wobble," a veritable, if slight, shaking of the axis of the Earth.

Most earthquakes have a localized shallow focus and originate within the crust, at or above the Moho discontinuity which may be regarded in quantavolutionary theory as the boundary of the Earth's shell and as the line of catastrophic slippage of the crust on several past occasions; but the Moho boundary itself was born of quantavolution, we maintain. It is both conventional finding, and quantavolutionary theory, "that some overall global factor, rather than conditions localized in the hypocenters themselves, is responsible for. generating terrestrial seismicity."[6]

The source of an earthquake varies. The seismograph stations of the world draw a fix upon a certain point that appears to be the focus of the earthquake, its epicenter within the Earth. The mantle of the Earth is a hot dense liquid. It does not lend itself to earthquake manufacture by simple mechanical thrusts and fractures. Are there substances in the mantle that are escaping and causing disturbances in the overhanging rock or crust of the continental and oceanic bottoms?

---

[6] Cook, *op. cit.*, citing Benioff (1955).

"Yes," says the up-to-date scientist. Chemical elements are decaying in the mantle and crust. They escape upwards and set up convection currents. These currents actually amount to so much force that, like the rising heat of a boiling soup, they can move the surface of the soup off to the side and down. But the forces of convection required to move ocean bottoms and continents is tremendous and many persons, including this author, believe that they cannot be assembled.[7] Earthquakes and earth movements are basically mechanical, and do not result from chemical or nuclear forces, as Cook has shown.

Still, the theory is fetching. For if one examines again the map of the rifts, earthquake zones, and volcanic regions of the world, one can see that there is an order or pattern to them all. They cut up the globe, and the pieces can be called plates. Some of the plates can be measured as moving very slightly; and it can be seen that lands that are now far apart fit together as if they once were of one piece. Since no other force can be imagined by our up-to-date scientist, the convection current force, upwelling and moving out laterally beneath the rocks, must account for rifts, seismism, and volcanos. But this accepted theory, it develops, may be incorrect, and we shall return to the issue of convection currents in a later chapter on continental drift.

An ominous kind of movement has always been the "conjunction," when two or more celestial bodies line up, especially the Sun and planets with the Earth. Earlier we mentioned the Gribbin-Plagemann phrase, the "Jupiter Effect."[8]. They chose to plot their scenario along the 600-mile-long San Andreas fault, part of the East Pacific Ridge system actually. Hence, the San Francisco Bay Area and many other thickly settled communities found themselves wondering when the "Jupiter Effect" will occur. "1982" or thereabouts, said the writers. At this time, which is passing as this book goes to press, Jupiter and Saturn were to line up with the Sun, Moon, and Earth and exert

---

[7] See Chapter 24 below.

[8] See Chapter 6, fn. 13.

an electrical gravitational tidal force upon the Earth sufficient to upset the delicate juxtaposition of rock surfaces along the San Andreas fault. The Moon is small, and 239,000 miles away on the average. Yet it affects the waters of the world with its tidal pull, daily and twice a month or every 14.8 days.

The same writers go a certain distance into history, where a few records are to be found, and are able to discover devastation by earthquake close to the time of past conjunctions, specifically in 1800-03. The timing is a bit off, the disaster by no means a catastrophe, but the evidence points to the "Jupiter Effect" as the culprit. In a close encounter with a large celestial body, the earthquakes would be immeasurably worse.

Before Gribbin and Plagemann, Charles Davison examined the same celestial motions to relate them to earthquakes. He found increases in seismism at full moon, 14.8 days, and 19 years and also found a sunspot period every eleven years: when the spots were particularly active, rains and earthquakes increased.

Davison's periodicities may thus be added to the planetary "Jupiter Effect." They show how sensitive are the shell and rock layers of the earth, in their fractured condition, to impulses from the outside. They are clearly tidal, i. e. cyclical.

Davison also discovered that atmospheric pressure could be correlated with earthquakes. Here there were two cycles: a daily one and an annual cycle. Midwinter midnight and midsummer noontime were seismic favorites. Perhaps the atmospheric phenomenon may be connected with the vast diffuse sky lights that occur before earthquakes, arising probably out of a discharge of electricity.

If changes of atmospheric pressure trigger quakes because they represent a "true dead weight" of the atmosphere above a certain shifting point of focus, then this too is a tidal effect. If it is itself produced by electrical changes, then the direct cause must be assigned to whatever assembles atmospheric potentials.

Sunspots have been increasingly blamed for climate and earthquakes. Recently a 70-year gap in the sunspot record between 1645 and 1715 A.D. was rediscovered and called the "Maunder minimum."[9] It was a time when the Northern Lights hardly appeared; when the Sun's corona was relaxed and clear of disturbances; when C14 was increased because solar particles were not blocking in their usual way the cosmic particles that cause the C14 in the atmosphere; when tree rings became irregular and thin; and when the climate was called a "Little Ice Age." John Eddy, in announcing some of these findings, declared, "We've finally broken a block that held us back - uniformitarianism. It was an assumption we took as fact." And "We've shattered the Principle of Uniformitarianism for the sun." As yet a negative correlation with earthquakes has not been plotted; earthquakes should have declined in number and intensity.

The possibility also arises that some earthquakes are responses to increases in the amount of ice contained in the polar caps. This may be true today and also of any prehistoric ice-caps. Cook and A. Brown develop this line of thought.[10] Cook points to a correspondence between total annual seismic energy and a seeming accumulated energy in the growing ice of the caps. The huge vertical and radial pressures exerted on the earth's rocks by the caps may be taken up by the elasticity of the shell, or, on the other hand, and at least occasionally, the pressures may be alleviated by a shearing or refracturing of rocks even quite far away from the perimeter of the ice.[11] We stated above that seismic origins are in global overall forces rather than in local areas of earthquakes themselves.

However, quantavolutionary theory leads us to suspect that, not the present ice caps, but rather the effects of the great catastrophic periods are still felt. Earthquakes are seismic

[9] John A. Eddy, "The Maunder Minimum," 192 *Science* (18 June 1976), 1189-1202.

[10] Hugh A. Brown, *Cataclysms of the Earth* (NY: Twayne, 1967).

[11] P. Jordan, *op. cit.*

memorials to ancient disorders.[12] The rocks of the Earth will not rest in place until their very gradual tailing-off consequences end.

The major source of present-day earthquakes is to be sought along the lines of the global fracture. The fractures will be discussed later on; here they must be mentioned because of their connection with earthquakes. Taking up first the north-south Atlantic rift and following it from the Arctic to the Antarctic, one observes intense seismism throughout its length but largely in the middle of the Atlantic and little on both sides of the Atlantic Basin. The fracture, like an almost healed wound, throbs, festers and drips a little, pushing the continents left and right almost unnoticeably. Perhaps the rocks of the Atlantic Basin are lagging or stretching behind the Pacific rocks, which are being pushed into the basin of the lunar genesis. No theory is yet adequate to explain the difference in intensity and frequency between the Atlantic and Pacific seismism.

Wherever the fracture moves - into the Indian Ocean, across Asia, and laterally across the Southern Pacific and up the East Pacific - it bears with it seismic strains that develop as earthquakes of shallow focus. Quakes of deeper focus take place along a belt that circles the Pacific area of erupted crust, from New Zealand north and east up to Siberia, across the north Pacific and down the west coast of the Americas all the way to Antarctica. It is famous as "the Ring of Fire."

A second belt of shallow and deep focus earthquakes pursues a route along the old Tethyan equatorial region. It begins in mid-Atlantic, pushes through the Mediterranean and the Near and Middle East, shifts to follow the Himalayans where these break upon the Asian heartland, and swings down and across the south Asian seas. Here, where it overlaps the "Ring of Fire," it is intensively active.

But the Tethyan belt does not appear to cross the Pacific basin. It would, of course, have been erased if the Moon had

---

[12] As is argued by Velikovsky in *Worlds in Collision*, Chapter 8.

erupted from the region. There thousands of seamounts stretch up from the ocean bottom, and long transverse faults occur.[13] Rather, it resumes off of Central America where, indeed, there is a meeting of all four great earthquake belts - the globe-girdling rift, the Pacific "Ring of Fire," the westwardly moved Americas, and the old Tethyan belt.

Afterwards it proceeds into the Caribbean which may have been once coupled with the Mediterranean. It ends with the outlying islands some hundreds of miles south of its connecting link, from which I began here to trace its around-the-world movement. Geographers have matched Spain with the West Caribbean region; it is to be expected therefore that the Tethyan fracture of the south would tie into a transverse fracture to the north, thus circumnavigating the globe.

Earthquakes are seismic memorials, it was said. Today there are precipitators, but not important new causes, of seismism. The old causes that regularly occur are themselves significant reminders of a time when the heavenly bodies were much more active. Just as most religious holidays around the world celebrate or re-enact the terror of primeval catastrophe and the relief of survival, the rocks of the world move from time to time in reenactment of their ancient catastrophic motions, prodded by the ancient forces when these are stimulated by recurrent anniversaries. [14]

But some still say that earthquakes go back in time without an increase in frequency or intensity. N. N. Ambraseys, a seismological engineer of the London Imperial College, concluded, after prolonged study of Near East documents, that the 3000 large and small earthquakes of which he found evidence in the period 1 to 1900 A.D. did not in this period show a decline of frequency and intensity.

His evidence is piecemeal, localized, undefined in regard to intensity, and barely usable. Only if there were enormously

---

[13] *Nature* (16 Aug. 1971), 375-9.

[14] In *Chaos and Creation* (1981) and *Worlds in Collision* (1950).

worse earthquakes early or late in the period could a conclusion be drawn.

By the first century A.D. the world was already seven hundred years past the last general catastrophe, as described elsewhere, and the skies had been tranquillized. Still, in the several hundred years before Christ, many accounts of severe seismism were handed down. The Spartans, most doughty of warriors, were so deadly afraid of earthquakes that if the land shook in the middle of a war, they would quit and retreat home; this kind of terror suggests a legendary experience recent to their times.[15] Ellen Churchill Semple, writing of ancient Mediterranean geography, admits the profuse claims of risings, sinkings, chasms, and upheavals both in legends and in the scientific accounts of such illustrious reporters as Aristotle and Strabo. Not to mention Seneca, who declared that "Tyre is as regularly shaken by earthquakes as it is washed by the waves..." But she simply puts them down as exaggerations and furthermore "they erred as to the time element in the problem," for they did not employ the million or so years that she gave to the geological order of the Mediterranean. (We see, though, that *her* Mediterranean is only Quaternary!) Yet who can deny Pliny, the natural historian, when he claims 57 earthquakes to have occurred in a single year at Rome in 217 B.C.?[16]

As we move back in time, the earthquakes increase in severity. Velikovsky points out that in the eighth and seventh centuries earthquakes were so numerous that when they occurred they were mentioned in a bare line of the astrological tablets of Nineveh and Babylon.[17] Nevertheless, "reports concerning earthquakes in Mesopotamia in the eighth and seventh centuries are very numerous, and they are dated. Nothing comparable is known in modern times." He quotes from a tablet of Babylonia,

---

[15] E. C. Semple, *The Geography of the Mediterranean Region: Its Relation to Ancient History* (NY: Holt, 1931), Chapter 3

[16] II *Natural History* 86.

[17] *Worlds in Collision*, 274-8.

"The earth shook; a collapsing catastrophe was all over the country; Nergal [Mars] strangles the country." Further, "references to breaches in houses, large palaces, and small dwellings are very numerous in the [Hebrew] prophets of the eighth century." Neglecting such sources, a historian could claim that "the earthquake held a place in the religious conception of the Israelites quite out of proportion to its slight and relatively rare occurrence in Palestine." Obviously, some literalness has to be restored to the language of the Bible, as well as to many ancient voices, if a better natural history is to be written.

A destroyed city may leave no records of its destruction; a sunken land leaves only an outsider's report and a myth. A lifetime (1937-1975) of work was dedicated by S. Marinatos before the archaeological and geological world came to realize, perhaps too enthusiastically, what earthquakes and explosions befell the island of Thera in the Aegean Sea some 3100 years ago.[18]

Velikovsky's research is especially thorough on the "tenth plague" of the Exodus, which he places at about 1450 B.C.[19] "At midnight, there was not one house where there was not one dead" in Egypt, says *Exodus*. All the houses were destroyed. It was the unlucky 13th day of the month. "The thirteenth day of the month Thout (is) a very bad day. Thou shalt not do anything on this day," according to an Egyptian myth. Why should a single event be frozen into all behavior unless it was far more frightful than other earthquakes, no matter how severe? "The children of princes are dashed against the walls" and "cast out in the streets," wrote Ipuwer, an Egyptian scribe of those days; "the prison is ruined;" again, "the residence is overturned in a minute."

It would seem that in those days the Earth shuddered and cities collapsed across the world from Mesoamerica through the Mediterranean, the Near East, Middle East, India and China.[20] The greatest modern earthquake becomes insignificant by

---

[18] *Chaos and Creation*, 233-4.

[19] *Ages in Chaos* (NY: Doubleday, 1952) and *Worlds in Collision, op. cit.*

[20] Velikovsky, *Worlds in Collision*, and Schaeffer, *op. cit.*

comparison with the disasters of the Exodus period. Even so, that is not the earliest period of catastrophic earthquake known to archaeology.

Claude Schaeffer systematically combed the files of all excavations in the Near and Middle East that were connected with the period from some 3000 to 5000 years ago. His conclusions are sharp: all known sites suffered multiple destruction; most of the time the destruction was by earthquake, often with fire, sometimes by unknown causes. In the city that he himself excavated in part, Ras Shamra-Ugarit, at least eight heavy disastrous discontinuities were discovered in the period 2400 to 1000 B.C., by his dating.

At five points in time a general destruction of the whole Near East occurred. Small earthquakes, that must have been very common, are of course not considered. They are hardly detectable in excavations. After practically all of these disasters, many years passed before a culture could renew itself or be resettled by survivors from other areas.

Schaeffer plotted the destroyed settlements on a modern seismic map that shows areas where earthquakes of intensities 6, 7, and 9 of the Mercalli scale are typically found. A number of the repeatedly destroyed settlements were located in regions of lower magnitude earthquakes. As noted earlier, this is true of Rome and Palestine, too. They are no longer so prone to earthquakes as they were then.

The destruction was so total in many of the cases which Schaeffer studied, and had such peculiar features - heavy combustion, for instance, and in the case of Troy II, "the Burnt city," which I too studied, both deep calcination and yet enough time for the population to escape - that the investigator is led to consider even exoterrestrial hypotheses. Invading troops, volcanos known to exist, and hurricanes acting by themselves are inadequate hypotheses. Deep ash falls might apply in some cases; unfortunately, archaeologists before World War II paid little attention to levels of destruction; anyhow, where would the ash come from? Once again, the lack of data frustrates theoretical reconstruction; moreover, the less severe modern experience of earthquakes had led to simplistic and negligent judgements even on the part of groups which spent years on site.

Were the quantavolutionary hypothesis to be increasingly applied, the contrast between the past and present would become more marked. Systematic review of the field work of the past two hundred years is needed, as is also a thoroughly objective analysis of ancient legends and records. Too, technical awareness and application of new paleo-chemical techniques are needed in further field investigations.

We can conclude that earthquakes were greater in early history and pre-history than they are today. Further, the seismic experience of the past century is not adequate to assure us that earthquakes a thousand times worse in their effects are no longer possible. They then approach a new level of destruction wherein fire, flood, fall-out, avalanches, diastrophism and other effects assume major roles. Under such conditions the seismism itself tends to become a relatively minor feature and even to lose its name to much greater movements of the land, sea and air. The earthquake is supremely prominent today because the rocks replay more of the history of catastrophe than the atmosphere, the hydrosphere and the biosphere. No people has recalled total cultural destruction by shaking but perhaps all recollect its destruction by fire, winds and water.

There are parts of the world where the rocks, seeming so firm to the naive eye and touch, are crisscrossed by what must have been an interminable succession of surges and shakes. Cores of the earth under Athens were drilled lately in the planning of a new subway; most of them pulled up cylinders of the so-called "Athens schist," a rock formation that is a mass of small chaotic fractures. It is conceivable that millions of years of erosion caused the cracking; it is perhaps more readily conceivable that the schist was macerated in a period of continual trembling. Plato reports that Athens suffered severe earthquakes in its earlier history; springs on the acropolis were stopped and cliffs were toppled. According to Plato, the Attica of old was practically unrecognizable by his own time, which seismically is our own time, the flattened end of the seismic curve.[21]

---

[21] Ph. Negris, *Plissements et Dislocations de l'Ecorce Terrestre en Grèce, leurs Rapports avec les Phénomènes Glaciaires et les Effondrements dans l'Océan Atlantique* (Athens, 1901).

CHAPTER SEVENTEEN

VOLCANISM

Five hundred volcanos of the Massif Central in France, now defunct, were erupting 12,000 years ago, or less, during the Magdalenian Upper Paleolithic culture. So maintained Escalon de Fonton of Montpellier University. The spheres of the Earth were once so active that humans must have been encouraged to a pan-animism, an omnidirectional feeling which would have dominated all religion and culture if there had not appeared some immense and forceful sky bodies that focused attention upon themselves. Mother Earth, now a picturesque name, was devoutly and literally supplicated by the ancients even in the millennia of the great sky gods between 13,000 and 2700 B.P. She was often married to the greatest of the gods, and it was generally believed that her nuptial ties explained much of the animism of the Earth. "A theory of volcanicity" must not only be "taking into account the whole range of geodynamic processes," as Rittmann says in his classic work on volcanos, but also the whole range of cosmodynamics.

The great movements have gone, but a restlessness remains, erupting locally; volcanos erupt solo, almost never performing duets. The volcanos of the world adhere to the world-girdling fracture system. The system organizes the world's volcanos. The volcanos of the land, active and extinct, follow the great fracture lines that pass underground as for instance in the Tethyan shear sub-system of the Caribbean-Mediterranean-Middle East, or beneath the Pacific coastal states of America. The same is true of the volcanic belts off the East Asian continent. The oceanic volcanos string along with most of the fracture system.

Isolated volcanos such as the Hawaiian Islands require special explanations; if the general theory here that seamounts (guyots) are fossil short-lived mantle taffy is correct, the isolated volcanos can have originated at the same time, "the same, but more so." The difference may be explainable by measurements made by Preston in 1893: "The lower half of Mauna Kea is of a very much greater density than the upper. The former gives a value of 3.7 and the latter 2.1, the mean density of the whole mountain being 2.9,"[1] for the height above sea level. Thus, like a seamount, Mauna Kea stretched in a taffy bubble until finally it burst and began operating as a typical volcano.

More puzzling is the absence of clear connections between volcanism and astroblemes. Why should not a deep shocking crater give rise to a volcano? That a meteoroid often makes a melt of a kind is undisputed, but where is the persisting volcanism? Obviously one must seek for deeper roots of the world's volcanos.

Volcanism takes the form of cones and fissures. It is also beneath swellings and bubblings of surface features. Most of the igneous basaltic surface of the world, including the ocean bottoms, was created by fissure volcanism. As occurs still in Iceland, fissure volcanos produce lava copiously. "During recorded history more lava has poured forth above the sea in Iceland than in all the rest of the earth's volcanic belts combined. Yet... Iceland's volcanic belt comprises less than one-half of one per cent of the total length of the world-encircling rift."[2] Beaumont points out that 40,000 square miles of the British Isles afford plateaus of basalt in sheets; though nowhere are cones or vents to be found, till and clay accompany the basalt.[3] Rampant fissure volcanism is today observable on planet Venus. "Recent first-class Pioneer photographs of Venus show that the planet is rent with fissures, and most remarkably has been described as 'the

---

[1] CXLV *Am. J. Sci.* (1893), 256.

[2] Heezen and Hollister, *op. cit.*, 557.

[3] Comyns Beaumont, *The Mysterious Comet*, (London: Rider, 1945), 197.

most volcanic planet' in the solar system."[4] By "most remarkably" the writer implies the theory that Venus is a very young planet and has been losing its heat of eruption from Jupiter only slowly.

When the Earth had to erupt magma on a large scale, from far down, because of a loss of crust and an expansion of crust, fissure volcanism had to be the means. The deep ocean ridges of today still supply lava for paving the abyssal surface; the process has assumed a certain orderliness. On the other hand, viewing the Pacific Basin one must conjecture that a very large surface was once removed and a deep wound was left exposed that repaired itself *in situ*. The concept of cone and fissure volcanism fails, then and there, and one must speak of sheet volcanism, creating its own hard skin.

Fissure volcanism stands for extensive catastrophic venting; if there is so little of it today, the reason occurs in the general global settling. Cone or tube volcanos represent a moderate 'need to erupt.' Volcanic fields denote an interconnected set of tubes with a number of outlets. Volcanic outlets are spaced apart in relation to the thickness of the lithosphere; thinner rock invites closer spacing.[5]

When dormant or extinct, all of these suggest either that a local rock crisis has been settled or that the global volcanic system has been shutting down its ramifications and further extensions. Many hills and uplifts, whence gases and lava have never escaped, are in the same fossil status. A major exoterrestrial encounter, the only event that can excite general volcanism, would reinvigorate the pattern of prehistoric and present volcanism insofar as the force vectors of the encounter prescribe, and would excite new volcanism wherever new stresses were imposed. Ultimately, geophysics should be able to locate as a set of overlays the total historical series of exoterrestrial encounters in fossil and live volcanism and go so far as to discover or substantiate the

[4] R. D. Mac Kinnon, 3 *S. I. S. Workshop* 1 (July 1980), 7.

[5] P. J. Smith, 265 *Nature* (1977), 206; Vogt, 21 *Earth Planet. Sci. Let.* (1974), 235.

detection of their avenues of approach, their duration, and their energy.

Neat surveys of past volcanism are not to be had. Rampino, Self and Fairbridge collected "known volcanic eruptions of large magnitude within the last 100,000 years."[6] Their interest lay in associations between volcanism and climate, and a shaky correlation was established, with climatic change apparently preceding eruptions, suggesting to this author exoterrestrial issues. Presently germane, however, is the possibility that the statistics will confirm or deny a greater incidence of volcanism in the past. No help is forthcoming, because of the inadequacy of the data: the dating methods are perforce questionable; the bias toward known historical instances is heavy (12 of 28 cases occur in the past 5000 years, one twentieth of the period studied); and there is no uniformity of occurrence over time (implying, if anything) that heavy volcanism is aroused by global events. Because fossil volcanism is generally assigned even older dates, most scholars do see very heavy volcanism in periods beyond 100,000 years; Australopithecus, for example, is often tramping in volcanic ash, but 'three million years ago and more. '

Some 13 ash layers have been already discovered in the Central East Pacific Ocean, none blanketing the entire region. There is a great discrepancy in dating between the argon radiometric and biostratigraphic methods, about half a million years within the single million years of total assigned time. The argon technique is faulted for atmospheric contamination and incomplete outgassing of lava containing radiogenic argon. (But is this not an inevitable occurrence, then, in all catastrophism, where atmospheric 'pollution" is inevitable?) Even so, both methods are faulted when it appears that preclassical Mayan artifacts are found under the 500,000 y argon-dated (or 50,000 y biostratigraphic-dated) so-called "D" (or Worzel) layer of ash in the region.

The explosion of the island of Thera about 3000 years ago sent about 40 km^3 material into the atmosphere. The seas were

---

[6]   206 *Science* (16 Nov. 1979), 826.

covered with pumice, some of which was driven ashore. Marinos and Melidonis plotted the story of one such incident at the small island of Anafi to the east of Thera. Two pumice deposits were noted. The one at Vounia is notable.

> On the base of a natural profile of soil, we observe the following sequence: lowermost schists of the basement (bed rock), on this a bed of earth and pieces of schists of alluvium and slope debris. On this the mentioned bed of pumice and on the pumice a younger bed of soil and small stones of the surrounding rock with the usual cementing of lime carbonate. The general dip of these strata is gentle (about 10°) to the bottom of the valley. The lower part of the pumice bed consists of broken pumice, though the upper one consists of almost powdered pumice mixed with small pieces of pumice, irregularly rounded, of some millimeters to a few centimeters. We cannot give any other explanation about the formation of the above pumice bed except the transportation and deposition of this material by the tidal tsunami wave following some terrible phase of the catastrophe on Santorin (Thera).[7]

The height of the foaming wave increased after rushing into the funnel opening of the narrow deep valley. It ascended, achieving 250 meters, and then retreated, leaving the pumice.

The authors do not comment on the heavy, late diastrophism evidenced: the absence of low-lying pumice beds, the abrupt cut-off of the bed, as drawn by them and the layering of *ca* 2 meters of alluvion talus atop the pumice bed. I have observed the same deep bedding of semi-consolidated rock over pumice in Thera-Santorini itself. Possibly there occurred subsequent explosions of rock and soil, or violent quakes that shook down hill-tops. The investigation of cases such as Vounia and Thera where the dating is relatively secure may enable us to reconstruct a larger and/or later sudden deposition of non-

---

[7] G. Marinos and N. Melidonis, "On the Strength of Seaquakes (Tsunamis) During the Prehistoric Eruptions of Santorin," reprint from *Acta* (see fn. 10), 280.

volcanic material. Without the historical dating here, one would be inclined to assign very old ages (as was the case here before Marinatos discovered Late Bronze Age artifacts in the ruins of Akrotiri) in order to account for the superposition of heavy 'erosional' deposits and then a slow landscaping.

Today, volcanism of all kinds may be remanent. Fascinating and destructive as it may be, it is as nothing compared with the volcanism of times past. The Soviet geologist, A. P. Pavlov, declared in 1936: "At the present time, only a residual, negligible manifestation of volcanic activity is observed on the earth; formerly, this activity was perhaps the most typical and almost universal phenomenon in the life of the planet."[8] Probably the phenomenon is correct, but the volcanism, like astroblemes, may have happened during only several immense exoterrestrial encounters.

The greatest eruption of modern times, some say (incorrectly) of all history, was the 1883 eruption of Krakatoa. The total volume of erupted material has been estimated at 18 to 21 km^3. "When compared with prehistoric ignimbrite-forming events ranging in volume up to 10^3 km^3 .... the volume of the Krakatoa eruption was very modest."[9] So declare S. Self and Rampino.

Thera's volcano (Aegean Sea) blew away most of a large, high island and its culture three thousand years ago.[10] Ilopango (El Salvador) destroyed a cultured Mayan area of thousands of square miles in an explosion of 1800 years ago.[11] The volcano of Tamboro on Sumbawa Island in the East Indies emerged from the waters in 1812. Within three years it grew the awesome height of 12000 feet, some three miles tall. Then it exploded. Approximately

---

[8] Quoted in S. K. Vsekhsviaskii, "Indications of the Eruptive Evolution of Planetary Bodies," (Kiev: unpubl. paper, *ca* 1973), 7.

[9] "The 1883 Eruption of Krakatoa," 294 *Nature* (24 Dec. 1981), 699-704.

[10] Acta, First Int'l Cong on Volcano of Thera, 1969 (Athens, 1971) J. Keller, D. L. Page, and C. and D. Vitaliano, eds.

[11] *N Y Times*, 101 Jan. 1977, quoting Payson Sheets.

100 cubic kilometers of material shot into the atmosphere. About 100,000 people were killed, many more than died in the Anglo-American War of 1812 being fought at the same time across the world.

Hawaii arises eleven miles from the bottom of the sea. It is the world's tallest mountain. It appears to be stable. Yet it ends a long fracture out of Mexico and begins an arc of seamounts that strikes Siberia. The scene of volcanism today is the pallid termination of the scenario of quantavolution. There is nothing objectionable in present theory; it is just not historical.

> Volcanic activity serves as a mechanism to release thermal energy from the Earth's interior. Thus, we can view the Earth as a boiler and the inactive volcano or vent as a sealed valve. Conversion of tidal energy to thermal energy by friction is concentrated at plate boundaries, where almost all active volcanos are found. Thus tidal energy helps heat up the boilers and increase the pressure, while tidal stresses weaken and break the seals. Both of these triggering effects increase during periods of increasing peak tidal stress... Once a volcano has erupted, its susceptibility to triggering remains low for a longer period of time and then increases rapidly following a hyperbolic or exponential stress.[12]

Now we turn to Rittmann for additional theory:

> Volcanic activity is caused by the loss of gases from magmas, a process which takes place wherever magmas can ascend from the depths and come into regions of lower pressure. This ascent of magma is, however, only possible if the earth's crust is stretched and fractured through tectonic forces. The existence of volcanos is thus closely connected genetically with orogenesis and epeirogenesis. We then attempted to explain these genetic connexions on the principle of the causal chain of disturbed equilibria, and so to place volcanicity in its correct

---

[12] R. G. Roosen, "Earth Tides, Volcanos and Climatic Change," 261 *Nature* (24 June 1976), 680.

position in the overall picture of geodynamic processes. The interpretation of a wide variety of observed facts led us to the conclusion that magmas could originate in two ways, and that we could distinguish between primary magmas having their origin in a subcrustal zone encircling the earth, and secondary magmas formed by the anatexis of sialic rocks within the earth's crust.[13]

One notes here, besides the requirement of a stretching and tearing of the crust, the origination of volcanic magma from the "subcrustal zone encircling the earth" and anatexis, or regurgitation of surficial rock. This region occurs some 15 to 30 miles below the land surface and about 5 miles below the oceanic bottoms. This layer corresponds not only to the Moho discontinuity, as I have mentioned in connection with the base of seismism, but also with the volume of "missing sial" from the ocean basins, which roughly approximates the volume of the Moon. Volcanism, then, like seismism, reflects the level at which, all over the globe, the still landed crust moved in reaction to the eruption of the Moon. Whether or not the mantle on which this lunar boundary level rides jostling is solid or liquid, in the years of its fast movement it would have heated, liquefied, and expanded. The volcanos are probably still draining the liquid.

Studies of volcanic eruptions arrive at correlations between the moment of major eruption and the tidal forces exerted upon the Earth by the Sun and the Moon. Similar correlations have been detected between tides and seismism. In this regard, volcanism and earthquakes reveal themselves as close relatives.

G. Beccaria (1716-81) with Stokeley, Franklin and others, set the stage early for a systematic approach to electricity in connection with earthquakes, cyclones, and volcanos, but the promised scientific drama has never been enacted.[14] As early as

---

[13] Rittmann, *op. cit.*, 267.

[14] *Artifical and Natural Electricity*. See Heilbron.

June 21, 1902, Elmer G. Still published his observations of the volcano-solar-lunar relationship:[15]

> The writer has for several years been observing this relation between the positions of the heavenly bodies and seismic, volcanic, and electrical disturbances, and is forced to the conclusion that the latter are caused in part by the conjunctions, oppositions, perihelions (or perigees) and equinoxes of the moon, earth, and seven other planets, especially when several of these occur at once.

He warned that such disturbances do not always occur at these times and that the relative position of the heavenly bodies have to be combined with local causes to produce volcanism and seismism. After all, he commented, if solar storms (sun spots) are excited by perihelion with Jupiter, why would not earthquakes and sun spots be transactive?

A second article in the same year stressed that "the influence of the Moon and planets in causing and intensifying seismic and volcanic disturbances is not altogether tidal action - gravitational; it is partly, or mostly, electrical, and seismic and volcanic action is an electrical disturbance."[16]

Once more in 1902, the same author, E. Still, continued his prescient argument, now declaring that gravitational tides of the Moon were quite inadequate as explanations of many terrestrial disturbances. "We know [Still was seventy years ahead of the field] that magnetic earth currents (which interfere with telegraphing), brilliant auroras, severe thunderstorms, violent storms of many kinds, and also earthquakes and volcanic activity accompany sun spots. All these are electrical disturbances, and the eruption of Mount Vesuvius and numerous seismic shocks which

---

[15] 86 Sci. Amer (21 June 1902), 433.

[16] 87 *Sci. Amer.* (26 July 1902), 54.

occurred at the time of the last large sunspots -about September 15, 1898 -were no doubt electrically caused by them."[17]

We are not surprised at these statements, in view of Chapters 4 and 5 earlier on in this book, where electricity was allowed a broad scope among geological effects. The electrical volcanism of Io, satellite of Jupiter, will be recalled, where ejecta speed at 2000 miles per hour from 60 to 160 miles above the surface. A number of factors operate holistically in terrestrial volcanism; electricity may sometimes take up center-stage; mechanical heat and pressure are probably the chief actors in late historical times. Yet the electric and the mechanical are always working together: no rock can be squeezed without emitting electricity; no electric charge can pass without heating rock.

Recently, Johnston and Mauk examined the unusually complete records of Mount Stromboli (Italy) over a 72-year period and related 33 major eruptions to the amplitude of tidal forces operating upon the Earth.[18] A distinct pattern emerged. Some ten days after the tidal peak is the significantly likely moment for the eruption. The eruptions concentrate in the days between full moons.

Roosen used oxygen isotope ratios in cores of the Greenland ice cap as an indication of mean temperatures between 1200 and 1976 A.D.

> Variations in tidal stresses on the Earth caused by the Sun and Moon cause changes in the stratospheric dust produced by volcanic activity; this in turn changes the thickness of the stratospheric dust veil and hence the atmospheric radiation balance. At least some significant fraction of the dust occurs at peaks of tidal stress. The tides measured vary over long periods. There is a peak of stress at approximately 179.3 years period. This period actually shows up in a (significant) correlation of 0.37 between the stress periods and the temperature curve.[19]

---

[17] 87 *Sci. Amer.* (27 Sep. 1902), 203.

[18] M. J. S. Johnston and F. J. Mauk, 239 *Nature* (29 Sept. 1972), 266-7.

[19] *Op. cit.* 682.

The relevance of such studies here is that tidal stresses and volcanism correlate; hence, great tidal stresses of the past must have excited great volcanism; conversely, evidence of heavy past volcanism denotes heavy past tidal stresses.

In the present placid astronomical order of the world, there is scarcely a place to look for such tidal forces. A mere 500 active volcanos occupy the world landscape, compared with the 500 of the Massif Central of few thousand years ago. Flying high over southern Italy, one may luckily see Vesuvius, Stromboli, and Etna all smoking at the same moment. Arriving in sight of the famous seven hills of Rome, there is a grandeur of culture, not nature. Yet Breislak in 1801 was arguing that the seven hills were debris amidst a large volcanic caldera, and Cuvier for one approved the idea. When the oldest hominids, human in some ways, walked the Earth at Afar (E. Africa), some ten nearby volcanos were active.

A great many dormant volcanos exist and an enormous number of extinct volcanos. If the belts of inactive fissures and the unnumbered thousands of seamounts are added, the Earth has undergone periods of the most intense exoterrestrial stress. Or else, one will have to parcel out these millions of volcanos and 'volcano equivalents' over exceedingly long stretches of time.

But if volcanism even in the stable "solarian" period of the past 2000 years exhibits a 'grouping' tendency in response to exoterrestrial tides, then pre-historic volcanism must have exhibited grouping, too. Once more, we force the question: quantavolution, yes, but could it not happen at widely spaced intervals over time?

Even with fossil and radiochronometric data that give, I think, ages too "old," the ocean volcanos and ridges are geologically young, under 80 million years. Is there some reason to believe that land volcanos should not be also as "young»? Probably not, inasmuch as most of the land volcanos are tied into the ocean ridges, into great faults, and into the ring of fire that bounds the Pacific Basin.

If there exist extinct volcanos and fissures that belie this statement by extruding from the surface far from the zones of present activity, these, it will turn out, are aligned with expired branches or special fractures of the Earth's crust. That is, it is plausible to assign all volcanos to the same geological time, and a young age; "where are the volcanos of yesteryear?"

If the continental and oceanic plates break up and drift apart, as the prevailing theory will have it, touring the globe every 200 million years, forming new combinations, where are the extinct volcanos that should dot the world like pine trees? That is, so far as volcanos are concerned, history ends recently. Presumably, before then, lands broke up and plates travelled without their fiery boundary-markers; this is implausible.

The innumerable seamounts are a standing reproach to opponents of quantavolution. I have mentioned their origins as pulled mantle taffy in cosmic encounters. They are an impossibility for tectonic plate theory for there the continents move on plates, not through them, and seamounts appear abundantly around the Moon Basin of the Pacific, with a solitary but impressive chain of hundreds off the New England Coast.[20] If the Moon were erupted from the now Pacific region, the seamounts could be visualized as pulled taffy drop-backs that could not follow the Moon into space. But the Atlantic Ocean off New England would only then have opened its abyss and "New England" would have been retreating westwards. To explain this particular "taffy-full" we must conjecture a prolonged explosiveness or subsequent passes of an attractive exoterrestrial body in order to assist their generation.

Morphological comparison of Atlantic and Pacific seamounts may be of use in deciding the sequence of events. One study of the former finds shallow water fossils, including coral and the algae Melobesia, at 3000 meters, and suggests that somehow the seamounts subsided that much. More in order is our

---

[20] J. R. Heirtsler *et al.*, 65 *Amer. Sci.* (1977), 466-72.

hypothesis that the sea did not fill the basin until recently; similar phenomena are discoverable in the Pacific seamount areas.

I would be loath to leave the subject of volcanism before tightening its awesome connection with the birth of the Moon in the parturition of Earth. In 1907, William Pickering was continuing George Darwin's effort, begun in 1879, to establish that the Moon fissioned from the Earth's present Pacific Basin. He called it "The Volcanic Problem."[21]   He alluded to spectroscopic binaries as examples of fission in the Universe.

He argued that when the Moon fissioned, "the Earth was in much the same condition that we find it at present, except that it was hotter." It was supposed to be rotating in only several hours (so as to provide the centrifugal force for whipping out the Moon). He matched the continents at the Atlantic to show the breaking away occasioned by the need to fill the emptied basin; he mentions "North America during its transit across the fiery ocean, in obedience to the pull of the Moon." (Thus he preceded Wegener with the idea of continental drift.)

Geologists generally abandoned the search for proof of Moon fission, even though they could choose their own time and state of the Earth to accomplish the feat. Thus they might afford a gaseous fission, or a thin crust, or a hot and molten body and they had no care for the biosphere or atmosphere or even stratified rocks. It is surprising that under such easy conditions for speculation, they could reject the theory. A reader of this book will surmise that an ideological block against any immense catastrophic event would account for the rejection of fission. Rather should the Moon come sailing in nicely and moor itself above the Earth. The catastrophic implications of capture were not generally pursued, except by Hoerbiger and the maverick mythologist Bellamy. Nevertheless, many establishment scholars looked benignly upon the fission theory, allowing that the event

---

[21] "Place of Origin of the Moon: The Volcanic Problem," 15 *J. Geol.* (1907), 23-38.

was to have occurred eons ago. Also, of course, exoterrestrial inducements to fission were taboo.

D. U. Wise, more rationalistically, attributes non-acceptance of the fission theory to calculation problems. "The traditional and seemingly insurmountable obstacle to all fission hypotheses has been the discrepancy of approximately 400% between the present angular momentum of the earth-moon system and the values calculated as being necessary for the last stable configuration before fission."[22]   That is, an incredibly flattened obloid would have to drop its end like ash off a cigar. After disposing of several types of calculations, he is satisfied that "the basic problem of excessive angular momentum in fission hypotheses may have a solution in volatilization and escape of a silicate atmosphere generated by dissipation of lunar tidal energy in a high-temperature early earth."[23]

The eruption of the Moon certainly extends beyond the conventional concept of volcanism, although Vsekhsvyatskii claims that planets and comets originated in volcanic episodes, especially involving escapes from Jupiter. Explosion is of course a fission; rocks are transformed; gases and electricity are part of the process, and so on. Also, exoterrestrial influences are connected with volcanism, both as to origins and to triggering activity. These influences are provable in our own time by correlations of volcanism with tides, electricity and seismism. They are provable for ancient times by the patterned system of volcanism in the world and the obvious function of volcanism in relieving stresses according to a pattern highly suggestive of transactions in outer space. Withal there is a uniqueness to the lunar event; the dimensions of the event soar almost beyond comparison with ordinary disaster and even all other catastrophes. But the theory of the fission is greatly simplified if it is conceived to occur through the passing intervention of a large body in space.

---

[22] "Origin of the Moon from the Earth: Some New Mechanisms and Comparisons," 74 *J. Geophys Res.* (15 Nov. 1969), 6038.
[23] *Ibid.*, 6044.

Furthermore, it is well to mention, as a postscript, that should the Moon have erupted from the Earth and all ocean bodies are young, then the eruption must have occurred recently. The basins are dated at under 100 million years. Thus the Moon episode, so incredibly destructive, would have occurred with the full realization of life on Earth, including many thousands of existing species and with most Earth rocks still present. If those species could survive, so even could homo sapiens.

Therefore, one must accept the possibility of the Moon originating by eruption. The evidence is that such occurred. The evidence is that it occurred recently relative to geological convention. The evidence is that it occurred without total destruction of the Earth's surface or its occupants.

CHAPTER EIGHTEEN

SINKING AND RISING LANDS

Vita-Finzi remarks that we cannot tell whether, over the past century or even now, the shorelines are sinking or rising.[1] Furthermore, there is much greater complexity and much less data when making such determinations for the longer past. The Earth has demonstrated a capability for moving up and down here and there leaving scarcely a clue as to the causes. The wisest path may be to pursue a general theory, such as the Ice Ages or, I think, a great lunar eruption, and build hypotheses and information upon it.

The legendary voices are worth an audience. Alexander Kondratov, a Soviet linguist and compiler of legendary and geological evidence of the sinking of lands, writes:[2]

> China's oldest myths tell of a war between the god of fire and the god of water 'at the beginning of the world. ' The mountains erupted fire, the earth quaked and the sea attacked the land. When the fire god was defeated he decided to commit suicide and struck his head against the highest mountain in the west. The frightful blow drove the land into the sea in the east like the prow of a boat, while in the west it flew into the air like a boat's stern. Since then all the rivers in China have flowed eastwards.

---

[1] *Op. cit.* 55, 59.

[2] *The Riddles of Three Oceans* (Moscow: Progress Publ., 1974) 101.

Kondratov inquired of geologist Yuri Reshetov concerning this myth and received the following in reply:

> Geological, geophysical, paleontological, archaeological and anthropological studies have shown that up until at least the middle of the last Ice Age the Japanese Islands and Indonesia were Asian peninsulas. During the second half of the Ice Age (from 40,000 to 20,000 years ago), vast areas of land subsided into the sea and were replaced by what are the Sea of Japan and the south China Sea. The sinking was accompanied by powerful volcanism and by earthquakes. At about the same time, that is, towards the end of the Ice Age, the ranges of Indo-China and the mountains of Central Asia rose another 2,000 meters. Many generations of Chinese must have witnessed the gigantic geological changes in south-east Asia. It is these events that the myths about the struggle between the gods of fire and water evidently reflect.

This is macro-geography, indeed. It speaks of a quarter of the world. Part of the world rose and part of it sank. The events described are probably much more recent, the 20,000-year figure reading 10,000 years in other sources.

Many Europeans still speak, as they have from the dawn of history, of a civilized continent of Atlantis that sank in a day.

The legend of the Lost Continent of Atlantis is a hardy tale; billions of words have been written about the few words of the legend. It is quite incorrect of F. M. Cornford, for example, to write that "serious scholars now agree that Atlantis probably owed its existence entirely to Plato's imagination." If Plato lied in his tale of Atlantis, there would be little truth in him generally; for Plato repeatedly insisted that his story be considered seriously and literally: the Atlantean culture did exist across a water barrier to the west; it had relations with the ancestors of the Athenians and Greeks; it did sink abruptly in an earthquake. Plato's date would place the event at about 11,500 years ago. I attribute this date to a confusion with the lunar catastrophe and assign it instead to the time of the Noachian Deluge, that is, about 6000 B.P. as described in *Chaos and Creation*.

An ancient document, the Oera Linda manuscript, which was written in Frisian with runic characters and whose age and authenticity is much disputed, claims a general Atlantis-type sinking of a prosperous civilization of the Fryas between the North Sea and the Baltic, where frost was rare and fruit trees blossomed. There came a summer of darkness, great earthquakes, a spitting of fire from newly bursting mountains, a general holocaust, an obliteration of rivers, and huge floods that advanced to cover most of the land. Whole islands were newly formed by the bones of dead cows and sand (one is reminded of the Siberian islands formed of mammoth bones). The survivors were subjected by invading Finnish bands (just as the Hyksos invaded Egypt after the Exodus).[3]

The Caribbean peoples talk of an "Antilla," now sunk beneath the ocean. The Pacific Ocean and American peoples of the Southern Hemisphere say that once a continent existed where now stand a few islands amidst a great deep sea. The perplexing books of Churchwarden concern this continent of "Mu." Legends of the Greeks speak of a drowned Aegean Sea, and the ancients believed the Mediterranean Sea was recently arisen.

In the Pacific Ocean of the North, there is supposed to have been a Beringia where now stands the Arctic Ocean on one side and on the other side the northern half of the "arc of fire" bordering the great Ocean.

The East Indian peoples and Indian Ocean people offer legends of the sunken continent of "Lemuria," whence came world civilization. T. Huxley and F. Engels were famous supporters of the theory over a century ago. And the islands of the South Seas, where Indonesia stretches out, are reputed to have been of a single piece before the waters rose or the land sank. The

---

[3] Unpubl. miss. communicated to author by René Roussel of Ablon, France, Apr. 19, 1974; *cf.* discussion by J. Bimson, *S. I. S. Workshop*, Feb, 1979, P. M. Hughes, *ibid.*, Sep. 1981, and 35-6 editor).

Dutch geologist Bartstva claims that a landbridge connected the Celebes and Philippines until Holocene times. In August 1982, Alan Thorne announced the discovery of Chinese human remains in North Australia with an estimated age of at least 10,000 years.[4] A map in *Chaos and Creation* outlines in the most general way all of these mythical lands that are said to have existed in human times.

If one is to believe legend, every large expanse of ocean once had its land mass. A form of quantavolutionary reasoning could proceed as follows: the ocean basins are new, created in the time of man; before the time of man, there was Pangea, a globe covered by continental crust that carried shallow freshwater seas, especially in the then equatorial area, which area, now greatly tortured, is still recognizable in the fabled Tethyan Sea remnants of the Mediterranean area and the "belt of fire" that girdles the world longitudinally.

The awesome depths to which the land has sunk or from which the crust has been removed should not halt the argument. If the Andes, the Alps, and the Himalayas can rise miles high, Lemuria and Atlantis can slump miles deep. If the sial debris of sunken lands cannot be scooped up by dredges or pierced by the few meters of core drills, that too is not surprising; the ocean basins were opened up and repaved recently with basalt; where the land was not exploded away, it was covered over by lava working furiously and fast under the catalysis of falling and flooding waters.

Where the continental fragments do not remain to be fitted obviously together, then the intervening land was blasted away or sunk. Continental sial has been extracted on occasion from the deep bottoms in the Atlantic and Pacific Ocean; this is surprising given the major discovery of recent oceanography that the ocean bottoms are covered everywhere with lava. Metamorphic rocks typical of the nearby islands and Italy were found 3000 meters below sea level in the central Tyrrhenian Sea, as reported by Heezen.[5] Fragments of black carbonaceous

---

[4] *Melbourne Sun*, Aug 14, 1982.

[5] 229 *Nature* (Jan. 29, 1971), 327-9.

sandstone were found on the Rocksall Plateau and Orphan Knoll, between Greenland and North America.[6]

Some legends have been confirmed by geology; many might be confirmed; most are not, because they are vague or misleading. It would be well to examine closely the myths that have proved quite accurate to see in what mythical form they found expression and then to proceed systematically to the translation of similar myths around the world.

The aboriginal Australians who live around MacDonnell Bay say that an angry witch once stirred up the waters and flooded the beautiful land to make the Bay. Geologists confirm that the land was high in the ice ages and recently sank to form the Bay. The image of the witch should not be discounted; Velikovsky has described how European and Chinese alike have an image of a witch riding a broomstick, which he traces to cometary images of 3500 years ago. Indians of the area of Crater Lake recalled in their oral history what geologists later confirmed - that a great volcanic explosion fashioned the beautiful basin in the mountains that has since collected rainwaters.

These were not the only risings and sinkings, but they were by far the major ones. Kondratov, for example, mentions that Bulgarian researchers have compiled a detailed map of underwater archaeological finds, dating from the eighth to the fifth centuries B.C., discovered along a large section of their country's Black Sea. Irish Celts were in America in this period, according to several recent studies of history, archaeology and linguistics; they were perhaps driven to explore and immigrate by a further sinking of their homeland coasts.

The age of the comet-god Athena-Venus preceded these episodes of the age of the god Mars by under a thousand years. The Gulf of Mexico may have been sunk at this time, for the peoples of the Mexican Gulf Coast were not long afterwards lamenting the destruction of their previous civilization by the jaguar-god (a Venus symbol) and storm-god Hurracan, and telling

---

[6] By the "Glomar Challenger," cf. 227 *Nature* (Aug. 22, 1970), 767-8.

of how they were taught their arts by a few people who came from the east. Kelly and Dachille wrote that the Gulf of Mexico has the superficial appearance of a meteoritic impact crater. In Cook's reconstruction of the area prior to continental movements, the Spanish peninsula is fit like a socket into the Gulf but a gap, possibly a crater gap, remains.

These several speculations treat of events of 11,500 years ago, or at the latest 7000 years ago, not of 3500 years ago -unless, of course, everyone is right: that is, the breakup of the area occurred and "western Europe" rifted outwards; the flood of Saturn deluged the shallow gulf areas; a fragment of the Venus tail spilled petroleum in the area and impacted.

The Caribbean area generally is rife with myths of disaster and immigration. The timetable is chaotic. Archaeologist Cyrus Gordon has described convincingly Mediterranean materials that originated between Phoenician and Roman times and that were uncovered in spots so far apart as the Brazilian Coast and Tennessee (U.S.A.).[7] Sanders and Price in 1968 set up a convincing case for direct Asiatic influences upon the New World. East Indian contacts with the Americas can be traced as well.

At maximum age, none of the materials would go back to before 1500 B.C. That leaves a great prior gap of culture, untilled save by indistinct legend. Brasseur de Bourbourg was one of many early European scholars who felt that, in these myths of white-skinned, technically competent people coming from the East, there were visitors from or survivors of a great continent of Atlantis.

Interest in East-West contacts has increased recently among scholars. That ancient "Japanese" had cultural contacts with at least "Ecuador" is a distinct possibility. That unusual blood types appear among villagers in settlements of the Andes is demonstrable. Also, the ancient Meso-Americans, as judged by sculptures and drawings, seem to be a population in which

---

[7] Before *Columbus* (NY: Crown, 1971); *Riddles in History* (NY: Crown, 1974).

African-Negroid and Tethyan-Caucasoid (Semitic) types were mingled with Mongolian-Sinyan-Amerindian populations.

John L. Sorenson, citing Kroeber and others, examines 200 basic, defined cultural features of the "Old World Oikoumene."[8] What would be called the "common heritage" of the peoples of the Near East. Of these features one in eight is found in Meso-America definitely. He believes that another tenth would be added to the New World list when checked out through the whole body of information; thus about 18 percent of the Old World basic culture traits are shared with the New World. The statistical probability that this percentage of correspondence would occur by accident is low. It suggests land bridges of past ages.

Perhaps it was around 1500 B.C. as well when Thule vanished into the Faroe Rise. Thule is famous in Northern European myth and is referred to in many books and accounts with tantalizing brevity. Russian geographer N. Zhirov argues this theory, citing evidence that Thule was near Iceland, that many islands were mentioned thereabouts, that it was in a warm oceanic current, and its people grew grain and other crops, (We are reminded of the Oera Linda manuscript.)

Indeed, the birth of the North Sea may have come so late as 1500 B.C. The famous amber of the North and Baltic Seas is conventionally dated at seventy million years; it comes from submerged pine forests that are assigned that date. Recently geologists have begun to stress the youngness of the area, prodded by archaeologists. Drowned settlements have been found at the bottom of the Baltic Sea, the North Sea, and off the British Isles. These are not to be confused with the sunken settlements of later time - Slavic Vineta in the Baltic by a tidal wave of 1100 A.D., many places off the mouth of the Rhine in 864 A.D. and so on. We are writing of the whole of these seas. "Europe was inhabited when the North Sea did not exist, when England and Ireland were

---

[8] In J. C. Riley *et al. Man Across the Sea: Problems of Pre-Columbian Contacts* (Austin, Texas: U. of T., 1971).

not islands and Jutland and Scandinavia were not peninsulas but were all parts of a single land mass." Thus writes Roy MacKinnon, who gives us a fix on these great submergences.[9]

Aristotle wrote in his book *Of the Earth,* "Inroads and withdrawals of the sea have often converted dry land into sea and sea into dry land." And Strabo, the most reliable geographer of the classical age, declared that "extensive submergence of the land, as well as minor submergence, has been known."

Reviewing these and other ancient writers, Professor Ellen Churchill Semple wrote in 1931 that they "attributed the straits and sounds of the Mediterranean and the formation of many islands to convulsions of nature. They found evidences of previous land connections in the similarity of relief and rock structure on both sides of the intervening channels, as do modern geographers, but they erred as to the time element in the problem" That is, she would accept what they said of sinkings and risings, the Mediterranean Sea, the Black Sea, the Red Sea connection, the Sicilian-Italian-Tunisian bridge, and so forth, but simply dismissed any short-time reckoning for the events. She is not alone in thinking that the ancient sense of time was palpably and *prima facie* stunted.

The same authority speaks airily of the Mediterranean Sea being of Quaternary origin or less (perhaps a million years); whereas now the *Scientific American* publishes maps of the Mediterranean as it was supposed to be half a billion years ago, a discrepancy of some 50,000 percent. Not to be outdone, Heezen and Ewing, two of the best contemporary oceanographers, found continental land far beneath the Tyrrhenian waves, even while the "oldest" parts of the seven seas are credited with a mere 200 million years. (Many say less.) That is, 200 million years ago would represent the time a continent was lofted by its convection cell currents over the oldest spot of the oceanic abyss, erasing its sediments and boiling it.

---

[9] "Cenomanian Sync., "I *S. I. S. Rev.* 2 (Spring, 1976).

If so, the Mediterranean could hardly resist for such vast lengths of time the passage of land masses over it, while land rocks betook themselves into its depths. One is attracted once more to the ancient idea of the Tethyan Sea, a shallow home for innumerable species until the new oceans were created to house them.

Geographers have long known of this mythical Sea of Tethys of which the ancients spoke. They appropriated the term for a Tethys Geosyncline or trough which they traced around the Old Worlds - from Gibraltar to Indo-China. The Mediterranean Sea is regarded as descended from it. I use the term for an equatorial belt and shallow seas circumscribing the original Pangean globe. With the new theory of continental drift and splitting of the Old World from the New by the Atlantic Ocean, Carey changed the concept of the Tethyan geosyncline. "The Mediterranean shear system links up *en échelon* with the Caribbean system to form part of a global sinistral shear system which I have called the Tethyan Shear System."[10]

The lands (and shallow seas) were wrenched apart between North America/South America and Europe/Africa. and Asia/South Asia. Then Africa rotated sinistrally and Asia dextrally. The Asian continent encountered land masses moving sinistrally from the South, Arabia and India primarily.

So incomplete is the understanding of great movements of land that, where one encyclopedia, the *Britannica,* is merely out of date, another, the *Americana,* assigns to the Mediterranean a Tethys origin that runs far to the north -taking the Black Sea route to the Caspian Sea, the Sea of Aral, and into Mongolia. In myth this is incorrect Actually this is a third and temporary great ocean of Tethys that may be called the Gobi Sea. It replaced the Tethys geosyncline and the remnant of old Mediterranean which are more plausible successors to the ancient mythical Sea of Tethys. It gathered waters in the great basin that is now the Gobi Desert or

---

[10] S. Warren Carey, *The Tectonic Approach to Continental Drift* (U. of Tasmania, 1958); *The Expanding Earth* (Amsterdam: Elsevier, 1976).

"the Sea of Sand," as the Chinese call it. Like the extinct Sahara Sea, the Gobi Sea lasted long enough to attract many human settlements to its shores. Then it was emptied in a great flood and its cultures disappeared, as described earlier.

Thus there were perhaps three Seas of Tethys, the latest being the Mediterranean Sea of recorded history. The first would be the Pangean shallow sea that carried the vast majority of marine species and supported a thriving population of plants and terrestrial animals, including australopithecines. This area, like the rest of the world, was severely buffeted in Uranian times but became known to the first modern humans. They were called pro-Selenians because the Moon was absent from the sky, and were the prototypical Tethyans of generalized Mediterranean race.

A second sea would be produced from the Lunarian catastrophes and be deepened by transverse cleavages of the world-girdling fracture system; it is discoverable today as the Balearic, Ionian and Eastern Mediterranean basins. It may have been a major locale of recovery for humans and their cultures. But the recovery was far from peaceful. Africa slammed into Europe at several points, raising the Alpine ridges and the mountains of Anatolia, then withdrew after a short interval.

The third Sea of Tethys was formed by the flood waters of the evacuated Gobi Sea basin, four thousand miles away. It was at first huge in expanse, then in a short while diminished to the earliest Mediterranean Sea known to history. The Sea of Azov and the Black Sea basins would have been filled and connected with the Caspian Sea. The Tyrrhenian area was flooded beyond its present level and survivors occupied the high islands of the Western Mediterranean.

Ancient history saw many more risings and sinkings of land and towns than have occurred over the past two thousand years. Extensive research would probably be able to distinguish the sinkings of what we have been calling sometimes the Solarian, Martian, Venusian, and Saturnian ages. They are all part of legend, of some remaining historical fragments and, unfortunately, of an age that knew writing and had a complex culture, but whose achievements are inadequately identified because of the great destruction and the unwillingness of scholars to entertain even a hypothesis of the events.

The names of the places supposedly sunk or serving as havens for survivors read like a roster of geography and mythology. Attica; many places of the Aegean Sea; numerous places around the Sea of Azov, a ring of towns around the Black Sea; the whole Adriatic basin (this was probably the location of the predecessor of the Po River, the mythical deep river of Eridanus, that used as its channel an arm of the global cleavage that forked from the Red Sea clear up through the Rhine), the Gulf of Taranto. The Straits of Messina and the Sicilian-African straits, the lands around Corsica and Sardinia, the coast of ancient Etruria, the Cyrenaican coast of Libya, Jerba in Tunisia, towns of Crete, the Gulf of St. Gervais off of Marseilles, the straits of Gibraltar, the Isthmus of Suez. One can only guess that the Sahara Sea (Sea of Triton of myth and ancient reports) was created during the Saturnian deluges. If so, it probably was emptied into the Atlantic Ocean and its cultures destroyed during the cometary intrusion of about 3500 years ago.

Scholars of every science have pondered the many tantalizing indications of shared history in the southern regions of the globe. Kondratov exclaims at one point:

> The most surprising part of it is that a study of the world's earliest civilizations reveals a whole series of riddles that can be solved only by using the hypothesis of Lemuria, a large land mass in the Indian Ocean that was inhabited not just by lemurs and not even by Pithecanthropi, but by human beings who had reached a high level of civilization. (p. 131)

And later he says:

> Lemuria... is connected with sciences that range from marine geology to the deciphering of ancient scripts, and geographically, from the Indian Ocean to the Himalayan

mountains and the Buryat steppes. It may be that Australia and Australian studies are also linked up with Lemuria.

One can conceive of the original extent of Austroafrica or Lemuria by noting that Africa, South America, Australia, India, and Antarctica were once intimately connected. Moreover, in the South Pacific a huge amount of shelf area exists beneath the waters and a great amount of continental crust is missing.

The Americas were heavily reconstituted by natural disaster. It is reasonable to presume that humans occupied these continents prior to the great catastrophes. Conventional anthropology and archaeology would do well to drop the theory that all Americans are descended from some few who made the passage across Bering Strait a few thousand years ago -some say 20,000, some only 12,000. They assume that the continents were in their present positions; only a bridge of land sank and rose. Even among believers in the possibility of contact from the Pacific islands by sea, a recent occupation is credited. A few think it more likely that the people of Tierra del Fuego and other southern stretches came from "down under," that is, Tasmania or other islands of the South Pacific. I think that it will not be long before some human remains of Uranian or pre-catastrophic times are discovered or rediscovered.

The same will be true of Antarctica. This huge continent, nearly twice the size of Australia, gives many indications of recent tropical climate, and produces many types of fossil animals and plants including those associated farther north with human occupations. Kondratov writes that "we do not know when the Antarctic region became covered with ice. Some glaciologists think that it cannot have been more than nine or ten thousand years ago."

Two maps have appeared in recent years after four centuries of gathering dust. One is the Piri Reis map that depicts the true un-iced coasts of Antarctica with considerable accuracy, another, the Orontius Fineus map, that carries interior topography with considerable accuracy. By conventional theory, mapping of a land mass of Antarctica could not have occurred until the middle of our present Twentieth Century because of the ice cover as well as the great difficulties in moving about without planes and snow vehicles.

It seems likely, then, in accord with the general theory of this book as well as such evidence, that African peoples occupied Antarctica during Pangea and Urania, and were decimated by the Lunarian disasters, especially by electrical and atmospheric ravaging; that they recovered somewhat during the Saturnian period, and then died out in the icy climate that descended in the age of Jupiter. Somewhere in the interior their remains will be found

Moving north from the frozen continent to the micro-continent of New-Zealand, largely buried under water, and to Australia, the situation is not too different. There few people were living before the European immigration. These few are supposed according to conventional wisdom to have come from the northern and western islands of the Indian Ocean some 20,000 years ago. Another theory says that this was impossible because there was open water that could not be crossed and that there would have to be land bridges. Presently, the geologists of the area have gotten together with the anthropologists, to the extent of saying that the land bridges existed for the movements of people. (Admittedly, the people of New Zealand, standing across a deep sea, would be difficult to account for by a shallow sea land bridge.)

So the theory goes back and forth in a way to satisfy now theorists of the bridges and then again theorists of the clever navigators. The theory which we employ is that the land masses of New Zealand and Australia were sliced away from Antarctica by the now quite evident earth cleavage and sent rafting along with other lands towards the excavated crustal areas, north and east. On the rafts were Austroafrican survivors.

Australia rafted mostly to the east; India moved mostly northwards and to the east; the Asian continent moved east and south nosing under the waters in places, and ultimately (after the Saturnian deluge) with a large section of its underside underwater as ocean shelf and slope. It is probable that the Indian Ocean was an excavated basin forming part of the great Pacific basin and then was closed in upon by Asia veering southwards and Australia going north.

India itself, it is agreed, became detached from Africa and Madagascar and rafted north to lodge itself into Asia. Half the crust of the earth was gone and the earth was expanding somewhat so that there was plenty of room for maneuver and titanic forces to propel the rafts.

Now to examine the human record in the southern regions. It is becoming ever more plain that the oldest surviving large-scale culture in the world is African, exemplified in the Tamil culture of India. For thousands of years we have heard claims that this south Indian culture was a survival of a great sunken culture. Ancient writers even asserted that India had been connected with Africa. Probably the first modern man to consider the evidence of the common roots of the Dravidians of Tamil Culture of Southern India with the natives of Australia. and then to connect this idea with the notion of continental drift, and hence continental drift in recent times, was the Soviet ethnographer, A. Zolotaryov. He was deeply influenced by Wegener's book and presented his synthesis in 1931.

Before Zolotaryov, the Tamil (Dravidian) legends and the many ancient commentators had impressed others. Thomas Huxley, the apostle of Darwinism, wrote that mankind had originated on the now sunken continent, Lemuria. Frederick Engels, the intimate cohort of Karl Marx, and a believer in Darwin's theory of evolution, wrote that a "particularly highly developed race of anthropoid apes lived somewhere in the tropical zone -probably on a great continent that has now sunk to the bottom of the Indian Ocean." Ernst Haeckel, German biologist, named the proto-human "pithecanthropus," and assigned its origins to Lemuria; he said it migrated from there to India, Africa and South-East Asia; indeed, in all three places pithecanthropus was shortly found.

The Dravidians, who are among the darkest in skin of the Indians and who had generalized features which could be called Negroid but by the same token primordial human features, are located principally in Southern India today. Their culture is called the Tamil and is now reputed among scholars to be the oldest in India, predating by far the Indo-European culture of the Aryan immigrants of the mid-second millennium B.C., not to mention the medieval culture brought in by Muslim invaders.

The Tamil scholars look back not only to a sunken Lemuria, but to a sunken larger continent called Gondwana. And it is this "Gondwanaland" that has given geologists the name for their conception of a united land mass of the southern hemisphere that split apart in the breaking up of the continents an alleged hundred million years or so ago, long before the age assigned to the primates. (I may note here the interchangeability in the context of this discussion of the words "sinking" and "drifting apart." One must be prepared mentally to think of sinking whenever rifting occurs, both because a cleavage is seen by terrified observers to be a sinking of the opposite lands and because flooding and sinking actually occurs in most areas of rifting.)

However, the number of species whose remains have been found in separate areas where there was once Gondwanaland (that is, around the world in the southern and tropical regions), increases from year to year. Some are alive. Earthworms of the same species are found in Australia, India and Ceylon. Pouched mammals or marsupials are found in the Americas and Australia and nowhere else. (In 1982 fossil marsupials were uncovered in Antarctica.) Old world and new world monkeys exist. So also, identical as well as related fossil species, of horse, elephant, tiger, camel and rhinoceros. So, too, both living and fossil plants.

From Kondratov's summaries, it appears that Soviet scientists have been most active in tracing the ethnic movements of pre-history from the Lemurian homeland. Surprising developments have occurred one after another, building up the case espoused by the old Tamil scholars. In the first place, and using "Dravidian" as the term for the basic generalized Negroid

(Australoid) race, the Dravidian language has been compared with and found to be related at some remote period to the language of Madagascar, thus supporting floral and faunal resemblances and geophysical similarities previously uncovered by other scientists from several nations.

Further, the Dravidian roots have been traced up through the Indian sub-continent to the proto-Indian high civilizations of the Indus valley and indeed up and across the whole north of India. Computer analysis of proto-Indian and a number of other writings indicated the Dravidian affinity.

Moreover, Soviet scholars contend that the proto-Indian, hence Dravidian influences, move up the Persian Gulf and into the very foundations of what were to become the Sumerian and other Mesopotamian civilizations. These have long been thought to be the rock-bottom, independently developed civilizations of the old world. This earliest pre-Sumerian culture has been termed the Ubaid. Kondratov makes clear that it is not alone a matter of trade and other intercultural relations; for the pre-Sumerians or Ubaids were part of the proto-Indian, hence, Dravidian complex. Place-names, language roots, religious images, god-names, and forms of building construction are similar if not the same.

Far to the East now is the present Khuzistan, Iran, once called Elam. The Soviet linguist I. Dyakonov has said that "the only hypothesis supported by a few indicative facts," in a comparison of Elamite with other writings, "is that of an Elamo-Dravidian relationship." Further, "tribes related by language to the Elamites and the Dravidians were scattered throughout Iran, or at any rate, throughout southern Iran, in the fourth and third millennia B.C. and perhaps later as well." Traces of the Dravidian race have been noted since then in various places in southern Iran.

Far to the north, recent Soviet archaeology has been uncovering a South Turkmenian civilization of the third and second millennia B.C. Again, statuettes, symbols and skeletal and cranial analysis point to close relationships to the Elamites, then the Ubaids, then the proto-Indian, that is, the Dravidian, and

THE LATELY TORTURED EARTH – PART IV.

ultimately to the sunken or rafted continent of Lemuria-Gondwanaland.

Kondratov does not leave his discussion of the Lemurian cradleland without elaborating two further items of significance. The origins of Egyptian high culture, following the Neolithic, have puzzled many scientists. Suddenly, upon the Neolithic, a high culture seems to have been imposed. I believe that it came from the Tethyan movement eastwards from the Atlantis-Mediterranean centers. Kondratov suggests that a Dravidian north-west thrust may have brought it in.

The earliest Egyptian writings are estimated at five thousand years of age. They are not primitive; they are classical, that is, developed and complex. Perhaps Dravidian India was the source. Indian archaeologist S. R. Rao has analyzed rock drawings of early Egypt found along the Red Sea coast and sees in their high-prowed, high-sterned boats portrayed there the vessels of Dravidian India. I find no contradiction, but actually two early post-diluvian civilizations encountering each other in Egypt.

The Dravidians, or perhaps more properly, the Austroafricans or the fundamental negroid race, did not cease their travels to the East until they reached the farthest islands. African blacks, Dravidians, and the Melanesians that reach across the southern islands of the Pacific to New Zealand relate to a basic African race that was not greatly different from the Tethyan and Sinyan groups during the Uranian age. (Racial differences develop rapidly in isolation and under conditions of inbreeding.) Now it appears that the languages of the Dravidian and Australian peoples - both of which, incidentally, throw the boomerang - are cognate. The Australian scholars J. C. Pritchard and William Bleek argued the case a century ago. In 1963 Swedish linguist, N. M. Holmer, systematized the grammatical and phonetic coincidences of the two languages. Kondratov continues:

> In the last century philologists discovered a remarkable similarity among the languages spoken over the vast area that extends from Madagascar, near the shores of Africa, to Easter Island in the eastern part of the Pacific. It has now been

demonstrated that the similarity is not accidental. The languages spoken on Madagascar and on Easter Island which, along with those of the Hawaiians, Maoris and other inhabitants of Polynesia, belong to the Polynesian group, the languages of the Micronesians, living on islands in the North-West Pacific, those of the Melanesians, inhabiting islands in the South-West Pacific, the languages of the Indonesian Archipelago, and those of the indigenous population of Taiwan all come from a single root and constitute the Austronesian (" southern islands") family of languages.

In view of all of the foregoing, which has relied heavily upon Kondratov, it might be reasoned that the whole southern hemisphere of the world and perhaps a very large belt moving north above India belonged once to a great African grouping and was catastrophized and separated during the lunar fission. Any Antarctic survivors were removed by the new ice age. It may be that the same is true of South America, but with flood, not ice, as the destroyer.

The scientific roots of catastrophism are more extensive than ordinarily believed. Alfred Wallace, co-inventor of evolutionary theory with Charles Darwin, believed that a single oceanic race had inhabited a great island in the Pacific Ocean which had then been sunk. So did Darwin's disciple, Thomas Huxley, and Darwin probably agreed with him. No injustice is done to Darwin by regarding his work as a great model of natural history, or "simply a theory" as some critics like to say. He had many doubts and made many "anomalous" observations about vast sudden catastrophes of species, of mountain building, and, when he experienced a now-forgotten earthquake off the coast of Peru, he was appalled by the high energy displayed, noting in his *Journals* that the surface of the stricken island was changed more in a day than in a century of uniformitarian processes.

Lesser known scientists developed more elaborate theories of the sinking of Pacific lands: a century ago, Dumont d'Urville, naval officer and explorer, Moerenhout, folklorist, both French; then earlier in this century, J.M. Brown, ethnographer, and M. Menzbir, Russian zoo-geographer. Others might be also named. All brought forward evidence of a great continent joining the Americas to Asia and of human cultures flourishing upon it.

What kinds of evidence of this theory might be advanced? Again, as with the Indian Ocean, the material is geographic, ethnographic, zoological, and mythological. Again the chronological problems are perplexing. Kondratov, whose work was passed by a high-level interdisciplinary committee of Soviet scientists, can therefore only hint at the possible resolutions:

> It used to be thought that the earth sciences possessed indisputable data. However, oceanography and geology are both developing so rapidly today that many seemingly settled questions are being revised. Substantial changes may soon take place in one of the cardinal questions of geology and oceanography -the dating of events that have changed the face of our planet.

Relating to the geographic is our general conception of the Pacific area as an exploded basin, filled promptly with water. The famous "ring of fire" is an effect like a fractured earth that is cauterizing the wounded edges of the continents. Repeated catastrophes irritated and reopened the wounds. The famous arcs of islands and their associated trenches were left in an advanced position when the Asian continent was forced back by the Indian collision and an elastic withdrawal after the continent had been pushed to its maximum.

Japan is rising out of the water. Eastern Siberia is also, as evidenced in a progression of shell mounds of shellfish-eaters marching inland from the coast where the food was taken and eaten. Is Eastern Asia still pulling back from its farthest advance? But southeastern Asia is still subsiding. Is the continent still moving southwards? Experts may be found to date these events anytime from the Tertiary Age to the end of the last Ice Age. As with the ice caps and climate, the rising and sinking of continents is difficult to measure, much more difficult to interpret in terms of localized theory, and always hard to time.

If only people had kept off of the hundreds of Pacific Islands, geologists of long-term persuasion might rest easily. But some surprising human developments have been going on throughout the vast region. Related to the great Sinyan race of the Asian continent are the Malaysians to the southwest and the Polynesians to the south and east. Farther south and mingled with these groupings in some places are the Negroid or Australoid types to which reference has been made earlier. Nor should one neglect the Negritos and pygmies who are found in the middle of Negroid regions but are reputed to have dwelt practically everywhere. "The little people" are a universal subject of folklore. Wherever found they are designated as a very old, perhaps aboriginal, type of mankind; they are said usually to be more clever and have a richer mythology than the peoples around them.

Not only are there peoples on the Pacific Islands, but also the peoples have cultural complexities and have exercised technologies beyond their recent capacities. Picture writing is found on a number of islands, the *kohau rongo-rongo* tablets of Easter Island and the Woleai Island script, for instance. Monumental sculpture, comparable to "Old Bronze Age" achievements of the Middle East, existed on Easter Island, Ponape, and the islands of the Caroline Archipelago. Brown found Easter Island sculptural forms in many islands: Hawaii; Pitcairn; the Marquesas; Christmas; Malden; Tinian; and Ponape. There are no two sculptures alike; hence the contacts were not recent and even originally the peoples must have been of diverse sub-cultures.

And everywhere, including the tiniest atolls, the peoples have myths of large populations, greater lands, of sinking lands, and of past ages of glory.

An Easter Island legend is typical. It is translated by Kondratov from Easter Island writings brought back by Thor Heyerdahl, the Norwegian archaeologist-explorer:

> The Youth Teea Waka said: 'Our country was once a big land, a very big land. '

Kuukuu asked him: 'Why did the country grow small? '
Teea Waka answered: 'Uwoke lowered his staff on it.
He lowered his staff at Ohiro. The waves rose, and the land
became small.
People began to call it Te Pito o te Henua. [Navel of the
Universe]
Uwoke's staff broke against Mount Puku Puhipuhi." A later
arrival on the island, Chief Hotu Matua is told the story.
It is added that "When Uwoke's staff was big, the land fell
into an abyss. The chief corrects the report:
"That was not the staff of Uwoke, my friend," said chief Hotu
Matua. "That was the lightning of the god Makemake."

The parallels here to the Phaeton and Typhon myths of
Greece and the Near East seem to be beyond mythical fantasy.
The comet (staff) of the god (cf. Uwoke, Yahweh, Ea, Yahou,
Hermes), the marine tidal upheaval, the near approach of the huge
comet, the sinking of the land into the abyss, the stroke of cosmic
lightning that broke off the comet's tail, and the resulting "navel
of the world," a sacred place like Delos Island in the Aegean Sea,
which was called by the same term.

But now we are given pause. The Phaeton incident was
of 3500 years ago, not 11,500 years. How explain the discrepancy
in time between the Lunarian fragmentation of continents and the
Venusian cometary catastrophe?

The question is actually an opportunity to advance the
theory. Perhaps the most perplexing of the problems enmeshed in
the multifarious evidence of grandiose Pacific happenings is this:
the debris of the Pangean continental breakup is scattered around
the Pacific as its fundamental morphology; yet reports of more
recent disasters occur. Many a geologist has dismissed offhand all
evidence of recent happenings because he knows how removed in
time were the major events; meanwhile the ethnologists are fixated
upon the evidence of a human history unfolding in the midst of
disaster. If the Pacific continents sank once, how could they be
there to sink again even in the past three thousand years, even,
indeed, in the nineteenth century, when reputable navigators

swore to the presence of islands near Easter Island and elsewhere that are no longer there.

Some indications fit different periods. One may conjecture that, in the Lunarian episode, small pieces of land survived the chaos, or disengaged from the nearest continent and floated into the vortex. But, as for Europe, Africa, and Asia, so for Oceania. There was no end to catastrophe. Considerable populations and cultures could still be built up, only to be drastically reduced by subsequent lesser catastrophes. The Earth has not yet achieved equilibrium, particularly in the regions that were most heavily damaged. Igneous islands, such as the Hawaiian chain, must be considered as the tallest of seamounts. Coral islands and atolls may be considered as debris of Pangean sea bottoms and as new growth, accelerated by heat and by being adaptable to a quick rate of bottom sinking.

Rare igneous bits of rock, such as rhyolite, are of continental origin and found on Easter Island. The Soviet geologist, V. Belousov, maintains that a large zone off of Western south America had once been continental sial. Moreover, the sea bottom of this southeastern sector of the Pacific rests upon a crust 20 to 30 kilometers thick; this is characteristic of continental crust, not of oceanic crust, which is only three to five kilometers thick.

It is possible that the area is continental sial, and even was once populated land, but that the stripping of crust by the moon eruption brought on lateral avalanching to the north and west and a sinking generally in this sector. Then as the world cleaved, the rift here overran the land that was to sink. The setting contrasts with western North America where the rift was overthrust by the continent. One may expect to find oceanic basalt or sima beneath Easter Island, which extruded from the rift to pave over the land. Just as in the northwestern United States, the rift extruded lava on top of the land in wave upon wave.

Regrettably, judgment cannot yet be passed on the origins of Tiahuanaco, in the Bolivian highlands, or upon its related areas of culture in Peru, Colombia, and Ecuador. It may well be connected with Polynesian settlements in mid-ocean. The

Galapagos Islands, once thought to be an isolated laboratory of plant and animal evolution, now gives up 2000 pieces of pottery and implements of human manufacture, as well as continental species of flora and fauna.

If Tiahuanaco last rose high when the Sierra Nevadas of California did, and when the area around Easter Island sank, then this event and a new phase of existence maybe placed in the second millennium B.C., rejecting its present dating of around 400 A.D. and moving backwards part of the way to Bellamy's 11,500 dating.

Soviet opinion on the American Indians, like that of most Europeans, assigns earlier dates to their arrival in the Americas. Kondratov works on the baseline of 30,000 to 40,000 years ago, and of 25,000 years for the Indians of the U. S. A. In Kamchatka Peninsula, sites are dated at 14,000 to 15,000 years. Red ochre, arrowheads, and beads and pendants like American Indian *wampum* were unearthed there. (Perhaps the question should be not "How so early?" but rather "Why so early?") Some 14,000 years of dealing in the same monetary exchange appears extraordinary in view of the fleeting career of historical monies. There are drawings in the American southwest of man and dinosaur; also footprints of man and dinosaur are linked; it is hopeless to calm the heated objections to these finds here; they are not impossible; then either the dinosaur survived until very late, or Americans are extremely old. Evidence is difficult to come by, but quantavolutionary theory may find profit in considering a humankind in America who was primordial with humans everywhere, who was almost annihilated in the subsequent catastrophe, who was Tethyan (Mediterranean-Atlantean), Melanesian (African) and Sinyan (Mongolian) - all three - and who was later reinforced by way of the Aleutians and the Bering Strait region; there is no gainsaying the supposition.

Perhaps by this time the reader has already noticed the magical phrase which conventional science uses to deal with recent catastrophes of all kinds: "the end of the ice ages." It is a useful way of saying what is uncertain, without admitting that it is uncertain, or that scientists are even in agreement on when the ice ages ended. It could be anywhere between 5000 and 25,000 years ago, with most scientists centering upon the date that I have assigned to the Earth cleavage and Moon eruption, about 11,500 B.P. Thus, Vladimir Obruschev places "the sinking of the land in the region of Easter Island at the time of the glacial epoch" when the ice melted and waters rose. Or "Bering Land began to sink at the end of the last glacial period, between 10,000 and 12,000 years ago." Probably in any sample of books and articles on quaternary geology, paleontology, evolutionary biology, and archaeology, most authors will be found to use the "end of the ice age" as a general synonym for catastrophe. From pole to pole and all around the world "the Pleistocene ended in disaster." The reader might examine the two contrasting hypothetical calendars that follow after the text of this book.

To claim a known sinking for a known time invites error. The help that one can get from geologists and prehistorians is mainly inadvertent. The calendar of events and dates could be readily improved were a quota of careful scientific attention granted to quantavolutionary hypotheses. Even conventional geologists of the holocene period have complained that their colleagues turn their backs on any phenomena that are recent.

Geology has traditionally opposed or ignored the interjection of legendary history and anthropology into its concerns, especially insofar as revisions of time scales are stated or implied. How then has geology coped with the rise and fall of land masses? After shaking off the idea that Noah's flood had covered the world (and the Deluge became a bogeyman to them, obliterating ancient human voices and behavior), geologists were possessed by the need to explain why marine fossils are found in lofty and protected enclaves of the continents. It seemed natural to resort to risings; then just as naturally, the land beneath the sea had taken part in sinkings. Blessed with the gift of time, they could assign to every parcel of land its turn above and below the sea. The mechanism for the many freight elevators was unfortunately almost as mysterious as the "Hand of the Almighty," and is to this

day. Furthermore, the mystery has in the past generation been
enhanced by the discovery that most land beneath the wave is a
stranger to the subaerial land. 'Sial is sial, and sima is sima, and
never the twain shall meet. '

J. Tuzo Wilson pioneered the theory of the destruction
and remaking of present ocean floors every couple of hundred
million years: so much for sunken lands; they are stuffed down
and run over by drifting tectonic plates. The rises are another
matter.

> The uplift of the continents is by the rise of flat domes of a
> variety of sizes, which have been called shields, cratons,
> batholiths and smaller domes... There have been intermittent
> uplifts involving the rise of land areas of the size of shields, or
> even of whole continents. Uplifts are followed by erosion and
> flooding of continents by the sea, each cycle requiring
> something like a hundred million years... Next smaller in size
> are *cratonic* uplifts of which Southern Rhodesia affords a fine
> example... Much of the shield of Southern Africa is underlain
> by a series of about a dozen cratons, each roughly circular and
> a few hundred miles across. These cratons are uplifted more
> actively than the shield as a whole... Smaller again are *batholithic
> intrusives*... Each craton was formed of a hundred or more
> batholithic uplifts... Those formed during a period of a few
> million years in Jurassic-Cretaceous time in the western
> Cordillera exceed in area by a factor of 1,000 all those formed
> during the rest of the half billion years since the close of the
> Precambrian eras. Once considered to have been intruded while
> molten, batholiths are now widely considered to have more
> likely resulted from plastic deformation with recrystallization
> and partial melting of piles of pre-existing sediments. They are
> often approximately circular and those showing the strongest
> evidence of recrystallization and igneous activity grade into
> uplifts of similar size that were clearly intruded while cold and
> in a solid state."[11]   Even smaller uplifts are very many in
> number.

---

[11] In Beals, et at., *Theories of the Origins of Hudson Bay, op. cit.,* 37-40.

Wilson's statements are descriptive: the mechanism is here presumed. Too, the language itself is non-operational and Aristotelian in undue proportion. Noteworthy in our view is the assignment of uplift to practically all land above the sea. It is thus that the marine sediments occur in all regions. The uplifts are circular, but not meteoritic; they seem like aborted volcanos, whether great or small.

The total impression is of immense uplifts from pre-existing sea beds, accompanied by smaller uplifts, then smaller, and finally quite small rises, a bloated skin with many thousands of protuberant patches. There appears also to be a heavy concentration of these rises in an age that concluded with worldwide biosphere extinctions, the Cretaceous. Further, the subterranean force involved a heat whose temperatures might begin by melting rocks and end in slight metamorphic deformation of rocks whose top levels were in fact pushed up in a cold state.

Might this whole worldwide process have occurred mostly in a single quantavolution? Some regions, even large parts of continents, would have been lifted hundreds or thousands of meters higher than others. Shallow marine sediments would be raised. Many sediments would be reworked in the heat, pressure, and churning of the uneven general uplifts. Erosion would be heavy in such an event, from mechanical disruption, uneven heating, electrical and gaseous outbursts, precipitated vapors, and winds. A great many inter-lift depressions and fractures, laying the groundwork for gullies, streams, and valleys, would develop.

The superpositioning of fossilized sediments according to age would be preserved, even as these were raised. High in the plateaus of Africa, Tibet, and Bolivia, fossils from shallow seas and swamps would be stretched out in their original beds. The Earth would have a largely new surface, uneven, less neat, and confusing to the eye of the beholder. Too, with all this swelling, could not one speak of a general expansion of the Earth? Again, we go in search of a mechanism.

Let us turn to another admirable geologist, whose work unwittingly has helped us to generate the theory of quantavolution. Shelion explains the modern theory of crustal movements of the Earth -diastrophism, in a word.[12]

> Most geologists look inside the earth for the ultimate driving force of diastrophism; no known exterior forces are sufficiently versatile to account for the variety of deformation we see... Plastic creep, perhaps in the upper part of the mantle, is the active element, and the brittle crust on which we live is passively tiding on this very slow flow. Of course, discernible forces arise from the rotation of the earth, from the tides, and from gravity acting differentially on irregularities in the crust and its surface topography, but these influences probably can do no more than modify and locally complicate what is probably the essential mechanism of crustal deformation - very slow plastic movements at about the level of the upper mantle.

One notices an absolute indifference to exoterrestrial forces and to their high energy expressions of an electrical, atmospheric, aquatic, and lithic kind. Shelton proceeds:

> This concept is attractive for many reasons. By postulating different directions of flow in the upper mantle, it is possible to imagine many different kinds of stress being imparted to the lower side of the comparatively passive crust. If the flow involves circulation in three dimensions it must include rising currents in some areas and sinking currents in neighboring ones hundreds or thousands of miles away, as well as horizontal transfer from the first type to the second.

One notes the speculative terms: "attractive," "postulating," "imagine," "must include." There can be no objection to speculation, especially in so excellent a volume as

---

[12] The material to follow is contained in Shelton's *Geology Illustrated*, 423-4.

Shelton's, but neither should geology claim to be a "hard science," fighting off speculators.

Shelton, perhaps embarrassed by the weakness of conduction currents, suggests that the rising heat of the deep mantle is so great as "to require the actual rise of masses of rock from hotter regions deeper in the earth." And he concludes that "some kind of very slow thermal convection -the rise of relatively warm columns and sinking of relatively cool ones -is a favored hypothesis for the ultimate cause of diastrophism." Then in two final paragraphs he reverts to basic questions, asking, too, for the essential information needed to answer the questions. He doubts finally that the information at hand is more than enough to tell one rock from another, and certainly not adequate enough that "a hypothesis of thermal convection currents in the upper mantle can even be formulated, let alone tested..."

Sometimes, when asked why he does not sufficiently quote "creation scientists" - George McCready Price, Donald Patten, Byron C. Nelson, Alfred Rehwinkel, to name a few -the present author answers that he has only a limited perspective, an individuated paradigm, which cannot move too far if it is to remain intact. Moreover, he cannot assimilate theoretically the instrumentation of some secular catastrophists such as Hoerbiger and Beaumont, whereas he feels comfortable in the modes of thought of such as Boulanger, Donnelly, Bellamy, Kelly, Dachille, Velikovsky, and a number of very recent historians and catastrophists. But finally he must confess that he feels more inspired by the contradictions displayed within the evolutionary and geological literature as it marches in fine array through the catalogues and journals of science. It profits science and pleases him more to tell the latter writers that he agrees with what they are saying but that they do not realize the full meaning of what they are saying.

CHAPTER NINETEEN

EXPANSION AND CONTRACTION

Mankind has been impressed by many lands sinking like Atlantis and Lemuria, and by others, such as the Atlas, the Cascades and the Chilean Cordillera rising. The movements, all legends insist, were sudden. And, of course, since it is the human who speaks, the movements were recent.

L.C. Stecchini, historian of ancient measures, maintained[1] that the Babylonians, calculating the diameter of the Earth subsequent to Egyptian measurements, arrived at a larger figure. Some of man's early obsession with geometrical measurements of Earth and sky were motivated by perceptions of terrific effects and of changes still then occurring or feared.

Geologists prefer to think of lands sinking in one place while rising in another. I doubt that ancient man would argue the point. The geologist may call the total process isostasy, by which is meant the belief-not necessarily a fact that the mantle around the world so acts as to stabilize the crustal surface. The mechanism of isostasy is questionable, but, since it is only a question-begging term, it is less questionable than the mechanisms for pushing up and pulling down the crust, which may be a non-existent practical fiction.

What would provide an intelligible mechanism? One such possibility is the expansion of the Earth as a whole.

---

[1] In conversation with author. His yet unpublished manuscripts may cast light upon the matter.

When the remarkable past changes of the globe first assembled themselves in my mind, I imagined them to have occurred solely as a result of the expansion of the Earth under the influence of exoterrestrial forces. Then the theory of lunar eruption appeared more convincing than a very large expansion, and finally I settled upon a combination of loss of mass and expansion of volume.

Whatever can explode can expand. Worlds explode. Radio astronomy and even visual observation on rare occasions, confirm this. The asteroid belt between Mars and Jupiter was probably a planet until recently. There are some small indications that it may have been identified with the Greek god Phoebus Apollo, hence be so close in time.

The Earth can explode. Therefore, it can expand. It is more difficult to construct a model of expansion than a model of explosion. Both layman and expert can readily conjure up an image of "more than enough" energy to explode any body. In so imagining, they may skip over the crucial problem of how much it takes exactly to explode the body. An explosion can be defined as a rapidly accelerating expansion that has achieved a specified rate where a set of effects occurs that is called "explosion."

The conservation of angular momentum of a rotating body depends upon its retaining the sum of its mass, its velocity and its radius. The radius is the distance from the center of rotation to the direction of its motion along the axis of rotation. Expansion signifies a change in radius.

The concept of radius describes a relationship of objects. It is not itself a force or an entity. Therefore, the expression "increase in radius" must signify a changed spatial relation between things that determine the radius. Once more, the salient question points at electro-mechanics, determinants of mass that might act to increase the radius.

Expansion of a rotating body then must be associated with a change in velocity or mass. In the case of the Earth during a lunar eruption, the loss of mass consists of half the crust and most of a dense atmosphere, altogether no more than 2% of the mass of the Earth. An interruption of rotation imposes an abrupt decline of spin velocity upon the Earth. This then requires an increase in radius and expansion in order to maintain angular momentum.

At the same time, the conservation of angular momentum does not occur in an isolated system. In the present case, energy representing the angular momentum is transferable to other external systems: the space plasma, the proto-Moon, the Sun and planets, and cometary bodies.

A body such as Earth will expand when it is freed from an external pressure. Possession of a dense atmosphere of the type of Venus would have limited the Earth's figure; if removed, the Earth would have expanded. Its outer surface will even spring back, that is, exhibit an acceleration and a counterpressure that causes it to "take off" from its base. Irregularities found in a number of places around the globe may be fossil expansions, if not fossils of impact explosions and massive eruptions.

There are reasons to believe such events can occur and have occurred. In anticipation of stating them, we may suggest why land has sunk; for the two behaviors of expansion and sinking are not independent, although they may occur at different places and lithospheric levels. Lands have sunk by collapse into new basins, by flooding, and by their contents disappearing in explosive clouds of debris. If the force that explodes the land expands the Earth, then we have sinking and rising in a new formula, one which contains its mechanism, and furthermore may be true.

The Earth was not pre-ordained to its present volume or density. No two planets have the same size or density. Earth's mass density differs considerably from that of the inner planets and much from that of the outer planets. So does its volume. It could once have been denser and smaller. That its mass and volume have been constant through long ages is 1) an ideological dogma and *idée fixe* 2) a mistaken simplism regarding the "hardness of rock" and the innateness of volume 3) a mistaken reading of natural history 4) a psychological denial of an undesired state 5) a practical fiction, or 6) a fact. The first five possibilities might be demonstrated without much difficulty, but will be left to such evidence as the reader may cull from this and related studies. If they are so, then the sixth may be in doubt and the contrary may be considered, namely that the Earth's volume has fluctuated or at least been subject to expansion. Such a consideration is the purpose of this chapter.

A number of theories have given the Earth different sizes in the past. A number of means of expansion are available. A number of reasons lead one to a probable opinion that the Earth was once smaller and has recently expanded in volume.

Pickering long ago realized the necessity of Earth expansion. "A rising region... must evidently be increasing its volume. This increase may occur either with or without an increase of mass. In the latter case, the increase must be due to a rise in temperature. It has been shown that, if a part of the Earth's crust fifty miles in thickness were to have its temperature raised 200 ° F, its surface would be raised to the extent of 1,000 to 1,500 feet. The Bolivian plateau has an elevation of two and a half miles. That of the Himalayas is about a mile higher. It is improbable that these elevations are due to this cause."[2] He finds that an increase in mass is impossible. He then turns for an explanation from a simple temperature rise to the possible pressure of water and steam, and since he was unaware of the lack of water over the rock of ocean beds, and since he presumed the Moon eruption and the catastrophic period to be very ancient, he called upon a still watery mantle to produce the necessary thermodynamics for expansion. Even were the age to be granted, the mechanism would be hopeless for the task. No exploding steam engine could blow the material of the Moon basin into space.

Carey and Jordan have devoted books to the subject of Earth expansion, and were cited earlier. Both see the process as very gradual. Carey estimates a 20% radial expansion and uses the projected expansion as a mechanism to account for continental

---

[2] 15 *J. Geol.* (1907), 34.

drift. T.M. Cook, in criticism, finds Carey's theory short in energy supply, and argues that the required release of chemical bonding of molecules would melt the Earth. Jordan, following Dirac, claims a relaxation of the gravitational constant over time. As gravitational attraction declines, matter expands. The application of Dirac's theory to Earth expansion would logically follow, but Jordan is unable to provide convincing geological evidence, even when presented with a long Earth history.

R.H. Dicke and C.H. Brans also predicted a slow drop in the force of gravity, and Dicke estimated that Earth gained from this source 15% in volume over 3.25 billion years. When the Atlantic basin was shown to be young, Dicke ceased to credit its widening as support for his theory, because it apparently had grown 300 times faster than his theoretical rate would allow.[3]

Egyed's theory of Earth expansion, based upon paleogeographical data showing a modest inverse correlation between the quantity of ocean waters and the passage of time seems vulnerable both because a uniform quantity of water is assumed and because the time periods, though conventional, can be challenged. Egyed cites Cox and Doell, further, in claiming an increase in the Earth's radius of between 0.5 and 1.0 millimeters per annum.[4] This would amount to 10 6 meters in a billion years, about one-seventh of the Earth's radius.

In comparison, my estimate of the radial expansion which accompanied the fission of the Moon is about 9%, less than one-tenth of the total radius; the estimate is the result merely of topographical scrutiny. The expansion in volume represented is about 20%, much less than Carey's estimate. Carey's expansion took place over many millions of years; the process here discussed would have occurred in perhaps three thousand years. Again, we rely upon a uniquely great exoterrestrial encounter to compress

---

[3] W. Sullivan, *Continents in Motion*, 50-6.

[4] "The expanding Earth," 197 *Nature* (16 Mar. 1963), 1059-60; see also P. S. Wesson, "Does Gravity Change with Time?" 33 *Physics Today* (July 1980), 32-7.

time, accomplishing in centuries what the aforesaid scientists have allocated as the task of very many millions of years. Any evidence at present of an expanding Earth, we would accredit to the extended uniformitarian tail of the exponential curve of quantavolution.

Of the several attempts at demonstrating expansion, Meservy's appears most clear and valid. He shows that "the separation and movements of the continents in the last 150 million years cannot be explained by continental drift on the surface of the present-sized earth."[5] This he does topographically. Following the Bullard and Hurley reconstruction of the supposed original supercontinent before its continental elements drifted apart, he retrojects the present continental map as it must have drifted and shows that the present arrangement could not have emerged from the reconstruction.

In order for the supercontinent of one time to fit the map of the continents of today, the continents of today would have to come from a smaller globe. "It seems highly improbable that the area enclosed by the perimeter [of the Pacific] was ever as large as half the earth's present area in the last 150 million years." Furthermore, he claims that his "argument is not very sensitive to the exact time scale or to variations in the rate of ocean-floor spreading, as long as these were reasonably monotonic in the period in question." That is, the solution of a smaller Earth would emerge even if time were foreshortened and ocean-floor growth were rapid. "The most direct interpretation of the evidence... seems to be that a large expansion of the earth's interior has taken place in the last 150 million years. The nature of the physical process that could have led to such an expansion is highly conjectured, but such a process cannot be excluded on the basis of present physical knowledge."

By what means could the Earth have expanded at the time of or subsequent to the breakup of the original super-continent?

---

[5] "Topological Inconsistency of Continental Drift on the Present Size Earth," 166 *Science* (31 Oct. 1969), 609-11.

Six means can be suggested, none of them excluding all others and in fact all six could be simultaneously operative to produce a concurrent breakup of the continental mass and an expansion of the globe. Meservy does not consider a sudden loss of over half the Earth's crust, as by Moon fission, but significantly the occurrence of such a loss, concentrated within the Pacific perimeter, only serves to strengthen his topographical demonstration.

An abruptly slowed rotation of the Earth over days of time, never to be restored, would reduce the centripetal force of the globe and tend to expand its volume. This would be especially prominent if the body causing the slowdown were electro-gravitationally attractive. The Lorentz-Fitzgerald (1893) equations assert that all matter contracts in the direction of its motion and the amount of the contraction increases with the rate of motion. The Earth rotates with a kinetic energy of 2.6×10^{36} ergs. If an interruption by an external body depresses its rotation by 35% and shortly thereafter the rotation assumes the level of a 20% reduction, an energy of some 10^{36} ergs is available, along with a large electrical, gravitational, and axial torque energy, to push the continents and expand the volume of the Earth. This heat of rotational slowdown is sufficient in theory to unleash 50 billion Krakatoa's. That volcanic eruption, one of the worst in history, released about 2×10^{25} ergs.

The conditions for expansion of the Earth were probably present, but they approached the conditions for a complete melting of the crust of the Earth. They approached, beyond that, the conditions for the explosion of the Earth. Nevertheless, in the end, the sphericity of the globe was maintained, half of Pangea was preserved, and small numbers of most flora and fauna, including *homo sapiens*, survived. The fall of cold water on the continents helped to preserve their structures against heat from below while the same waters moving into the oceans and the falling waters there catalyzed the expansion process.

The sudden acquisition of a huge heat presented problems of storage and prompt use, if the Earth were not to explode. The Intruder's pass-by and the forces it exercised upon the globe would begin some days before the moment of maximum impact and continue for several days thereafter. Thus the heat would not be applied all at once; by the time the critical moment arrived, the Earth was committed to partial explosion and expansion. The loosening of the Moon-making crust and the cleaving of the globe would take place quickly; then immediately the heat would be drawn upon for the reconstruction of the Earth.

Also, a great proportion of the heated matter would be exploded into space. The global fracture system would help to handle the venting of enough heat and material to cool, pave and expand the Earth's surface. Moreover, it would develop the capacity to do so within the required time. And the density of the Earth's interior would be originally sufficiently high to provide the material.

A decline in atmospheric pressure by the temporary and permanent removal of atmosphere, especially a heavily vaporized one, would also contribute to an expansion of the Earth. So too, of course, would the actual removal of crustal material of low temperature. It is not necessary that the rising magma be less dense than the escaping crust but only that temporarily it be in a molten state, mixing with gases and water as well, and hence capable of freezing into a solid at a higher level or over a larger expanse of surface. H.J. Binje said once that "the driving force of rising magma lies in change of the nuclear structure of the magma itself."[6]

Water added to a silicate solution reduces its melting point. The lower crust and mantle boundary might melt at as low a temperature as 500 ° C under water saturation. The water itself would be provided by old surface waters and incoming deluges of rain, snow, and ice.

The upper crust on which the biosphere and sediments rested would be shielded from the abyssal heat by thousands of volcanic vents penetrating its surface and by the cyclonic venting

---

[6] Quoted by Jordan, *op. cit.*, 121.

of heat into space over the immense flayed crater of the Moon. Still the thermal pressures throughout the globe would be heavy and accompanied by rises in temperature that would increase the expansion.

The globe would fracture throughout. Pictured as scraped of its biosphere and surficial sediments, the globe today presents a thoroughly fractured appearance. Nowhere on Earth is one very far from a great fissure that would have been involved in expanding the globe. Perhaps one of the reasons for the discontinuity and absence of expected sediments in so many places is the underlying expansion by igneous intrusions that once occurred. Furthermore, the very 'success' of the globe-girdling fractures in producing ocean beds of lava and pushing away the continents is that they were engaged in expanding the volume of the Earth.

The sial continents that remained obviously were not destroyed in the process of partial explosion and expansion. However, they were penetrated at many points by expanding lava. The sial could be lifted by less force than would be required to dissolve it. Given over half the surface as a direct outlet, and a huge fracture network for disgorging heat and magma, there would be less occasion to obliterate the many large areas of sial overhang.

Willis once wrote that "it is established by observations on rocks that the chemical compounds of which they consist can adjust themselves to changes of pressure or of temperature or of both by changes of volume as well as by alterations of form. Larger volume would result if a mass of rock were heated and at the same time relieved of some of the load resting upon it."[7] He even went so far as to say that erosion can cause underlying rocks to expand their volume. Rock crystals respond to new conditions, not even highly thermal, by reorganization of their structure.

---

[7] B. Willis, *East African Plateaus and Rift Valleys* (Wash. D. C., Carnegie Inst., 1936), no. 470; 306, 309.

"Crystals are almost human in that they always seek the easiest way out... Where crystals grow vertically, continents rise." A sudden and massive change in crystallization may have occurred in many rocks. Now we might claim that the lunar explosion may have been the chief factor in expanding the Earth and producing the granites of the continents whose origins we had been wondering about in an earlier chapter.

A definition of stability and even of structure is that the defined complex resists electro-gravitational dissolution. If a complex, say of rock, is stretched in a lowered gravitational field, that is, attracted by another field, and obtains a revised structure, then, after release from the second field, it will tend to retain the form temporarily assumed. This may be a factor to be considered in relation to an expanded Earth. An analogy suggests itself: rock under conditions of the assumed encounter would behave like oil shale when it is processed. The rock that is mined expands its volume by 20% or more.[8]

Seismic signals experimentally transmitted through the Earth produce more or less sudden changes in velocity, indicating "boundaries" at six radial distances before reaching the center: the Moho discontinuity, and at 400+, 950+, 2900+, 4800+, and 5100+ kilometers of depth. There seems to be little explanation for these seismic transitions unless they represent levels of response to an historical torque. The interruption of a rotational motion of a mass must be perceived by the whole body. At some ratios of density to torque, indications of a phase shift should occur. These indicators would be erased by a huge expansion, but by the same token, will remain vivid under conditions of moderate expansion.

In another work, I asserted briefly, and probably in error, that the Earth would lose electrical charge in a grave encounter such as would remove the lunar material: "loss of electrical charge may also have decreased the density of the Earth."[9] This was

---

[8] *Encyclo. Britannica yearbook*, 1976, 289.

[9] *Chaos and Creation*, 154.

based on the assumption that piezoelectricity from rock turbulence and electrostatic charges would be lost into space to the larger intruding body; then matter hitherto bonded electrically would be unbonded and take up more volume. However, after discussions with E.R. Milton, I became persuaded that the intruder would have carried a heavier charge, since it was transporting charge from the outer solar system toward the Sun; it was also much larger than the Earth; therefore, it would have deposited charge upon Earth. The charge would then be incorporated by the Earth's molecules and cause the stretching of their internal atomic bonding. Hence expansion. But where the charges would accumulate is critical, whether on the continental surfaces, or diffused in the mantle, etc.

"It is not generally known that the volume of a Leyden jar is increased by charging the jar and diminished by discharging it, wholly or partially. The crust of the earth resembles a Leyden jar, of which the coatings are represented by the liquid core and the enveloping atmosphere." So wrote Abbé Moreaux in 1909.[10] He envisioned a daily expansion and contraction of the crust. Possibly the same charging phenomenon would effect a larger and more enduring expansion of the Earth.

The almost non-existent evidence, and the complexities of the electrical phenomena accompanying such an encounter, make all reasoning highly speculative. It is possible that both processes occurred, a gain and loss of charges, with the gain predominating.

Yet another set of phenomena may be connected with Earth expansion, rather than simply the adjustment, unexplained, of coastal margins to which it is otherwise attributed. That is the tendency of continental margins to stretch out over the ocean basins. For instance, "as late as the beginning of the Quaternary period the land of Siberia reached much farther north and at the

---

[10] 68 *Sci. Amer.* Suppl.(24 July 1977), 3.

end of the last glacial epoch was broken up, large areas sinking into the sea."[11]

Elsewhere we read, "while exploring the seismic structure of the continental margin off France, Lucien Montadert, of the Institut Francais du Pétrole, noticed that the upper part of the continental crust of margins has been fractured into a remarkable pattern of narrow sedimentary basins bounded by listric faults, that is, faults that 'curve, ' being steep at the surface, becoming more horizontal with depth. He suggested that the continental crust at the margin was extended at the time of rifting by up to 20 per cent."[12] The listric faults are not found where internal basins, such as Lake Michigan, have been examined.

We are inclined to view this oceanic marginal fault system as a possible stretching to accommodate expansion. If the rift did not cleave cleanly, however, the stretching might be expected. As the Earth expanded, and radial pressures pushed upwards, blocks of rock would be broken off serially from the continental mass. The stretching might also, still in accord with our general theory, be a result of a differential speed of rafting, with 'France' here heading eastwards faster than the bottom of the basin could be paved with fresh lava.

When the Earth's surface is viewed from a detached intellectual perspective, it begins to appear as a thoroughly disorganized assemblage. Instead of its presenting logical conformities on a grand scale, its every feature becomes an anomaly. All of its real rules seem to have come from violating the rules of the earth sciences. When such a condition is manifest in human organizations, such as the factory, or the hospital, or the government, that is, when what is regularly done contrasts with the way things are supposed to be done, the usual recommendation is to change those rules that are inapplicable to reality. Unfortunately in the present case, as in many cases of social

---

[11] II *Catas. Geol.* 2 (Dec. 1977), 3.

[12] Tony Watts, "Plate Tectonics," *New Sci.* (6 Nov. 1980), 362.

organizations, new rules are not easy to write and, meanwhile, the old stable mixture of reality and pretense that has been managing the enterprise dissolves into fantasy and disorder.

PART FIVE

RIFTS, RAFTS AND BASINS

The globe is girdled by lines of fracture, and from these lines stretch perpendicularly hundreds of transverse fractures. The pattern of rifts and cleavages shows that they created basins (as in the Arctic and Atlantic) and exuded material to fill the Pacific Basin which was dug out by exoterrestrial force and assembled or dispersed in space. All of this happened fast. Continental drift is true, but not true enough.

Further, the major mechanism of drift – convection and subduction – is impossible. The major mechanism of sedimentation: uniform and gradual accretion by erosion and wind or water carriage – is likely impossible. Quantavolution readily supplies alternate short-term mechanisms that are adequate to the facts and that also educe new facts.

CHAPTER TWENTY

THRUSTING AND OROGENY

When nineteenth century geologists departed from their original simplistic uniformitarianism, they found it useful to identify in Earth history several points of great diastrophism (" turnabouts" in Greek) or revolution. Whence came the Laurentian, Algonkian, Killarney, Appalachian, Laramide and Cascadian Revolutions, each marked by profound unconformities in the rocks. Naturally a quantavolutionist will wonder why these never evolved into a new catastrophist geology. First, there was the obstacle of ideology, which a social psychologist can appreciate more than a natural scientist: the social atmosphere of the times, the breakaway from religion, the need of biology to pursue prolonged development periods, and the empirical fascination of studying the processes going on before one's very eyes -these acted to subdue diastrophism and revolutionism.

Long periods of slow changes were supplied until the revolutions themselves appeared as continual skirmishes of the elemental forces. The search for internal forces capable of sculpting the Earth's surface went so far as to conceive of the massive core of the Earth wobbling within the globe so as to push out or pull back crustal features. Some have thought of shrinkage, so that as the Earth aged it wrinkled (apparently not willing to move out upon the seabeds). Nowadays radioactive decay, rising from the rock deeps and engendering heat, has been called upon to push the continents slowly about. And this is said to crumple the colliding edges of continents into mountains and to stretch and reform the landscape.

A second reason why catastrophist geology could not evolve is also related to the ideological: geologists have refused to look into the skies for the forces needed to accomplish the revolutions that they perceived; without a mechanism, they were left with mere names -and questions, such as ones asked by K. Krauskopf;[1] "What are the irresistible forces which can twist and break the strongest rocks?" "Where do the forces originate which can raise and lower continental masses vertically? ... Why have not forces in the crust long since reached an equilibrium? With questions like these we have long since reached an impasse."

Since we have essayed answers to questions of vertical movements for the sake of this chapter, we may add: "How can kilometer-high sediments be pushed over thousands of kilometers of the surface of the Earth?" Every thrust that has occurred or might happen can be described by the same few variables. The permutations resulting in reality may be numerous but are still intelligible. Price details several thrusts; Cook and Velikovsky describe a number; Burdick, Brock and Engelder have produced case studies.

It should be possible to conceptualize thrusting. Suppose a thrust as any lateral motion of a definite mass. The mass will have an initial velocity and acceleration, a momentum and inertia, a direction. It will have a surface to ride upon, and the interface will have a characteristic viscosity. The mass need not be solitary, even though it is definable; a limestone may be riding on a schist, or rock upon oil or water slurry, and so forth; hence there will be another set of variables for each definable component in a complex thrust.

Melvin Cook and Charles Hapgood employ prior ice caps as a mechanism of sudden diastrophism. Accepting prior calculations and proof of the existence of towering ice caps at the poles in recent times, they weigh the ice and decide that enough mass is available to cause unbearable pressures laterally (Cook) and a lever effect (Hapgood). The ice mass avalanches upon the

---

[1] Quoted by Kelly and Dachille, *op. cit.*, 76.

world, perhaps in conjunction with the fracturing of the globe. The massive thrust of the ice bulldozes the surfaces of all sediments and biosphere in many areas; the fractured Atlantic region of Pangea, now the Americas, moves westward and the bows of the continents rise into high mountains as they plough through the oceanic crust. Hapgood adds a tilt to the Earth, product of the same event, and this permits him to add another string of disasters to that of the precipitating cause. I cannot criticize these works here. In general, to tie together the apparently interconnected Pacific Basin and continental movements I find a need for a more universal force.

Mountain ranges are folded. What is a fold and what is a thrust? There can be no fold without a thrust. Nor is there any major fold that comes from two opposite thrusts at the same time. There must be a source of the push that folds, and sometimes folds in two or three laps. And the push must be along a surface that is the base for itself and the fold. Conceivably an uplift might come from an expanding Earth or an attractive electrogravitational force above the Earth. In the latter case, however, irregular outbursts would occur, and the landscape afterwards would be volcanic or batholithic or like the seamounts of the ocean bottoms. In the former case the surface would crack, swell into circular rises of different sizes, cause gentle slopes, and also erupt in volcanism. There is no need to deny the ordinary idea of a fold as coming from a push.

Enough of the high mountain ranges of the world are poised at the edges of the continents to admit the possibility that they were pushed from behind by the moving continental mass. Their pitch, too, suggests a seaward thrust. If the thrust was initiated by ice blocks, they would ultimately take the form of a scow uplifted at the stern and bow. If in movement because of a forward electrogravitational slide and an upwelling and expanding lava flow from the rear, the bow would be much less pronounced than the stern. If the movement were accompanied by a swelling of the magma below, especially if the expansion were more pressing from the rear magmas, the scow would tend to nose down and come to its ultimate halt with towering mountains and deep roots. If the uplift were general beneath the thrusting mass, the prow and mass as a whole would lift itself, too, and ride more easily on the magma. Like a motorboat that rides higher as its speed increases, the continents would be elevated and move faster once in motion over a swelling magma. Unlike the motorboat, the continental blocks as a whole would not then sink; the supporting rock would be metamorphosized at a new density.

The emptiness of the Pacific Basin stands for an event quite capable of initiating global diastrophism once and for all. This would require the withdrawal from the Earth of a moon-sized body, in fact the Moon, an event that must call upon an enormous electro-gravitational attraction, which must come from a body even larger than the Earth that passed close enough to pull out over half the crust. And this event is described in *Chaos and Creation* and *Solaria Binaria*.

Thereupon all the terrestrial processes that Melvin Cook so well portrays proceed: the remaining crust fractures down to the mantle in an explosive network providing the globe-girdling rift and fault system. Orogeny occurs rapidly as the cut-apart continental blocks scramble for position. Cross-tides of water and wind race around the world. Rock and ice are in motion as great bulldozers, thrusting here and there. The immense number of faults, not only of the global girdles but practically everywhere, establish the infrastructure of the valleys and rivers of the world. The true ocean basins are created for the first time.

Under such circumstances and sequences of events, the vocabulary of science is strained. The most extreme case of thrust would be a force gripping or pushing the crust of the Earth like a shell so that it moves independently of the mantle and core. That such an idea may be rooted to some degree in reality is attested by studies proposing analogous movements in the Sun and Jupiter, and at least one suggestion that the Earth's core rotates out of step with the crust. The contacts of the crust with the plasmas of space and with its atmosphere may set up a continuous drag and eccentricity on the mantle, manifested for example in seismic and volcanic responses to heavy solar storms. Natural history may have witnessed, if not a complete and neat slippage of the crustal shell, some diastrophic approaches thereto.

I do not know where to place the finding of F.A. Vening-Meinesz: as related to lunar eruption, Earth expansion, rifts and fractures, landforms, or to thrusting? He studied the major topographic features of the globe in relation to the Earth's axis of rotation. Their pattern of shocking and shearing evidences a clockwise rotation of the crust in relation to the core of 70°. That is, as the Earth moved east, its landforms struck out south by east.[2] The unified nature of his finding suggests a single giant thrusting episode sequential to the evacuation of the southern hemisphere.

The continents move; this is a form of thrust. Often it is a thrust through water and basalt bottom; then, again, the Indian subcontinent thrusted upon Asia. Sedimentary rock layers are scraped and dumped over the sides of the awesome abysses; here is thrusting. Coal fields are forests bulldozed and deep buried: this is a form of thrust. Mountains are piled upon one another, again a thrusting action. Tides and winds lay down field upon field of debris, of vast extent; are these not thrusts, too? It is fruitless to argue over definitions. As with earthquakes, which are moving earth, and shaking, so with thrusting: beyond a certain intensity, the vocabulary is inadequate to the quantity of cleavage, the quantavolutions. Then, too, earthquakes and thrusting can come to a marriage; in a discussion of even the relatively mild seismism of our times, Frank Lane writes that "where an earthquake is concerned there is no such thing as an unmovable object, even mountains are moved." They are thrusted.

Yet it has been a long time since "the mountains skipped like rams," as the Biblical Psalm goes. Although the records of solarian geology are far from complete, we suspect that such a sight has not been seen in the past two millennia. The occasional spectacular rock avalanches and submarine mud avalanches that are presently recorded are not what the Psalmist had in mind. In

---

[2] Discussed in Velikovsky, *Earth in Upheaval*, 125-6.

an age that experienced earthquakes abundantly, he was celebrating and reporting Yahweh at a peak of power, probably in the centuries that remained vivid to him -the seventh to the fifteenth before Christ -reinforced by the cherished accounts stretching back to the breakdown of the pangean surface.

He was speaking for the hapless ones who watched the Alps rise up from the Tethyan geosyncline to be "shoved northwards distances of the order of 100 miles" where now are located Italy and Switzerland. The famous "nappes" of the Alps are but smaller thrusts laid upon great ones. The alpine massif smothered the long rift that once cut through the "Adriatic Sea" and "Rhine River Valley." Or the American cordillera, thousands of kilometers long, stretching from Alaska to Tierra del Fuego; there mountain uplifts amounting to thousands of meters have occurred, it is agreed by a range of authors from C. Darwin to I. Velikovsky, in absolutely modern times. The Sierra Nevadas of California are a single block, a thousand kilometers long, thrust up westwards. The Himalayas rose steeply in human times. "The highest mountains in the world are also the youngest," wrote Helm and Gausser.[3] But the Himalayas are also reasonably accredited to the crumpling of the "Indian" subcontinent against Asia with the vast inertial forces initiated in continental rafting. And probably the rising of the Tibetan and African plateaus occurred under lateral and subterranean pressures of the same time.

One after another, explorers and writers have expressed surprise at the youthfulness of the mountain ranges until at least and at last all that are spectacular have been moved up in time to the age of humans. Velikovsky published a brief survey of this evidence citing the geological works of R.A. Daly, G.M. Price, R. F. Flint, B. Willis, A. Heim and A. Gausser, H. de Terra and T.T. Paterson, and R. Finsterwalder. He offered general catastrophic forces as the cause operating most often in human times. Melvin Cook placed orogenesis in a single set of great earth movements

[3] *Ibid.*, 78.

of human times. The present work unites the recent risings, the great global faulting, and exoterrestrial forces mainly of the lunarian age.

Cook uses huge avalanching ice blocks convincingly as the bulldozer of many thrustal incidents in America and, in one case, in South Africa. The ice sheets push the sedimentary strata for many kilometers, melt and flow around them, crack through them, scatter mounds of debris in their path "like loads of loose, dry snow thrown ahead of a fast moving snow shovel."[4] Possibly the only alternative to his mechanism would be rapid continental movement toward the south, deceleration of the basal movement of the crust, a swelling of the Earth beneath the rear echelons, inertial continued movement in the same direction by weaker overlaying sedimentary strata, in some cases even overrunning the halted forward elements. Under this scenario, one would have numerous cases of inverted strata, older on top of younger, and hence the fossil inversions sometimes deemed a disproof of evolution. One would also expect then the occurrence of thrusts in regions of the world where no ice sheets were at work; in fact, the Alpine overthrusts, the Atlas Mountains and other overthrusted areas were not near to overpowering ice masses.

Thrusting played a large role in the formation of coal, lignite, and fusain deposits, which range in depth from the surface down to over a kilometer. The distribution of world coal deposits, Cook shows, follows in significant part the radial avalanching of the ice caps. Coal deposits radiate from cracking and thrust points of the old ice cap and shell-slip. "Most coal deposits are found apparently squeezed by crustal thrusts, between the ice cap depression zones and the concentric, flow-resisting mountain ranges."[5] Ice cap fragments moved outwards upon the biosphere with the scooping and scraping motions of a giant earth-moving machine, depositing it, often smoldering, often slurried with ice and sky waters, into heaps, folded and thrust them over and in-

---

[4] Cook, *op. cit.*, 188.

[5] *Ibid.*, 11.

between with thin sands, clay and gravel, and abandoned them in a state of thermal-retaining and heat-generating compression laterally and from above. Super-hurricanes, fast deep water tides, and typhoons can also scoop and pile up the total biosphere. Since the deep oceans did not exist during much of the quantavolutionary crises, the massive scoopers, scrapers, and in-folders might handle the marine life of shallow seas identically. Coal of different grades, and in thin beds, is interlarded with layers of ash, charcoal (fusain), clay, till, and pebble, that is, with all that goes before the blade of the bulldozer.

Velikovsky's summary of H. Nilsson's analysis of the lignite or brown coal of Geiseltal, Germany, is revealing. The original studies were the work of J. Weigelt and associates. There plants from contrasting climatic regions of the world are identifiable, as are insects, algae, fungi, reptiles, birds, and mammals, (including apes). "Plants are represented there from almost every part of the globe."[6] The material is well preserved: chlorophyll, colors, membranes, and nervature are in many cases apparent. The fossilization says Nilsson, happened lighting fast - "blitzschnell;" the catastrophic process is evident.

Nilsson explains the event by tidal waves moving in from around the world. The time is given as early Tertiary. Velikovsky is noncommittal. To us, more likely than tidal action would be cyclonic action: a great funnel of gases passed over a wide band of territory collecting the biosphere, macerating it, and finally dumping it. Little heat and pressure is needed to bake lignite. Carbon 14 would be low in coal deposits, not for the reason commonly given, that coal is an old deposit, but because it was not in a constant state of equilibrium and is, as Cook shows,[7] not now in equilibrium and, when the rate of growth of carbon 14 is projected backwards, it arrives at a zero state around 13,000 years

---

[6] *Supra*, 219-20. On Nilsson, further, see B. Gray, VII *Kronos* 4 (Summer 1982), 8-25.

[7] "Continental Drift," *Utah Alumnus* (Sept. 1964) 12, and see discussion, *ibid.,* Nov. 1963, Oct. and Nov. 1964.

ago -subject to much turbulence, of course, but pointing to a thoroughgoing reformation of the atmosphere around that time. Where is the thrusting and folding of the ocean bottoms? There is very little of it, unless, as we said, continental drifting is called thrusting. The seabeds are flat, save for the steep oceanic ridges, the great rises, and the innumerable seamounts.

Geophysicist Edward Bullard marks the contrast:

> The mountains of the oceans are nothing like the Alps or the Rockies, which are largely built from folded sediments. There is a world-encircling mountain range -the mid-ocean ridge -on the sea bottom, but it is built entirely of igneous rocks, of basalts that have emerged from the interior of the Earth. Although the undersea mountains have a covering of sediments in many places, they are not made of sediments, they are not folded and they have not been compressed.[8]

The last sentence points up an impossible predicament for conventional geophysics: a supposed situation in which the continental crust folds and thrusts and compresses into abundant mountains while the oceanic crust slides up and under and around without making mountains, having once and for all and by gradual processes made its igneous ridges and seamounts. That no continental mountains are to be found imbedded in oceanic basalts is remarkable. Considering how recently most of the mountain ranges of the Earth have formed, however, we surmise that the mountains came on the heels of the ocean basin creation or thereafter. But this points to the conclusion that the world has been flat until very lately. And this leads to the idea that quantavolutions of all kinds may have begun only recently.

The seamounts are igneous, and usually flat-topped. They came into notice during and since World War II. Their astonishing numbers point to a common and concurrent origin: almost all of them must have been both extruded and pulled up in the exoterrestrial engagement of the lunar fission period. There are no

---

[8] "Origins of the Oceans," *op. cit.,* 19.

substantial currents to erode their tops and anyhow erosion creates peaks and gradual slopes. They are not volcano fields, connected underground by a piping system for magma flow.

Sedimentation on them is slight. Some have "surprisingly young" fossil-impregnated rocks on their beveled tops, write Heezing and Hollister.[9] Some of the fossils are subaerial, not marine. Could sea levels have been 400 meters and more lower than today, ask the same authors. (Actually, subaerial fossil species have been found at 1000 m depths.) Or could the ocean bottoms have subsided by that amount? Neither hypothesis finds favor.

They probably stem directly from the lava pavements of the ocean floors. They probably lifted up into a maelstrom of air and water, rather than grew up underwater like some volcanos, even now, are observed to form. For a short period, they stood amidst a rising ocean of water. The water ceased to rise rapidly. Life took hold on some of them. After a couple of thousand years, an immense quantity of water was poured into the ocean. The seamounts now drowned.

The rhetoric of geology is overpowering in its stress upon time. It rolls along in the cadences of an epic poem, stressing eons of time like the pause at the end of the lines. But today has its poetry of the absurd, and this may drive the incessant echo for a moment from the mind. Consider, then, the absurd: that legitimate arguments can maintain, facing the geological world, an age of 10^9 years and an age of 10^4 years - ten billion against 10 thousand years.

The absurd, of course, is the theory of quantavolution: time is squeezed out of explanations of the Earth until only the minimal amount remains, like forcing the air out of a bottle until a nearly total vacuum is reached. The analogy is not so remote: some say that the Earth is losing its atmosphere, atom by atom, until one day, eons from now, it will move denuded of air in the vacuum of space. The same might be done in hours and days by

---

[9] *Op. cit.*, 521.

the near passage of a body sufficiently large and electrically attractive to suck up the atoms of the atmosphere.

The absurd in geology makes statements of a related type. All the igneous rock and its formations of the Earth's crust, could be brewed by sudden heat over 1500° C and pressures over 5000 atmospheres within a few years. Igneous rock is the greater part of all rock. All rock that is metamorphic needs less heat and pressure to form, and the same short time. Metamorphic rock is a small percentage of all rock. Sedimentary rock, least common but plentiful nonetheless, by definition never boiled or overheated or intensely pressurized, can be laid and formed as fast as material is provided, this consisting of biosphere products, fall-out, and erosion of other sedimentary, igneous and metamorphic rock.

Slower than all of these in forming are the biosphere products. Still, if upon the crust of the Earth were laid the seeds of plants and the eggs of animals, and these were enveloped in an electrified atmosphere, and souped up with nutrient minerals, a passage of several thousand years would find the crust blanketed kilometers deep in biotic debris. If one were intent upon preserving the evolution of species, species would mutate to their present forms in two to thirty leaps, say, and this would provide the varieties of today. It would, of course, require several thousand extra years. It is no secret, actually, that the fillip of evolution has supported geology's claim to time, rather than the contrary (except for radiochronometry); life takes longer than rocks, and fossils can be used by the theory of evolution to push back the age of the rocks.

The absurd idea still has not gone far enough; the Earth's surface and crust are a complicated mixture, of thin and thick pieces, of sliced and hacked out layers, and of dense and light materials, under different pressures and temperatures. How is it to be fashioned to bring order? Dispense promptly with the word "order". The natural order is largely in the mind. The "order" is a wish and illusion. Pursuing the absurd, the mixture of forms and materials of the Earth's crust are but the work of a clumsy chef, who shakes his pot, stirs it erratically, burns the bottom and adds ingredients to his strange tastes. Or, to assign no blame to a divinity, the same effects are achieved by forces born within the Earth and coming from outside of it, but great forces, of the kind that can form the materials. The force that can suddenly slow or change the world's motion can thrust and scatter about the formed materials, and concoct others. What can chop and grind and break the materials can inject all the heat and pressure to make them in the first place, and again and again.

What is more, in this absurd scenario of quantavolution, processes occur simultaneously. The chalk cliffs of Dover do not wait to form until the Anatolian chalk cliffs are made; nor does the mutation of species await a sunny "bowr of earthly blisse." While the Earth's crust is reforming into the Moon, a multitude of volcanos blaze, and deluges of water and debris fall upon the world. All the rocks everywhere are in movement, under pressure and exerting pressure; electricity exudes from every pore and catalyzes the already hard-working floods and vapors; radiation and adaptive saltations are differentiating many species and exterminating many more.

How does one argue against the absurd conception of natural history? One would draw books on the Grand Canyon of Colorado from the shelves showing "two billion years of history passing before one's eyes." But the quantavolutionary vision of the Grand Canyon springs readily to mind: the complex can be put together in a short time in uplift and cross-cutting floods, then cleaved, supplied with torrents, and finally quieted down to make it attractive for tourists. Should one appeal to radiochronometry to resolve the vision, it occurs that the radioactive isotopes might have been stopped or raced in the catastrophic maelstrom. We recall again some of the features of the Earth's surface previously discussed. One by one, it would appear, the morphological features of the world succumb to quantavolutionary explanation.

"Long distance overthrusting has occurred (a) for whole continents over the ocean crust where overthrusting has been several thousand miles (continental drift), and (b) for the superficial Cambrian and 'younger' sediments over the continuous, strong basement rock." (Cook) The greatest thrust and rift and the smallest rock-crack can be considered as "faults."

"Shields [the flat barely covered rock of Canada, Scandinavia, and elsewhere] are here interpreted as crustal rocks denuded of sediments by thrusts of their original sediments from beneath the ice caps driven by the hydrostatic pressure and the friction of the ice flow" (quoting Cook).

"Welts" define pre-Cambrian rocks (that is, with slight signs of life), exposed at the surface as a result of uplifts and crustal buckling.

Huge troughs such as the Mississippi Valley are the result of an immense flow of turbid ice-laden waters and tidal flooding, so recent that spectacular anomalies such as the great New Madrid earthquake can occur.

Countless rubble hills are dumped in place by floods and wind from rocks expanded and broken up by earthquake. Most of the rubble orogeny has occurred in times of quantavolution, not by evolution nor uniformly bit by bit.

Uprisings occur through collision of rock masses, undercutting, compression, heat expansion of undercrust, and cooling of quasi-exploded material. Here would be included the igneous mountains of the world, such as St. Helens or Vesuvius. Here also would be earth that did not escape upon explosion and appears as mounds or hills swollen up (not buckled). Here too would be broad plateaus caused by a heat-expanded crust that cooled in its expanded form at great heights.

Finally, closely related to the previous item, are the submarine ridges around the world and the myriad seamounts (guyots). The ridge mountains, the world's tallest, are igneous productions, still bubbling and bursting along their length. The seamounts, as noted earlier, are the taffy-like pullback, unexploded lava blisters of the lunarian outbursts.

Quantavolutionary theory, then, holds that any hill and mountain of the Earth can be explained by concepts such as these. All involve energies that erase millions, even hundreds of millions of assigned years of time. Nor is it difficult, either, to imagine a quantavolutionary definition of other features not before discussed.

Where a gorge, a rift, or a canyon is observed, we are traumatized into seeing faults, fissures and turbulent waters rushing to shape them.

Where others see placid lakes, long ago hollowed from rock and fed from melting ice, we see land sinks, quick filling with avalanching waters, now stranded and in all shortlived.

Where deep surface deposits of clay, pebbles, sand, till and their associated rocks occur, we see tidal catastrophes, cyclones, and exoterrestrial fall-out.

In lava fields are seen, not occasional flare-ups after long-prepared mantle heating, but the rivers of boiling rock forced up and out by large earth movements and expansion.

Fan deposits are not gradual accretions at the foot of a flow, but sudden dumps by turbulent currents, and the continental slopes are the largest of fans.

Catastrophic winds, tides, and floods form dunes and peneplains, abetted by seismism.

Basins are formed and erupted by catastrophic uplift, changed Earth motions, or meteoroid impact explosions.

What is left to mention in the lexicon of landforms? We still have to do justice in succeeding chapters to several major Earth features: the ocean basins; the rifts, canyons and channels; and the sediments, including the continental slopes. Otherwise one is driven into sub-classification. Faults, for example, can be classified into tilts, grabens, horsts, and troughs and each of these is divided into sub-categories; these are treated in textbooks and present no unsurmountable obstacle to quantavotutionary theory. Each of these pertains to its parent-category – faults -and cannot supply something which the parent lacks. Metamorphic rock is of many kinds -schists, gneiss, limestone, marble mycorites and migmatites -and a natural history museum will present an orderly array of them.

The world "order" occurs again. And again the order is in our minds. The several conditions of heat and pressure and the several minerals that altogether manufactured these rocks were a disordered composition baking inside a faulty oven. One is seduced by the vast quantities observed of each type into imagining orderly production. A tall mountain of sedimentary rocks appears orderly to us, but so does the simple snowflake under a microscope.

One is impressed also by the very many material compositions and forms. But this is an illusion arising from the many different combinations which a few conditions and chemical elements can create; a mere eight separate states of being, described in terms of a temperature, a pressure, and a chemical compound free to combine, can, after all, supply some 2 8 or 256 entities to contemplate. There is order in all things and alongside this order there is chaos in all things; that is, we can look at any event or thing as orderly or chaotic, just as Parmenides looked at the permanence of being and Heraclites at its eternal flux.

Where in the world is the remaining virgin land of Pangea? If one is to believe surveys of the presence around the world of all the conventional geological ages, the answer is "practically nowhere." Perhaps 2% of the world's land can claim a full geological column. The ages are either a fiction, or the victims of quantavolution.

Still, even at this early stage of quantavolutionism, when few minds -and even fewer resources -have been brought to bear on the issues, it appears that by employing only a modest increment of time, quantavolution can move from the absurd to some respectable level of probable validity. One can comfortably and scientifically operate given an Earth age of a million years, with a late resurfacing of the Earth accomplished during the past fourteen thousand years.

It might seem impossible to reconcile the 5000-times-greater time span of conventional geological theory. Actually it is not impossible. The processes reflected in the Grand Canyon profile could be temporarily collapsed by a factor of 5000, making every five million years become a thousand years, without scrambling ordinary explanations. The rules to reduce time are:

increase heat; increase pressure; add motions; introduce electric potentials; and look into the skies. Says the sage to the astronomer, writes Friedrich Nietzsche: "As long as you still experience the stars as something 'above you' you lack the eye of knowledge."[10]

---

[10] *Beyond Good and Evil*, Epigram 71.

CHAPTER TWENTY-ONE

OCEAN BASINS

The planet Venus, which has been shown to have had its share of astroblemes, lightning activity, melting, volcanos, plateaus, mountain ranges, great valleys, and closed depressions or basins, has a "curious dearth of great basins," whence we surmise that Venus never underwent the trauma of Earth, which resulted in most of the Earth being ocean basins. "The tectonic forces that have shaped the surface of Venus have raised only 5 percent of the surface into 'continental masses' and left only 15 to 20 percent of it as basins... "[1] Nor do we know whether these are "real" basins, that is, distinct from the continental material as they are on Earth.

The ocean's "trackless wastes" may be a nice metaphor for the 71% of the Earth's surface covered by water, but the ocean bottoms are marked by enough signs to revolutionize the earth sciences and natural history. Essentially the ocean basins are three in number, the Pacific, the Indian and the Atlantic. The Pacific Basin was the recent scene of the most awesome event ever to have befallen the Earth since its early times, the outburst of the Moon. The Indian Ocean appears to have been created at the same time by the migration of continental land driven to the scene of the disaster. The Atlantic Ocean was rather obviously originated from a great wedge that helped propel the continents east and west so as to distribute the mass, heat, and electrical

---

[1] Richard A. Kerr, "Venus...," 207 *Science* (18 Jan. 1980), 291.

charge rather more evenly in the expansion and filling initiated in the evacuated areas.

The only possible mechanism for the lunar outburst would involve an exoterrestrial body, to which I have alluded on several previous occasions. The clearest description of the event and the closest to our own theory was provided by Howard B. Baker in an obscurely published article of 1952.[2] He was an American geologist, who from 1909 worked on the problem and completed a manuscript in 1932 that was never published. Both works were discovered by the present author after the manuscript to *Chaos and Creation* was completed; no changes were needed as a result, except to credit Baker for his achievement.

Baker stipulated an eccentrically orbiting planet, "Pentheus," as the intruder, and illustrated

> how by perturbative increase of orbital eccentricity alone, without any alteration of mean distance ... an orbit of mean distance 3 (astronomical units) might be so displaced that perihelion would be tangent to the Earth's orbit and aphelion well into Jupiter's danger zone, that is, greater than Jupiter's perihelion distance, which is 4.95...

> The planetary disturber is conceived to have been broken up by gravitational encounter with Jupiter, as suggested by Jeans (1934), and much of its ocean water, frozen with sand, gravel, and other debris, continued on a cometary orbit. The Earth occasionally met with these showers during the Pleistocene glacial epoch.

> The Roche limit, as explained by Jeans (1934, p. 269), is 2.49 times the radius of the larger of two bodies in an expanded or a contracted state as computed to make the density the same as that of the smaller body. With equal densities, the volume and

---

[2] Baker was born in 1872. In 1932 he mimeographed *The Atlantic Rift and Its Meaning* in Detroit. Fortunately, a copy reached the library of Congress. The article is "The Earth Participates in the Evolution of the Solar System," Detroit Acad. Nat. Sci., 1954 (pamphlet).

THE LATELY TORTURED EARTH – PART V.

mass are both represented by the same figure and are proportioned to the cube of the diameter.

Thus Earth, with mass 1 and diameter 1, and a radius of about 4000 miles, would encounter Pentheus with a mass 27 times greater, a diameter 3 times longer, and a radius of 12,000 miles. In this case, the Roche limit would be 2.45 x 12,000 or 29,400 miles from center.

As Pentheus progressed in its orbit it occupied a path 58,800 miles wide, within which no body much smaller could survive. The earth is conceived to have been deeply touched on the Pacific side by the Roche limit of the larger planet at the latter's perihelion... with the result that the Moon was born... If Pentheus were a mass of 64, radius 16,000 miles, its Roche limit would be of 39,200 miles.

Baker's model path calls for a two hour passby between 10 PM and 12 PM, at a perihelion velocity of 23.5 miles per second. A bulge distending the Earth appears at 10 PM with a tide raising power 225,425 times that of the Moon. At 11 PM the Roche limit is almost tangential to the Earth with a power of 816,818. At midnight, the tide power is 1,170,701 times that of the Moon and the Roche limit embraces the whole outbursting section of Earth, which then escapes into space. The Earth has lost most of its crust, but has gained water and a fall-out of rock.

Baker does not use the electrical power that would also operate effectively to the same end as gravitation. The distance might be several times farther given the same masses, if the intruder had come from afar bearing an electrical potential much different from the Earth's charge. Also the model proposed by this author is of a more gaseous and heavily electrified body. Its detailed treatment is available in *Solaria Binaria* and it ought perhaps not be discussed further in these pages, whose subject is the bottom of the oceans.

The prevalent view of sea-floor spreading has molten material exuding from the great oceanic ridge volcanos, pushing into place as a strip and jostling the older strips that compose the floor to move further away from the ridges. Ocean floor chronology and drift theory are based upon observations that from one strip to another, every several "millions" of years, there occurs a magnetic field reversal.

However, besides the other problems, which I have recounted, one core (395 A) from the Atlantic ridge flank shows magnetic differences in depth; the upper 170 meters is normally magnetized, the next 310 meters is reversed: and the following 40 meters is again normal.[3] This is an unwelcome surprise to chronometry and the theory of convection currents.

Still, pursuant to our theory here, we should expect erratic magnetic effects to accompany the great outpourings of lava; as soon as a batch is dumped off the ridge it hardens with the magnetic orientation of the moment. Very soon, before it has moved away, another batch is dumped on top of it, then another, all occurring before the whole thickness of lava moves far enough to be free of additional burdening. If, as we think, the ocean basins could mostly be paved in a thousand years, during which time the Earth's field would be moving geographically and oscillating, the laminated magnetic structure of the floor must follow.

Allen Cox points out that "if sea-floor spreading has occurred at a constant rate, the marine magnetic profiles may be interpreted to yield a reversal time scale going back 75 million years. The apparent average duration of the polarity intervals was greater during the time $10.6 < t < 45$ million years than during the past 10.6 million years, and during the time $45 < t < 75$ million years the average length was still greater."[4] That is, periods between reversals of the Earth's magnetic field occupy ever broader stripes or bands on the ocean bottoms as we go back in time.

---

[3] "Testing Vine-Matthews," *Open Earth* 3 (Apr. 1979), 28-9.

[4] *Geo Rev.*, 244; and see A. Cox R. R. Doell, 189 *Nature(* 1956), 45 which contains summery of paleomagnetic tests; and "Geomagnetic Reversals," 163 *Science* (17 Jan. 1969), 237-44.

Cox realizes that this might be an effect of an inconstant rate but dismisses the idea. With our larger theory that negative exponential rates followed a catastrophic opening of the basins, we find this data supportive. The ocean basins opened fast and then ever more slowly, giving the appearance of a magnetic field that used to reverse more slowly than it does now.

The Arctic Ocean scarcely deserves the name.[5] The North Pole area is flatter than the lands to the south and some miles lower than the swollen equatorial belt. If it were not so, there would be no Arctic Ocean. By far the greatest pan of the Arctic Ocean floor is continental shelf, less than 300 meters below sea level. There are half a dozen abyssal plains with depths from 2700 meters to 5000 meters. The Mid-Atlantic ridge forks northwards around Greenland and the two prongs come close together north of Greenland, then move in parallel across the ocean bed sandwiching the North Pole abyssal plain between them. A third "Alpha Cordillera" meanders northwest from the North Greenland regroupment, with many seamounts. The three ridges enter the continental shelf of northeast Siberia. They seem to disappear. But the Nansen Cordillera moves into the continental shelf in a great "Sadko Trough" and, precisely in line with it, some 400 km on, there begins the delta of the Lena River and a great valley, probably a rift valley. This rift cuts down through Asia ultimately to join the Indian Ocean ridges.

Throughout the Arctic ocean bed, the continental mass rises abruptly above the abyssal plains. Sheer cliffs of over 2000 meters are the rule. Although, on the one hand, a defender of erosionary theory would offer in explanation that the solid ice cover has preserved the "original" morphology, it may be argued that the fractures are new, occurred when the ice cap avalanched in Lunarian times and then were covered up during the Saturnian-Jovian age-breaking events that included a new ice cover, the present one.

---

[5] See *National Geographic Magazine,* map, "Arctic Ocean Floor" (Wash. D. C., 1976).

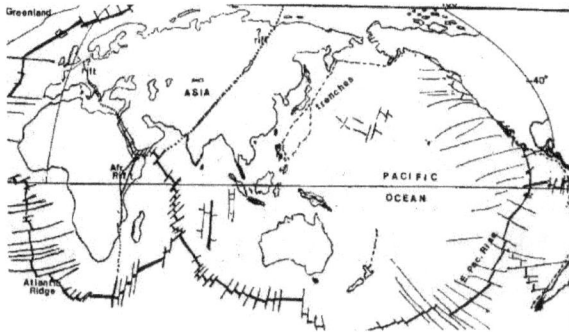

FIGURE A: Sketch of the main ridges and fractures of the Pacific Indian ocean bottom with main trenches. Possible Trans Asian and Trans-Euro-Mediterranean rifts are added to the drawing, which is adapted from O. G. Sorochtin, ed., *Geophysics of the Ocean (in Russian)*, vol. II, fig. 17. The lithosphere (crust) is everywhere shallowest beneath the ridge lines. Thousands of seamounts shooting up from the ocean bottoms are not drawn here.

FIGURE B: The Arctic Hemisphere, indicating the largely continental (rather than basaltic ocean-type) bottom; and the North Atlantic Ridge passing by the North Pole and proceeding towards Siberia, where possibly it becomes a land rift proceeding to the Indian Ocean via Lake Baikal. (Pages B to F are author's sketches. In all of them, the outlines of the full continents, including shallow shelves, are drawn.)

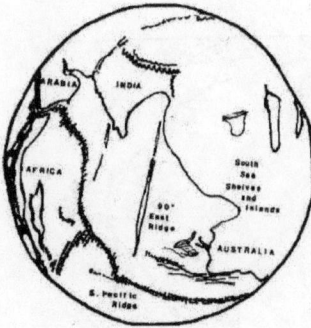

FIGURE C: The Indian Ocean Hemisphere, noting the African Rift on the extreme left, the East Ninety Degrees Ridge, and the largely continental rock platforms that underlie the vast Asia-Australia area.

FIGURE D: The American Hemisphere, noting how both the Atlantic and mid-Pacific Ridges follow the shape of South America at great distances. A world-circling Tethyan shallow sea belt may once have passed through Central America, the Mediterranean and the South Seas, but can hardly be discerned because the ocean bottom growth and expansion and crustal slippages have largely erased it.

FIGURE E: The Antarctic Hemisphere, showing how ridge-fracture cut the south polar Continent off completely from all land to the North, as by a circular saw. It would appear that the main fractures occurred before the main continental shift, (as in the Arctic Basin to the South), because there still is a semblance of order to their progression around Greenland and into Asia. Furthermore, Greenland adheres in shape to the North American continent and its neighboring western fracture does not seem to descend as deep as the eastern one. And on its East, Greenland seems conformable to the Scandinavian platform.

Semi-tropical, fully human cultures have been uncovered in islands only a few hundred kilometers from the North Pole. Iceland is apparently a high element along the North Atlantic (Reykjanes) ridge, volcanically produced.

It would appear that the main fractures occurred before the main continental shift (as in the Arctic Basin to the south), because there still is a semblance of order to their progression around Greenland and into Asia. Furthermore, Greenland adheres in shape to the North American continent and its neighboring western fracture does not seem to descend as deep as the eastern one. And on its East, Greenland seems conformable to the Scandinavian platform.

The simplest scenario for the mass movements that created the Arctic basin would call for a fracture, a swinging down of North America with a widening of the fracture valleys to create the abyssal plains. Northeastern North America was stretched out; Greenland and the many Canadian islands moved more slowly. Later Asia pushed northwards at one point in its generally southeast torque -the Yermak underseas Plateau -almost restoring contact with North America (Greenland) but letting the great ridge system pass through.

The total true ocean area created by and in consequence of the explosions and worldwide venting system amounts to 310 million km^2. Its depth averages 4 km. A floor of 1.24 billion km^3 would have been laid almost entirely in a period of about 2000 years. After that time, the activity of basin-evolution would begin quickly to subside. The basins would continue to evolve at a greatly reduced rate. Most of the vents would have become inoperative. The ridges and fissures are still expanding around the globe but at a scarcely discernible rate; like today, rarely would the oceanic surfaces be troubled by sea bottom volcanism and spreading.

A full life would have arisen in the warm oceans; the marine species of today originated in the shallow Tethyan waters. Men began navigating the oceanic surfaces now. Whereas ancient fossilized life-forms have been discovered on high mountains, they are absent from the bottom of the sea. It is presumptive, if not incorrect, for geological writings to state that the oceans have covered and uncovered the land on several lengthy occasions. The mountains have arisen from the shallow waters of Pangea, bearing the fossils, or the fossils have been laid down by flooding and tides, or they have been dropped by cyclones. The abysses of the ocean contain only species whose origins in shallower waters are patent. The oceans were born recently, and therefore hold only what has lived in these times. Heezen and Hollister, recounting the scarce record available of life on the ocean bottoms, conjecture that "either there was no abyss then, or the relics of these ancient seas have been completely destroyed. The deposits of earlier seas are found exclusively on the continents." To us it is clear that these earlier "seas" were the only "ancient seas" and were the shallow Tethyan seas and swamps.

The length of the oceanic fractures and their transverse fissures (transform faults) amounts to some 300,000 linear kilometers. The funnel volcanos number in the tens of thousands. The emission from a volcano cone can be given a value equivalent to 5 kilometers of fissure volcanism, and the number, never counted, can be set at 50,000. Then 550,000 kilometers of venting area was available to produce on the average 2,254.5 km 3 of ocean floor per venting kilometer within 2,000 years, or an average of 1,127 km 3 per year.

In two days in 1902, the Volcano of Santa Maria in Guatemala erupted and emitted 5.5 km 3 of material. A fissure of Laci, Iceland, part of the Atlantic northeast ridge forking, was quite active in 1783 and along a 25 km line emitted 15 km 3 of material in 4 1/ 2 months. In the tenth century an Icelandic fissure one year erupted 9 km 3 of lava alone along a 30 km trench. It is evident that if its activity were continuous at its full rate of eruption, the fissure of Laci would eject about 3,000 km 3 in 1000 years, 9000 km 3 in 3000 years, far more than its quota.

The figure used as a base requirement, 1.24 billion km^3, is twice as large as required. The underside of the ocean floor, comprising half the thickness of the floor, appears to be not a product of lava flow but a melting and cooling of basaltic rock in place. As the gaps widened, and the lava flowed to fill the chasm, the floor of the chasm at first softened from the heat all around it and from the waters, and then quickly hardened beneath the lava flows. This is but the cooled crust of the exposed magma of the mantle. When geologists declare, as does Shelton, that "... we cannot yet explain *why* magma exists where it does or seeks escape when it does,"[1]   they are not considering this kind of quantavolutionary and exponential solution.

The continents can be viewed as the rims of the ocean basins. They are steep-sided blocks, whether they plunge directly into the waters or have sea-covered shelves that then plunge down. The continental slopes, on the other hand, are water-covered moraines of continental debris laying on top of ocean abyssal basalt. They have a triangular profile, making nearly a right angle where continental block meets ocean floor; the hypotenuse is a lengthy stretch moving from the top of the shelf at an angle of 5 ° on the average. The declining rate of expansion of the ocean floor contributed to the profile of the slopes. By moving first rapidly, then ever more slowly, they heighten the illusion that a gradual off-flow of sediments has created the sloping figure. More likely exponentially declining rates of continental debris and sea bottom spread worked together to provide the profile. Deep river canyons extend hundreds of kilometers into them. Elephant teeth are found far out on the slopes at great depth; probably the slopes were laid down, occupied by terrestrial life forms, and then lately flooded. Deep turbidity currents, if they were to transport them, would bury them or destroy them. They lay near where the elephant died not long ago.

As told in the previous chapter, the continental slopes are free of continental mountains, as are the true ocean bottoms. The logical implications of this fact have evaded geology. If most great mountain ranges are new, whether by our chronological reckoning or by that of conventional geology, why have none appeared on

---

[1] *Op. cit.*, 69.

the continental slopes? The answer suggests itself: the mountains rear up at the edge of the precipices of the continents; they dump their debris into the abysses.

Immense floods and tides traversed the continents and poured off the miles-steep continental blocks into the ocean. The canyons occur where the blocks were fractured, and consequently where the waters poured out most heavily. The canyons, which will be treated soon in more detail, were not submerged beneath the oceans until the ocean basins stopped growing and their waters crept up upon the continental blocks and shelves. The seas do not come in and kidnap the land; they beat back the detritus and even build land. Thus the great slopes could not have formed under uniformitarian conditions or even underwater.

Prolonged, universal run-off of deluge and catastrophic tidal water produced slopes; the blocks were often towering water falls, dropping sheets of slurry into the abyss to form the slopes. The coarse gravel typical of the slopes far out to sea signals the impetuous rush and transporting power of the waters going to fill the basins. The scale would have dwarfed even the scene pictured by K. J. Hsh for the Mediterranean Sea (our dates and events differ, of course), "a giant bathtub, with the Straits of Gibraltar as the faucet. Seawater roared in from the Atlantic in a gigantic waterfall." If the falls delivered 10,000 cubic miles of seawater per year, they would have exceeded Niagara Falls 1000 times, and filled the Mediterranean basin in 100 years. "What a spectacle it must have been for the African ape-men, if any were lured by the thunderous roar."[2] The Mediterranean basin requires in its complexity an analysis that we cannot afford here. It appears to have been primordial, that is Pangean, and shallow. Then it may have suddenly closed and as suddenly opened, dry for a few years, and then overwhelmed by floods of water much greater than at present.[3]

---

[2] "When the Mediterranean Dried Up," *Sci. Amer.* (Dec. 1972), 33

[3] E. Smith 28 *Sea Frontiers* (1982), 66-74.

The ocean basins are composed of sima, rich in silicon and magnesium elements. They are of basalt. They are igneous, formed in red heat. They are thin. They are denser than the continental sial. The continents probably sit upon similar material, but much deeper, perhaps directly upon the upper mantle, save where the magma of the mantle may have expanded and intruded upon the continental granites.

The continents and the ocean basins are distinct formations that were produced at different times and by different mechanisms. The sial is old. The sima is new. The fact that the shell of the ocean bottoms is only one-tenth as thick as that of the continents in itself suggests that the ocean crust is the product of a melt, that the seas are new, and that the continents were somehow in a position to resist complete volcanism or explosion. The fact that ocean crust is more basic or less acid than the continental crust indicates that it separated from the primeval melt after the granitic crust; so says M. Cook.

The continents were produced by a cooling of the Earth's surface and by their own erosion and debris, and in direct contact with ultra-basic material of a heavier composition. Hence, the igneous marine floor does not cover a former continental surface, and density probes show this to be the case. Nevertheless, the floor probably contains continental debris in small amounts. With all the sinking of lands reported in legends, one would expect ocean-bottom drills to collect continental material here and there. Very little appears, leading one to suspect that most sinkings have occurred on the continental slopes or shelves.

The ocean basins are scarcely sediment; they hold only 1% of all sedimentary materials. Under uniform conditions, this would represent only 16 million years of runoff deposits amounting to 10 18 tons^3. Dissolved solids in the ocean waters compose 3.5% of their mass, far from making up the difference, nor can these solids be allocated to detritus removed from the continents.

Often the rocks are bare along the circumglobal ridges. They are 20 meters thick or less. The thickest ocean sediments are not on the basins proper but on the continental shelves and slopes. Further, next to these areas where the abyss begins, sedimentation is thicker and can reach 1000 meters in exceptional areas.

All of these oceanic sediments come either from cataclysmic off-pourings from the flooded continents, or from fall-outs, both volcanic and exoterrestrial. Material lagging at the end stream of the fission of the Moon might have dropped back to form islands of continental crust in mid-ocean. The time required for such sedimentation is calculable in a couple of thousand years or less under quantavolutionary conditions.

The character of oceanic sediments varies. It differs markedly from much continental sediment that is rock. It is clay and ooze. The shelves carry clay; the polar regions, the slopes, and some of the abyss carries ooze; and the deep abyss carries clay. The polar basins also carry sand and boulders.

Carbonates are heavy on the shelves and bottom oozes, but compose only from 2 to 10% of the clays (since they dissolve in the colder waters). Layers of distinct calcination and ash are interlarded with the oozes and clays in many parts of the world. An unknown proportion of additional ash has been incorporated chemically into the clay and ooze and remains to be distinguished. Much clay is igneous in origin, a product of volcanic tephra, volcanism, and cosmic fall-out. Much manganese has been precipitated onto rocks, pebbles, fish teeth, and bones over many areas, and pure manganese has been found on the bottom near the ridges.

The towering ridges that girdle the world have flanks that descend gradually. They present almost no underseascape for many hundreds of miles. There is no thickening of the ocean basin crust beneath the ridges, unlike the so-called isostatic thickening beneath the mountains of the continents, much of which is probably due to blunted thrusting. This occurs despite the fact that the ridges rise higher than the continental Alps. Thus they are distinct in origins, as was pointed out in the last chapter. The continental mountains were shaped by horizontal forces, with the intense, sporadic assistance of electro-gravitational forces from outer space. The ridges were formed by vertical forces from within the Earth, with similar assistance; unlike continental mountains, they lack rock roots, evidencing that they were not thrusted.

An impossible predicament is presented to conventional geophysics; how can uniformitarian forces produce this contrast? The continental crust folds and thrusts and compresses into abundant mountains; but the oceanic crust, having made its igneous ridges and seamounts once and for all, slides up and under and around without making mountains, but exudes lava in discrete amounts, and shakes seismically from time to time.

The Pacific Rise conforms generally to what one would expect from an exploded, as contrasted with a cleaved, basin such as the Atlantic. Worthy of quotation here is a passage from the *Encyclopedia Britannica* (my remarks in brackets):

> Fast spreading... as is characteristic of the Pacific [because the basin was already blasted out], produces a rise. Slow spreading... results in the formation of a ridge. Sea-floor spreading is a symmetrical process that accretes new ocean floor equally to both flanks of a rift; [The East Pacific basin obviously did not accrete symmetrically.] When a former landmass splits apart, the ridge maintains a median position as the newly created ocean basin increases in size. This phenomenon occurred in the Atlantic and Indian Oceans, but, in contrast, the rise in the Pacific did not rift a landmass when it was formed, and consequently there is no reason for it to be median. [Again, no land mass.]

A slash wound upon already swollen human flesh produces a swelling along the line, but a lower ridge than a single slash wound upon healthy flesh; so the Pacific rise is swollen high off the middle of the ocean bottom, and has a less marked ridge from the slash wound cutting it than the Atlantic basin has from its same slashing.

The Earth expanded as well as exploded; whatever can explode can expand: a chapter has been given over to this subject.

Although the Pacific Basin is concave, no one can examine a relief map of the Pacific Rise, for example, and say that the volume of the Earth remained unchanged thereby. Since this rise occurred, along with many other bulges, then a considerable expansion might be demonstrated by survey without resorting to theoretical physics. The globe has many slight bulges. Russian geophysicists have recently described its shape as formed by at least two geometric networks of lattices, a many-faceted figure.[4] So there may even be a pattern to the expansion of the global crystal. The latticework can be viewed as expansion joints; the total pattern makes the surface of the globe a set of convex plates rather than a perfect sphere.

Under the conditions imagined here, much of the expansion would be expressed simply in a hurrying of the basin-paving process, accelerated by inrushing waters. The salt of the dropping canopies would also promote magmatic melting. Molten lava takes up more volume than solidified basalt; wherever the crust was boiling, it would expand the surface of the globe. The tidal pulls of the Intruder, temporarily, and the new Moon, permanently, would draw the surface of the Earth outwards; there the surface would pause, *cool*, and harden.

The ridge mountain volcanos, and the ridge and transverse fault fissure volcanos, differ from tens of thousands of sea mounts, atolls and guyots that rose tall and slumped back upon the escape of the Moon from the Pacific Ocean Basin and smaller crustal material elements elsewhere. Some of these became instantly created volcanos and continued activity after the others had collapsed back. The Pacific seascape differs from the Atlantic by its incomparably more numerous holdings of seamounts. Morphological examination would indicate that the seamounts do not have the extensive piping systems of continental volcanos.

As the main blow struck and the fracture opened in North America, it drove that continent as a block southwestwards until it overrode the East Pacific Rise (fracture) that had just appeared

---

[4] Chris Bird, *New Age J.,* 36-41.

off what was now its west coast. Much of this western area promptly erupted into *volcanism* and was *covered* by huge lava flows and extensive, faulted desert plateaus and plains. The Asian and Australasian coasts and islands do not fit into the North American continent because vast spaces opened up and the whole arc from Alaska to Southern Asia broke away with the explosion of the Moon. A boundary ridge is not easily visible but extends down the Pacific basin on the West from Kamchatka Peninsula to the Campbell Plateau and ties into the Emperor Sea volcanic seamounts and the Line Island Ridge.

The Indian Ocean bottom, unlike the Pacific and Atlantic basins, appears to have been well-traveled. Antarctica has been shoved southward some hundreds of kilometers, and girdled by two great ridges. A newly discovered rift pierces the Waddell Sea, probably a transverse fault from the ridge to its north, and is lost under the great ice plateau hundreds of kilometers inland.

Australia has been ushered eastward by a fork of the same fracture that pushed India north and Antarctica south. Indeed, if one wishes an up-to-date definition of the continents of the world, useful for some purposes, one may say that a continent is a body of land surrounded by an oceanic cleavage. Even in the case of Europe and Asia, some believe the fracture to exist, going up from the Indian Ocean through the Persian Gulf, the Caspian Sea, the Ural Mountains and into the Arctic complex earlier described.

Contemporary geological theory has also traced the path of the Indian subcontinent from Southeast Africa to the Tibetan Plateau. "The vast Himalayan range was created when a plate of the earth's crust carrying the landmass of India collided with the plate carrying Asia some 45 million years ago, having travelled 5,000 kilometers nearly due north, across the expanse now occupied by the Indian Ocean."[5] The drift itself took much longer, since it occurred at the rate of 6 to 16 cm/year, if one were to accept the belts of magnetic reversals that mark the stretches of

---

[5] D. P. McKenzie and J. G. Schlater, "Evolution of the Indian" *Sci. Amer.* (May 1973), 63.

ocean bottom along the line of march and the dates given the lava from one belt to the next.

However, cores drilled into the Indian Ocean bed not far from the observed course produced gaps in dating of sediments by fossils of many millions of years, perhaps fifty millions in some cases. "Why these accumulations are missing," commented the directors of the survey, "is at present a mystery."[6] Fifty from a hundred million or so years is a big proportion. That "there are more gaps than record" is, of course, a familiar complaint among paleontologists on land as well as under the sea. In the other great basins, gaps of twenty million years in the fossil record are common. With sediments so thin, the gaps are not so important, say some -a turbidity current or two, and there you are. (Geophysicists and paleontologists can be catastrophists *à la minute,* when it is demanded of them.)

Still, when there is a gap in the fossil record of between 50 to 70 millions of years ago, we are speaking of late Cretaceous times and of the disastrous end of the dinosaurs and most marine species. A layer of unfossilized chert tiles the floor just above this zone, "as though some catastrophic development killed off most of or much marine life." One begins to suspect that the Cretaceous boundary may be considered as the primeval age of the ocean beds and that all which is found in the abyss arrived there afterwards; further, the finale of the Cretaceous may have been the end of Pangea and the outburst of the Moon, even if both are to be dated at a few thousand years ago. Almost all of the sea floor assigned to a date is Cretaceous or younger.[7]

We mentioned earlier that the Himalayas are agreed to have risen steeply within the last dozen thousand years. We called to the attention of Raikes and other students of the destruction of proto-Indian civilization that their "uplifts" were part of world-wide catastrophe. Today, the people of the southern Himalayas

---

[6] Sullivan, *op. cit.,* 172.

[7] See map in Sullivan, *op. cit.,* plate 22.

are suffering from a horrendous erosion of their soil. They are blamed for improper farming practices and overpopulation. This may be true enough, but, considering the youngness of this region, it is also fair to suggest that the Himalayan slopes have simply not existed long enough to have come sliding down on their own accord.

The cruxes of the internal activity of the Earth during the lunarian period occurred at two well-marked belts of discontinuity. One is the Moho discontinuity just below the shell in which the oceans and continents are fixed. Here the upper mantle boundary preserves an almost liquid character before it resumes a hotter but hardened condition farther down. This boundary of difference would scarcely be noticeable if it had not marked the torque and twist of the surface in the phases of shock and adjustment. For a simple unagitated melt produces, except in purely statistical terms, an undifferentiated transition of rocks up to the sedimentary level. The Moho boundary marks a breakdown of viscosity on a worldwide scale.

The second crux occurred at the 2900-kilometer-deep level of the lower mantle, some 50.0 kilometers before the upper core's boundary. Suddenly the density index, that had been moving at a fairly even rate of increase through the rocks after leaving the lighter crustal regions, leaps from 5.42 g/cm^3 to 9.91, a difference of 4.49 g/cm^3. This is about one-third of the total value of the scale, which begins at 3.31 and ends at 13.00 at the center of the Earth. No marked changes in pressure, gravitational intensity, or incompressibility are notable at this level. The largest secondary torque of the globe in reaction to axial displacement and rotational torque and retardation happened here. Lesser torques occurred at the 400, 1000, 5000 and 5100 kilometer depths where seismic discontinuities are observed.

Several points deserve stress in reviewing what has just been said and looking ahead to the next chapter.

An immense part of the Earth's shell is simply missing. It had nowhere to go except into space, for it cannot be decomposed, mixed with plutonic material, or shoveled under the sea bottoms.

A psychological fallacy pushes us to believe that the ocean basins were made by and for the primordial waters. That the basins exist is one accident; that waters fill the basins is another accident. The accidents added up to a "miracle" of good fortune for mankind. Better near extinction than a totally frozen or drowned globe. At first, the waters were below the rims of the basins; now they slop over the rims.

The East Atlantic Basin corresponds to the West Atlantic Basin. Their former juncture is plain. The Pacific Ocean is deeper. The East Pacific Basin is sharply marked where the southern, western and northern margins are arranged as a giant set of arcs detached from a blasted area.

All the ocean basins are young, thin, and scarcely sedimented. Oceanographers who recently discovered these facts were amazed; in a few years, the basins became four billion years younger. Only potassium-argon datings, which are vulnerable to catastrophic events, let the bottoms achieve even this young age. The basins are not 200 million as against 4,500 million years old. They maybe only aged a dozen millennia. The surprise is greater: not one-thirtieth as old, but one-ten-thousandth as old.

Three additional, related points are stressed in other works of the author and have only been mentioned in this book:

Despite the almost total destruction of the biosphere by heat, explosion, suffocation, and famine, many species survived. Marine life soon found vast new breeding grounds. So did plants and land animals. Even before they were drowned in the later deluges, the cooled seamounts harbored many forms of land life on their summits.

Horrified, stunned, fully human beings saw all of this happen. Wherever archaeology finds "Paleolithic" and "early Neolithic" sites, it finds not slow soil coverings but fast disaster coverings. Much legendary and physical evidence points to a newly emplaced Moon and a worldwide catastrophe about twelve thousand years ago.

The network of fractures around the world is unitary. Mechanically it must be considered as the effect of one and the same event. The Moho discontinuity recorded today beneath the Earth's shell at from 5 to 50 kilometers depth may denote where the shell rafted and where it was peeled off. The next two chapters deal explicitly with the fracture and rift system of the world.

CHAPTER TWENTY-TWO

FRACTURES AND CLEAVAGES

In the past few years, the public has become well aware of the revolution in oceanography, a major element of which was the uncovering of an immense integrated global fracture system. It is a kind of reverse harness which works from the inside instead of the outside to control natural behavior around the world. The question is whether the harness emerged from deep within or whether the globe was harnessed by an exoterrestrial force. Except in westernmost North America, in East Africa and the Near East, through Iceland and Central Asia, through the Adriatic-Rhine River rift, and beneath India, the fractures course below the sea, where they are rendered visible by the ridges running alongside of them. Many years ago, De Lapparent and Howard Baker had recognized the oceanic rifts and called them recent, while Heer had assigned the boundaries of the Mediterranean to the era of the drift.[1]

The system is worldwide. It may be said to begin in the arctic region, moving south from both sides of Greenland. It shoots down to the Antarctic region, forks west and east, and forks again north and east. The east fork traverses the South Pacific and rises northwards when it strikes South America, proceeding up to and around the North Pole where it is reconnected with the northward fork that has shot up through the Asian continent via the Persian-Indian coast, Lake Baikal, and northern Siberia. There it probably connects with the fork around

---

[1] Beaumont, *op. cit.,* 190, 197. For Baker see the preceding chapter.

Greenland, completing a circuitry of the globe. Less apparent is a worldwide rupture that carries through the East-Central Pacific, Caribbean, Mediterranean, and South Asian areas, possibly a fracture along the line of the old Tethyan Sea equatorial belt.

The present globe does not portray the original situation. A Pangean globe would show nothing but land and shallow seas. Today's named areas stood unbroken. The globe then was without ocean basins. Its main body of water was the Tethyan Sea, corresponding to the present Caribbean, Mediterranean, and trans-Pacific northern tropical region. This was the equatorial region. The South Pole was bounded by lands now disappeared, unless New Zealand and a few other continental areas are remnants of them. The continents of South America, Africa, Australia and Antarctica were far to the North, and part of the Pangean land mass.

The fracture originated at the old North Pole and proceeded rapidly towards the old South Pole, bending as the north geographical axis of the globe shifted to the northeast and as global rotation slowed and resumed. The globe must have jerked suddenly as the Atlantic cleavage passed through what are now the Brazilian and African humps, and then resumed its more direct southerly course.

The polar ice cap is said by Weyer to have shifted its position by 10 to 15 degrees along a line 60 degrees west and 120 degrees east.[2] Possibly the cap was cleaved and the rift began running; then almost immediately the Intruder began to cut its swath from the Pacific crust and staggered the Earth to a momentary pause, driving the rift eastwards in Mid-Atlantic.

When the fracture reached the South Pole, losing momentum but cleaving the Earth rapidly at the full depth of the continental crust, it veered sharply eastwards slicing through the then polar south region until it met with the westward shifting "American" continents, whereupon it veered northwards until it reached the northwestern fork of the north polar fracture. It

---

[2] V *S. I. S. R.* 2 (1980-1) Discussed by Warlow, 34-5.

skirted the eastern rim of the great pit of the Moon material that had been blasted up and away.

A secondary forking sent the fracture northwards shortly after the south polar fracture occurred, slicing through "Africa/ India," then, after crossing the Tethyan fracture, resumed in diminished depth its course across central Asia.

Meanwhile the initial point of rupture at the old North Pole sent a forking movement northeast and northwest, isolating Greenland. Both of these fractures joined the trans-Asiatic fracture at different points.

Earlier, as the main "Atlantic" fracture encountered the equatorial Tethyan area, it incited a trans-world secondary fracture, that moved more rapidly east than west. The western Tethyan fracture cut through the continental mass then occupying the Gulf of Mexico and lost itself in the inchoate molten mass occupying the blasted crater of the fissioned Moon material. It may scarcely be perceived to end at the West Pacific Rise (rupture). The eastern thrust moved, however, through the "Mediterranean" and "Near East" then through a blast area which soon was overrun by a jumble of lands moving southwards.

Finally, major rifts struck out from the Tethyan fracture north and south. On the south a Mediterranean and a Syrian fracture join the Red Sea rift and continue south across East Africa to join the proto-Indian fork. In proportion to a number of submarine fissures, this rift was a moderate addition to the world fracture system. Africans of the Rift countries retain legends of great structural changes in their land. To their stories are to be added similar Arab and Hebrew stories.

From the beginning to the end, the fracture system might have been the work of a day; geophysicist Cook speaks in terms of hours. It conceivably inspired the "Third Day", during which "God created the oceans" in Hebrew story. "And God said, 'Let the waters under the heavens be gathered together in one place, and let the dry land appear. '"[3]   At the end of the day, the

---

[3] Genesis I: 9 and fn *Oxford Annotated Bible* (NY 1965), 1.

continents had been carved out, many islands had been sliced along the Tethyan way, the Antarctic region, the arctic region and the "East Pacific" area. The continents were in motion. The Earth was girdled by chasms and ready to move and expand. Pangea was ended. The climax of chaos had passed.

Two characteristics of the world fracture system deserve much more attention than geophysicists have allowed them. Only Cook, to my knowledge, has frankly expressed what is so apparent, that the total system was the work of hours; perhaps he could utter the shocking sentences because he had won a Nobel prize for his work on explosives. That our precious globe could be treated so abruptly and cruelly is inconceivable to most people; it is like an innocent child coming upon the scene of an autopsy. Cook remarks that "there is evidence for the hexagonal structures characteristic of shock fracture..." This is no less than what many geologists have been trying to say in the "tectonic plate" school of thought and the Russian "crystal grid structure" theory that C. Bird has described, all hesitating to give voice to the necessary implications.

Cook goes on to add the clause, "but this evidence is by no means perfect." He may be saying this because he does not deal with the two essential components of the epoch-making event, the intervention of a great exoterrestrial body and the blasting of the Moon from the Pacific Basin. These elements of the scene tend to obscure what would otherwise appear as a more normal hammer fracture of a solid crystal globe in rotation.

The Antarctic continent (including the continental shelf of the Ross Sea) is steep-standing in its surrounding ocean. About half of it executes a remarkable circular tour, from 0 ° to 180 ° east. The other half presents a more jagged coastline, deeply retracted from the imaginary circumference of the eastern arc. Opposite the uniform half circle are the continental masses of the world. Opposite the retracted half of the continent occurs the South Pacific Ocean, between New Zealand and South America, where by our theory the Moon was drawn forth.

The Antarctic continent, we surmise, must have been located north and east, and its south and west side was the limit of the exploded crust. Its north and east portion was broken off from the neighboring continents by the forking of the Atlantic fracture, east and west along the circular arc, and had just been isolated to its west and south by the lunar explosions. Forced down by the fracture and up by the new abyss, it settled centrally over the new South Pole, contained there by lava flows from all directions. Its slopes are heavy with debris, indicating that the separation and explosion happened when the continent was ice-free and/ or that an ice cap, if there, melted catastrophically. The lack of fossils more recent than the Cretaceous in Antarctica seems to pose a challenge to short-term time reckoning in quantavolutionary theory. If the terminal Cretaceous was the time of lunar fission, however, the lack tends to confirm the theory. Thereafter Antarctica was isolated.

The puzzling fractures of the Pacific basin north of Antarctica invite puzzling quantavoluntory assumptions. The Nazca Ridge and its associated seamounts moving west off northwestern South America find their mirror image in the Tuamotu Archipelago far on the other side of the mid-Pacific Rise (Albatross Cordillera). Also the whole of the western coast of South America conforms in shape and fit to the same cleavage. The cleavage image is shifted southwest.

Are we seeing double? These features must have originated together. The Rise must have pulled away from South America faster than South America, impelled from the east by the Atlantic, moved to follow it. This is understandable if the Rise had no crust, but a yawning basin, to its own West; meanwhile it was being pushed reactively by its own east side lavas as these were blocked and pushed by South America.

Farther North, the Rise loses itself in the great transform fractures of what we call the Tethyan Belt and is then overriden by the North American continent which has been shifting southwest with the opening of the Arctic and North Atlantic Oceans.

A second matter calling for attention is the form of the fracture system: the ridges move rectilinearly with sidewise steps and with a great many perpendicular fissures. Perhaps the successive torques to which the globe was now being subjected shifted the main line of fracture. Every time there occurred a glitch in the crustal velocity of rotation, the main fracture line would shift to the East. At the new equatorial belt, a great shift to the East is observable. Several more 'glitch-points' occur before the fracture cuts through Africa-Antarctica and then, perhaps because the slowdown of rotation had terminated, the sidewise steps are no longer in evidence.

Nor are the transverse fissures any longer apparent. The long east-west fractures seem independent of the main ridge. Instead, passing now for the rest of its journey through evacuated surface, the major fracture, in its bifurcation, is accompanied by myriads outbursts of lava mountains, the seamounts. Seamounts occur in large numbers along the Atlantic ridge and in various evacuated regions of the basins. A close statistical analysis may ultimately use the seamounts as indicators of torque, time of fracture, velocity of the land masses, and other events, now quite obscure, of this period.

The striking conformity of the Mid-Pacific Ridge with the shape of South America and its passage beneath western North America persuades us that the original continental land on the east of the fracture is still there. But there exists no sial continent west of the Rise, that has any kind of morphological association with it. There is no well-defined boundary of the oceanic expansion to the west, nothing to compare with Euro-Africa on the other side of the Mid-Atlantic Ridge, no well-defined boundaries such as tie South America, Africa, and Australia to Antarctica. The "Circle of Fire," which marks an arc of volcanism and seismism from Southern Chile to the Aleutians and down through Japan, stops; so the Circle is not a circle and not a fitting image of the American side either. The morphology of the basin of the Pacific would be an incredible coincidence, a gross improbability, without our positing the disappearance of its entire crustal covering west of the Rise.

Attempts to produce a unified time-scale for the spreading away from the ocean ridges have not been successful. Heirtzler and his associates found that relative to the time scale for the spread of the South Pacific, the North Pacific time scale was in error by a factor of two. The principal technique employed has been potassium-argon radiochronometry. Nor do the spreading patterns moving from the ridges around the world agree on the location of the North Pole around which presumably they would evidence rotation. Still, because of similarities in the spreading pattern of widely separated regions, it is believed, and we think rightly, that the spreading of lava was a universally concurrent phenomenon. The similarities, where they show a discontinuous floor-laying off of one ridge show the same off of another; such similar discontinuities connote simultaneity.

Earlier Cook (1963) had advanced evidence for the recent rupture of the continental crust that would probably have erased most of the perplexities just evidenced.[4]

1.   The uplifts observable in Fennoscandia and Northeastern America *"began at the same time and followed essentially the same relaxation equation. This equation, derived by Vening-Meinesz, is an exponential rise equation characteristic only of a sudden unloading of the crust followed by a normal relaxation."*

2.   The maximum depression at the center of the ice cap was along the seashore where presently stand Baffin Bay, Davis Strait, and Labrador Sea.

3.   Without the missing land in these northern areas, no great ice masses would have collected: "the ice would simply have rolled off into the sea." The seas of the region could not have existed prior to 10,000 years ago.

4.   The uplift data fits into the extended fracture that thereupon moves down the Atlantic and around the world.

---

[4] 40 *Proc. U. A. S. A. L.*, part I, *op. cit.*, 74-7, also in *Prehistory...*

5.　　　The ice stored in the ice cap is calculated as equal to providing the water that would fill the Arctic and Atlantic basins.

6.　　　A "Great Arctic Magnetic Anomaly" defined by E.R. Hope from the magnetic remanence of crustal rocks exhibits "a *surface* dipole magnet in the North Pole region." One apparent pole is in Northern Baffin Island, the other offshore from Severnaya Zemlya in Siberia. These two apparent poles appear to be at one and the same location, if the two separated lands represented are pushed back together at the location of the former pole. "In other words, it would only be necessary to return all the land masses in the northern hemisphere to their original position by reversal of the process described by Du Toit [the splitting and rafting of the Arctic crust] in order to completely remove this magnetic anomaly."

Two sets of conditions governed the occurrence of the world-girdling fracture and the Earth's expansion. The first condition of fracture is an unevenly applied pressure on a shell. The shell is the Earth's surface down to a level which presently can be called the Moho discontinuity but which in the Age of Pangea was the point when the coolness of the Earth's surface disappeared into the mounting temperature of the crust and mantle, caused by primordial rising heat convection from the center, by pressure from the rocks above, by radioactive flow blocked from emission by the surface charge of the Earth, and by the greater centrifugal force of rotating material of greater density than the surface material.

The unevenly applied pressure consisted of ice caps rapidly formed in the thousand years before by falling ice and icy waters; these did not need to be melted at the equator, then raised by evaporation, then blown to the north, and then dropped again. They contributed directly to the ice cap, and to such an extent, that shortly there formed a tall mass of ice covering Pangea around its North and South Poles.

If the Earth had not had its magma sources opened up by fission, fracture and expansion, it might have been frozen completely over. Great depressions were formed in the rocks, depressions which have still not relaxed after 12,000 years. This pressure was a mechanical potential exercised around the circumference of the ice bowls. The fact that oceans did not exist permitted a much greater piling up of the ice caps, for a deep water basin cannot hold the same amount of ice.

A second condition of fracture is a formation that can be split. Millions of geological faults of the Earth attest to the potentiality of rocks for splitting and shearing. If the body to be split is spinning, the slightest delay in spin along a line of fault will drive the one side of the fault away from the other side. The centrifugal force in the Earth's rotation achieves this.

The setting up of a massive horizontal circular pressure against weaker rock and the resistance of denser and stronger rock below incline the potential event towards a split rather than an implosion or collapse. The buildup of ice will continue until the horizontal walls will give way through folding and thrusting. But the ice mountain does not thrust over because it is sunken in, with the form of a cap.

The horizontal strain to the depth of the ice cap causes a continual heat at its edges. It leaks water, but accumulates more ice than it loses. The heat augments below, too, from the pressure of the ice upon the non-basic sedimentary rock and granites below. These grease the cap undersurfaces.

If there were now to be a sharp blow upon the center of the cap, the cap would crack radially. In addition, the weakened crust beneath the cap would give way. The Earth's axis of rotation would be tilted to meet the first blow; the Earth's spin would take up a new figure with an axis towards the direction from which the blow came. The fracture would leap out of the blow and race around the globe in the manner described above.

All of these conditions were fulfilled. The blow struck. The hammer could have been a lightning bolt from an Intruder from the northeast. At a distance of a million kilometers, it began to agitate the space sheath of Earth. The axis of the Earth tilted to meet it. The bolt struck the ice cap and sent radial fractures in all directions. At the same time electro-gravitational force was applied, with particular stress upon the pole, wrenching the Earth by its cap against its rotational direction. Earth's rotational velocity slowed sharply.

All lines of weakness were stressed. The Globe shuddered from the blow and fractured deeply. The eastward rotation of the Earth sent the deep fracture rushing down the "Atlantic" and "Pacific" sides to the other end of the spin, the South Pole. The Intruder swooped closer and passed over the Southern Hemisphere, the "Pacific Basin," where it flayed the Earth of half its crust, and then passed on. The crustal debris shot up into space in pursuit. Most of it turned aside and became the Moon. Some fell back to Earth, now and in the succeeding years. After centuries of a ring of debris, the Moon was fully assembled. The Earth came to see the new great light and the Sun and other planets as well.

The globe was probably spinning east before and during its exoterrestrial encounter, and the Intruder apparently approached from the northeast. Thus a swath of crust was removed that began narrowly in the North, barreled out at the epicenter of the encounter in the Pacific Basin, and continued to explode for thousands of miles until it passed into farther space. The "crater of the Moon" was elliptic in form.

Because of its possibly being remembered and because of its continental geography, the great Rift Valley of East Africa might be recalled for discussion. Viewed from the south, it appears to begin where the Island of Madagascar was detached from the African continent, proceeds north, bifurcates, resumes a unified path and leaves the continent at the Afar Triangle, thence moving northwest below the Red Sea, bifurcating once more to pass up the valley containing the Dead Sea into Syria (where it loses itself in the jumble of mountains observing the burial of the old Tethyan Sea and Tethyan welt that is moving generally west and east; the western bifurcation is questionable, but is likely to pass across the Mediterranean, up the Adriatic, beneath the Alps, and out along the Rhine graben that ends far to the northwest beneath the North Sea.

Arabia fits cleanly into Africa across the Red Sea. Why the Rift should turn northwest at this point may be explained by the westward thrusting Gulf of Aden-Indian Ocean faults, which have sent out a powerful arm in this direction, thus reinforcing each other and cutting a neat right angle around the Arabian Peninsula. The narrows where the Gulf of Aden enters the Red Sea are called Bab-el-Mandeb, the straits of tears, after the legendary devastation the rupture caused. The Olduvai Gorge and Afar Triangle, whose hominid fossils have been assigned ages up to 3.5 million years, sit upon the Rift Valley, which is kilometers wide and houses its own world beneath the towering plateaus and mountains abutting it.

Opinions differ as to the age of the African rift. That it has been active in human times seems evident from legends and excavations. Its origins have been set as far back as 2.7 billion years, however, by R. B. M. McConnell, speaking most directly of the 4000-kilometer section from the Red Sea to the Zambezi River.[5] He speaks of "transcurrent movement" between more ancient shield rocks, but also of "perennial" reactivation. So eminent an authority as Flint accorded the Rift an origin in the late Pleistocene, well within our ken.[6]

If India and Madagascar were dissociated from the continent some 100 million years ago, as is currently believed, certainly the Rift would have been strongly activated then. Also, if Africa and the Americas had separated not long before that time, then, too, the Rift would have been agitated. The great platform that hovers above the Rift might represent the kind of worldwide swelling expounded earlier as an accompaniment of the general global cracking.

From the standpoint of this book, the arguments giving a long history to the Rift are worth no more or less than the arguments for long time-scales elsewhere in the world. In *Solaria Binaria*, which is primarily a work in astrophysics, the age of the Earth's rocks is put at less than a million years; in this work, which

---

[5] R. B. McConnell, 83 *Geol. Soc. Amer. Bull.* (Sep. 1972), 2549-72.

[6] *Glacial Geology*, 523; *Glacial and Quaternary Geology* (1971).

concentrates upon the recent reworking of the Earth's surface, we are not interested so much in the older rocks as in their recent upheavals.

In this context, we see the swift movement eastwards of the African continent and the lifting of its great southeastern plateau region as concurrent. The Rift had already happened; two masses were pulled apart in the global fracturing; but reactive pressures from the even larger fracture to the east, now below the sea, compressed the Rift and let the dropped rocks fall only a small distance before halting, trapped as they are today, covered with lakes, volcanic ash, and plains.

The Olduvai Gorge has been assigned 200 million years; it was then a late fault branching off the main faulting of the Rift. If it is so old, it becomes difficult to explain the hominid and mammal fossils protruding from its walls. They could not be cliff-dwellers; so the Gorge must be younger than they. How young they are is in question; the legends of heavy rift activity weigh upon our mind, and there is a variety of evidence that the hominids may be much younger, material that is treated by this written *Homo Schizo I*. The evidence extends to the Afar Triangle, a flat land-fill actually, born of the pull-out of Arabia, where related hominids are found. It also extends to the Palestinian portion of the Rift where Olduvai types of hominid sites are discoverable.

The Gulf of Aden and the Red Sea, we have said, seem to have been produced out of sharp lateral faulting shifting the end section of the Carlsberg Ridge of the Indian Ocean northward. This might indicate that the total area east of the Owen Fault Zone, including the African Rift-Gulf of Aden-Red Sea rift occurred at the time of or only a little later than the globe-girdling rift of which the Carlsberg Ridge forms part.

Further activities of the Rift advance into proto-historic times, particularly into the Bible. The occasion of the destruction of the Cities of the Plain, including the story of Sodom and Gomorrah (see below, Chapter 29), treat specifically of the same rif T.M. Blanckenhorn placed the age of the Syrian section of the

Rift in the early glacial period.[7] W. Irwin retrojected the influx of magnesium salts into the Dead Sea, on uniformitarian principles, and arrived at a 50,000-year approximation of its age.

Velikovsky gives several reasons for reducing this age drastically, and estimates both the Dead Sea and Jordan Valley have an age of 5000 years. In all the disastrous effects of the biblically described destruction, a sea is not mentioned; yet when the Israelites under Moses and Joshua arrived on the spot around 3450 years ago they encountered the Sea.

The Jordan River, argues Velikovsky, had changed the direction of its flow, too. "Prior to the Exodus, the Jordan Valley was on a higher level than the Mediterranean Sea. With the rupture of the tectonic structure along the river and the dropping of the Dead Sea chasm, many brooks in Southern Palestine which had been flowing to the south must have changed their direction and started to flow towards Palestine, emptying into the southern shore of the Dead Sea." Legendary references indicate that heavy bursts of lightning were involved in the production of fire, smoke, and sulphur, whether by cosmic stream injections in which the planet Jupiter (Marduk in Babylonian, Zedek in Hebrew) is insistently implicated, or by subterranean upheaval along the rift (by no means excluding an exoterrestrial prime mover).

Allowing therefore that some of the major rifting of the Earth occurred as late as several thousand years ago, we conclude this chapter. All of the great rifts of the world are connected in time and by cause. They form a system that harnesses the world to the recent fission of the Moon. The individual histories of the sections of the world fracture system are insignificant by comparison with the common historical experience of the whole. The system functioned to balance the world by redistributing the crust and by expansion and to vent gases and heat during the

---

[7] Velikovsky, "Destruction of Sodom and Gomorrah," VI *Kronos* 4 (1981) 49, and see the accompanying note by Frederick B. Jueneman; also J. E. Strickling, "Sodom and Gomorrah," 2 *S. I. S. Workshop* 4 (1979), 3-5.

process. The climactic event was tangibly sensed by the Pangean Earth days in advance; it occupied a day in establishing the new morphology of the Earth-Moon system; thousands of years were required for its major effects to devolve into the processes recognizable in the world today.

CHAPTER TWENTY-THREE

CHANNELS AND CANYONS

The model river channel combines the history of an earth fault, a catastrophic torrent, and an erosional runoff bed. Most large rivers, perhaps all of them, are children of Okeanos, whom the Greeks called "the Father of Rivers" who personified the sky waters before the first deluges, as we said in Chapter 13, and then came down to Earth. His children carried his waters into the new ocean beds.

Many myths appear to conjure rivers where none exist, and, of course, a great many dry river beds of once tremendous rivers are to be found around the world. During the Universal Flood of Deucalion, a small chasm was said to open in Athens into which the waters emptied. According to Lucian the people of Hieropolis (near Aleppo, Syria) "say that a great chasm opened in their country, and all the water of the flood ran away down it." Again, myths warrant hypotheses. In the Volta River Project (West Africa), a onetime shallow river bed was suggested by a deep river bed with a jagged bottom. Local legends spoke of upheavals in the now quiet area, and when the water was lowered prior to constructing a dam, several protuberances became islands, and at a depth of 35 feet revealed carvings whose age was estimated at 3000 years.[1]

The Po River is probably an extension of the African-Rift-Red Sea-Rhine rift valley that connects with a buried rift in the Adriatic Sea. It carried down the immense debris of the sudden

---

[1] Anon., 229 *Nature* (5 Feb. 1971), 371.

uplift of the Alps. It may be the ancient sacred river, the Eridanus, of Greco-Roman legend, long-lost because later a sea. The Po serves in truncated form to water and drain the Po Valley. The Rhine River picks up the graben northwest of the Alps, and moves it far out into the North Sea; not long ago, it shared its burden with a westward flowing river that was then naturally dammed so as to reduce the Loire River of France to more modest proportions.

The Colorado River may be a ramification of the East Pacific Ridge, that runs up the Bay of Lower California and strikes through the desert into the raised platforms of the southwestern states, abetting the disintegration of the Rocky Mountain uplift; once its tectonic work was done, it began its present work of erosion.

The great rivers of China flow in the direction they do, says a Chinese myth, because the goddess Niu-Kwa made the waters of the great flood stream off towards the southeast; the whole Earth had tilted and sunk into the sea there.[2] Most great rivers of the world understandably conform to the processes set into motion by the lunarian outburst. Many hasten along courses conveniently provided them and their tributaries by fractures, the Rhine, the Colorado, the Susquehanna, the Indus, the Congo and others.

In decoding the natural history of river beds, geologists fighting the ghosts of catastrophism have refrained from extremes. M.G. Wolman and J. P. Miller in 1960 essayed an analysis of the "Magnitude and Frequency of Forces in Geomorphic Processes."[3] Using mainly four rivers as their cases, they conclude that "dwarf" gradualist forces operate steadily to perform most transport of sediments, that "man-sized" moderate forces of brimming "bankfull" waters supplement the "dwarf" work in carving banks and valleys, and depositing sediments, thus accounting for perhaps 90% of the changes effectuated. The rare

---

[2] Bellamy, *M. M. M.*, 261-2.
[3] 68 *J. Geol.* (1960), 54-74.

work of "giants" make up the balance, including many switches off channels and movements of erratic boulders.

Unfortunately, they lack respectable data over time even for these "giant" events, which they estimate at 50-year intervals; yet they call them catastrophes. Like the experts on seismism, their extremes are historically confined to what no one doubts have been uniformitarian times. Of course, then, they must pass over with the weakest of scenarios the grand metamorphism and concentrate upon pygmy processes playing out recent history. They realize that they are dealing with exponential, logarithmic processes, but excise the peak curves. In the only concession to longer history, they murmur at one place about "materials inherited from a period of greater stream competence which possibly existed during glacial times." As we have noted, "the end of the ice ages" is a cover-up fiction of all that has happened to the lately tortured Earth.

Not alone of river channels do they speak but also of beaches and winds. With regard to beaches they introduce the commonly accepted concept of an "equilibrium profile." It is "an average form around which rapid fluctuations occur. Waves from storms may periodically destroy the equilibrium form, but over a period of years there is an average equilibrium profile by which the beach may be characterized." The more meaningful question is where does this profile come from in the first place -these millions of profiles, we should add, unique in themselves but in distribution worldwide? Where is the "supergiant's" place, that smashed out the profile to begin with, in the analysis and theory. As for the effects of winds upon river and beach morphology, many analyses, they say, "indicate that a log-normal frequency distribution of wind velocities is a general rule." The log-normal winds, like log-normal river flows and sea waves are what recent experience and the authors give as "log-normal"-curves that rise scarcely enough to make their uniformitarian hearts skip a beat.

Their last paragraph is naive, but so unconsciously significant as to be worth quoting:

Perhaps the state of knowledge as well as the geomorphic effects of small and moderate versus extreme events may be best illustrated by the following analogy. A dwarf, a man, and a huge giant are having a wood-cutting contest. Because of the metabolic peculiarities, individual chopping rates are roughly inverse to their size. The dwarf works steadily and is rarely seen to rest. However, his progress is slow, for even little trees take a long time, and there are many big ones which he cannot dent with his axe. The man is a strong fellow and a hard worker, but he takes a day off now and then. His vigorous and persistent labors are highly effective, but there are some trees that defy his best efforts. The giant is tremendously strong, but he spends most of his time sleeping. Whenever he is on the job, his actions are frequently capricious. Sometimes he throws away his axe and dashes wildly into the woods, where he breaks the trees or pulls them up by the roots. On the rare occasions when he encounters a tree too big for him, he ominously mentions his family of brothers -all bigger, and stronger and sleepier.

In their last sentence, they suggest the truth as in a dream. This should be the extreme dimension of their theory, accounting for the largest facts before their eyes. Thus the larger catastrophic origins of the morphology under examination are excluded.

A century ago, geologist Clarence King was describing the river system of the Pacific coastal area of the United States.[4]

A most interesting comparison of the character and rate of stream erosion may be obtained by studying in the western Cordilleras, the river work of three distinct periods. The geologist there finds preserved and wonderfully well exposed, first, Pliocene Tertiary river valleys, with their boulders, gravels and sands still lying undisturbed in the ancient beds; secondly, the system of profound canyons, from 2000 to 5000 feet deep, which score the flanks of the great mountain chains, and form such a fascinating object of study, and not less of wonder, because the gorges were altogether carved out since the beginning of the glacial period; thirdly the modern rivers, mere

---

[4] 311 *American Naturalist* (August 1877), 449-70.

echoes of their parent streams of the early Quaternary age. As between these three, the Early Quaternary rivers stand out vastly the most powerful and extensive. The... present rivers are utterly incapable, with infinite time, to perform the work of glacial torrents. So, too, the Pliocene streams, although of very great volume, were powerless to wear their way down into solid rock thousands of feet, at the rapid rate of the early Quaternary floods. Between these three systems of rivers is all the difference which separates a modern (uniformitarian) stream and a terrible catastrophic engine, the expression of a climate in which struggle for existence must have been something absolutely inconceivable when considered from the water precipitations, floods, torrents, and erosions of to-day.

Uniformitarians are fond of saying that give our present rivers time, plenty of time, and they can perform the feats of the past. It is mere nonsense in the case of the cañons of the Cordilleras. They could never have been carved by the pygmy rivers of this climate to the end of infinite time. And, as if the sections and profiles of the cañons were not enough to convince the most skeptical student, there are left hundreds of dry river-beds, within whose broad valleys, flanked by old steep banks and eloquent with proofs of once-powerful streams, there is not water enough to quench the thirst even of a uniformitarian. Those extinct rivers, dead from drought, in connection with the great cañon system, present perfectly overwhelming evidence that the general deposition of aerial water, the consequent floods and torrents, forming as they all do the distinct expression of a sharply-defined cycle of climate, as compared either with the water phenomena of the immediately preceding Pliocene age or with our own succeeding condition, constitute an age of water catastrophe whose destructive power we only now begin distantly to suspect.

These passages, according to the model for which we are groping, refer to the three phases of recent quantavolution. The Pliocene river beds represent a period of increasing disorder and deluge in the world for about two thousand years prior to the climactic lunar fission. The awesome dead rivers of the Early

Quaternary are relics of the phase of mountain thrusting, westward movement of the American continent and the deluges associated with it, which broke down and flushed away the elevated landscape onto the shelves and slopes along the Pacific scarp. The rivers of the American heartland do not exhibit so obviously the recent catastrophic forces. Still, in the late Pleistocene, both the Mississippi and the Ohio rivers changed their courses markedly along an east-west axis, provoked by great seismism,[5] and watched, most probably, by awestruck humans.

Today's third phase finds "pygmy" rivers, many in new channels, watering and draining the country. We group all three phases in the latest of holocene period of the past 14,000 years. "Nothing comparable" with the second phase river action, "ever now breaks the geologic calm," writes King. Then, with prescience of the concept of "collective amnesia," he adds that the idea of "catastrophism is therefore the survival of a terrible impression burned in upon the very substance of human memory."

Some rivers possess drowned deltas of enormous proportions. The collision of India with Asia produced, besides the Himalayas, two equally large-scale, if less visible, phenomena in the deltaic fans of the Indus and Ganges River. These stretch into the Indian Ocean, one to the west, the other to the east of the subcontinent, covering with detritus ocean basin areas together as large as India itself. Like the raging torrents of yesterday in North America, these great transporting systems are today inactive. Although the rivers still carry two of the largest flows among all of the world's rivers, they are, as King would say, "pygmies" compared with their ancestors, their "fathers," or "holy fathers" at that, because all of this work that conveyed the tumbling slurry from high places for hundreds and thousands of kilometers had to do with mountains and plateaus just created. There stand no millions of years behind these works of nature.

---

[5] . A. C. Johnston, "A Major Earthquake Zone on the Mississippi," 246 *Sci. Amer.* (Apr. 1982), 60-83.

THE LATELY TORTURED EARTH – PART V.

It would seem appropriate to pass from the subject of rivers to that of undersea canyons by way of the most famous of natural monuments, the Grand Canyon of the Colorado River. Grand Canyon is a monument also to deceased uniformitarian geology. It is so well-studied and rationalized, with long-time-term reckoning, that every geologist is expected to recite its history liturgically. Not so Cook, nor Kelly and Dachille, nor the present writer.

Conventionally, following Woodbury, Shelton, and Redfern, we commence with an age approaching two billion years ago. Radiochronometry supports the great ages found in the canyon. The canyon proper is allowed an age which Derek Ager, for example, sets at ten million years, but, pursuing a negative exponential principle, gives one million years to the mere latest fifty feet of erosion.[6] (That is, a practically catastrophic rate is seen to have occurred at times.)

The floor of the Grand Canyon complex is an unknown material supporting what is called Vishnu schist, composed of mud, sand and lava. Thereupon the miles of sediments begin to pile up, most of them now missing, and probably eroded, but today some three miles can be accounted for: one in the bottom and main canyon itself, a second mile from the brink of Grand Canyon to the top of Zion Canyon, and a third up the face of the higher plateaus to the top of Bryce Canyon. Wind and water bring in the sedimentation layers. Many in variety, several distinctive deep beds of schists, sandstone, limestone and shale compose the great bulk of deposits. Discoverable in the series are ten major unconformities and many minor ones, where intervening layers existed and were worn away before being covered by new deposits.

The area was uplifted and submerged a number of times with relation to the seas around. Some lapses in the record are so prolonged that whole mountain ranges on site could be worn down and planed off by erosion, succeeded by new tall deposits.

---

[6] *Op. cit.*, 48.

Fossils of algae, primitive and later vertebrates, fishes, and footprints of amphibians are discovered in ascending. Fossil trees, fishes and reptile tracks are found in higher Triassic rocks. The fossil record stops at the Eocene epoch of the early Cenozoic (recent) era. In the Cenozoic, the entire region was uplifted from near sea-level to the present elevation. During uplift periods the Colorado River system has washed away materials and cut the gorges. So goes the gradualist solution of the Grand Canyon scene.

The quantavolutionary view, as may be supposed, stresses high energy forces, fractures and quick deposition. "Many of the pools and rapids in the Grand Canyon are located where the river crosses regional and local fracture zones."[7] Cook points out that the Canyon is narrow at Supai Village and that the gorge appears to have ruptured open in a brittle fracture. The Grand Canyon, as was mentioned earlier, is perceived as a branch of the earth-girdling rift system; numerous other branches of the fracture system are observable north and south of Grand Canyon also. All of this occurred when the continent was thrust westward over the Pacific Ocean rift and the ocean rift fractured the continent. A number of orthogonal embayments of the Canyon are perpendicular to the main fracture or canyon, and these have been filled with debris from the outpouring of temporary great inland lakes known to have existed in the region.

The three miles of sediments, all heavily fractured, were products of overthrusts from afar and of great slurries that brought in and laid down beds of fossiliferous sand and mud. Speaking of the sediments of hundreds of feet, "if all this was a very slow process requiring millions upon millions of years, how did it happen that the rivers carried nothing but clay for millions of years and then suddenly changed to sand?" And "nowhere today do we find rivers producing deposits of such uniform

---

[7] R. Dolan, A. Howard and D. Trimble, "Structural Control of the Rapids and Pools of the Colorado River in the Grand Canyon," 202 *Science* (10 Nov. 1978), 629-31.

nature..."[8]   The erosion was generally prompted by heavy seismism. The fossils found in the beds would have quickly disappeared if they had not been buried in sudden local and general disasters. The radiochronometry employed is of dubious validity, or, let us say, requires a specific set of challenges going far beyond these rudimentary paragraphs. All may agree that in the deep non-marine but water-deposited Eocene limestones of Bryce Canyon may be found some excellent carvings.

Grand Canyon would be a minor feature of the continental slopes of the ocean and a minor canyon among submarine canyons. Even the Hudson River possesses one as awesome; it proceeds underseas for hundreds of kilometers, first cutting into the continental shelf, and then extending down the continental slope to the abyssal plain of the ocean, 4.5 kilometers below sea level. The difference is not that the one has grown sub-aerially and the others aquatically; both types have been sub-aerial for all their active lives. The seas encroached as the lunarian period created the sea basins, slopes, and canyons. Grand Canyon and several other such remarkable sub-aerial features are of the ilk; a comparison of a profile of Monterey Submarine Canyon (California) and of Grand Canyon[9] reveals very close similarities and indicates strongly a common ancestry.

Scores of impressive submarine canyons extend the courses of rivers around the world. The idea that they were once active as rivers was resisted for a generation. In 1936, Francis P. Shepard could formulate the predicament, which still stands unresolved:[10]

> Investigations of submarine canyons carried on for a number of years with the cooperation of the Coast and Geodetic Survey, the Geological Society of America, Scripps Institution and other organizations have revealed that these sea-floor

---

[8] Kelly and Dachille, *op. cit.*, 113.

[9] *Ibid.*, 81.

[10] 83 *Science* (May 22, 1936), 484.

canyons have all the characteristics of river canyons and are distinctly different from fault valleys. Also tests of the idea that the submarine canyons might be the product of currents have produced negative results so that they have evidently been cut by rivers. The significance of this sub-aerial erosion on the present sea-floor is particularly disturbing, since the submarine canyons extend out to depths of from 2,000 to as much as 10,000 feet and are found off practically every coast of the world. Also all available evidence favors a Pleistocene age for the canyons. Accordingly, there is the implication that the coasts of the world were greatly elevated above their present positions during the glacial period. That all the continental margins both off stable and unstable coasts could have been subjected to such movements in comparatively recent times is scarcely credible. The alternative that there have been sea-level changes connected with the cause seems much more reasonable. Such changes are indicated not only by the submarine canyons but also by many of the phenomena of coral reefs and by oceanographic data from various parts of the world. The only cause of sea-level change which does not meet with almost insurmountable objections is that of glacial control. It seems quite possible that the continental glaciers during some of the earlier glacial epochs may have been sufficiently thick and sufficiently extended to have allowed a lowering of 3,000 feet or more. While such a lowering was probably insufficient to account for the deeper canyons it is felt that it would have resulted in the development of a universal canyon system which, connecting with much older sunken canyons in some places and modified by subsequent sinking elsewhere, would account for the present situation.

The world would have to be a great ice mountain to provide such waters. The waters had to come from elsewhere, and be accompanied by great tectonism. We hold rivers to be based upon faults.

In the same year, geologists Harry H. Hess and Paul MacClintock presented a striking solution. They saw in the canyons evidence of recency, a late Pleistocene age, of suddenness of creation, and of worldwide simultaneity. Here are the three primary tests of quantavolution, all passed by the submarine valleys. Then they are compelled, with reluctance, apologies, and special consultation with H. N. Russell (who advised against it), to advance the quantavolutionary mechanism, exoterrestrial encounter. The passages deserve quotation:[11]

> The valley-cutting conditions resulted from a sudden change in the shape of the hydrosphere, depressing sea-level in low latitudes, raising it in high latitudes; in other words, a change in the ellipticity of the sea surface. At present we can think of no orthodox cause for this change... However, a speculation comes to mind; if a sudden decrease in the rate of rotation of the earth took place, the hydrosphere would respond by being drawn into polar latitudes. The solid body of the earth would less rapidly adjust itself into a new spheroid in equilibrium with the slower rotation, which adjustment, when complete, probably would restore sea level to approximately its present position. But during the adjustment, it is postulated that there would have been time enough to allow rivers to cut valleys on continental slopes. While of course we do not know what could have caused the sudden change in rotation, it is conceivable that a collision with a small extra-terrestrial body would be competent to produce the effect.

The authors then sought for evidence that the depths of the canyons would decrease from the equator to the poles, and, second, that there would be found high marine terraces in the northern latitudes where the shores would have been temporarily flooded. Indications of both were deemed favorable.

The failure of theory to move along such lines is unaccountable, except in terms of the psycho-sociology of science of which we speak in the *Velikovsky Affair* and The *Cosmic Heretics*. Many years later one reads in a study by Landes approvingly:[12]

---

[11] 83 *Science* (1936), 332-4.

[12] Reprinted in W. Corliss, compiler, *Strange Planet* (Glen Arm, Md: Sourcebook Project) vol. El, Doc. ETS-002).

I claim that the finding of graded clastics and misplaced (shallow-water) faunas deep beneath the sea is *not prima facie* evidence that they were carried there by turbidity currents: that the finding of cobbles does not prove that they were transported by submarine landslides; and that photographs of ripple marks lying at a depth of 4,500 feet do not necessarily mean that they resulted from current action operating at depth... I likewise believe that deep-sea- floor current ripples, like the truncated seamounts, are relics of shallower water.

At this point, Landes should be looking into the ancestral skies. Instead he suggests that the deep ocean basins might once have been over 20,000 feet deeper. Even this idea might lead somewhere, but, instead, the *ad hoc* argumentation that so often passes for geological theory obtrudes; when in trouble, call upon isostasy, diastrophism, time, lifting, and, as here, sinking, and thus by name-calling the problem is solved and the matter ends; the data are not pushed to their ultimate meaning.

Landes writes: "What manner of logic allows us to accept evidence, such as marine strata, of a sea-level far above present datum of 25,000 feet, but causes us to run from evidence of a sea-level depression of 25,000 feet?... What is so sacrosanct about current sea level?" The trouble here is that the logic is not good enough. One ought not to have indulged in the notion of a sea-level 25,000 feet higher because of the marine fossils up there, especially while he was laughing over Noah's Ark. Furthermore, the present sea bottoms and therefore sea-levels can be depressed by another 25,000 feet, but again no mechanism is perceived.

He, and others, should be asking the deeper questions: "What are these deluges that humanity has been clamoring about since the dawn of history?" "Must every drop of water bear the holy stamp, 'Made on Earth'?" "How long does it take a pre-designed fracture trough to make a river channel, complete with fractured and non-fractured meanders? .... What is so sacrosanct about the ocean basins having always been filled with water?" I think that we have progressed far enough along in this book to dispose readily of the submarine canyon problem. The canyons were instantly created great river courses that rushed down, first, precipices, then, steep slopes, then gradual slopes, into the ocean basins that were only partly filled with water. Drainage of the water-logged continents and successive deluges filled the ocean basins to overflowing. As the seas encroached upon the rivers, the rivers were also receiving far less water to give to the sea. The underseas box-like, sluice-like channels ended their careers as turbulent rivers within perhaps two thousand years.

They have not filled with sediments. Gross, in his *Oceanography,* says that submarine canyons would soon fill up if they were not being emptied by turbidity currents. Geology has invented some bizarre mechanisms to circumvent catastrophism and here is one of them: turbidity currents. They have never been actually observed; they are "intermittent;" they are caused by earthquakes; they have speeds of 20 km/hr; they account for anomalous continental sand and fossils found on the ocean floor. A rare study assigns them credit for having broken a trans-Atlantic bottom cable. (Still, no one denies seismism.) Would not such currents act as bulldozers instead of sweepers, and fill, rather than clean out the canyons? Our quantavolutionary theory is adequate for all that bespeaks turbidity currents, including the oceanic sands and fossils.

A question remains to perplex: if the continental blocks were meanwhile rafting over long distances, would they not have left behind their detrital slopes? The slopes would then be flat and spread over the abysses. A logical answer is available here, too. We have but to recall that the continents travelled because they were both pulled and pushed. If they had been only pulled they would have left their ocean moraines behind. But they were standing on a kind of conveyor belt, as has been said by Harry Hess and others, and their slopes moved right along behind them; the belt was being pushed by the lava currents issuing from the ridges, fissures, and volcanos. Anyhow, the canyons were working rivers after the continents ceased to move rapidly, and before new ocean waters drowned them.

In concluding the chapter, a few words may be in order on the more puzzling problem of the deep sea trenches. These deep, narrow and often long slits in the crust are found in various regions but are especially prominent around the Pacific. There they gash the sea floor off of South America, Central America, the Aleutians, Kuriles, Japan, the Philippines, Java, and various island fronts, including a long stretch north of New Zealand.

In a typical large trench, a depth of ten kilometers is precipitously achieved, with a slant toward the continental rock against which it is emplaced. Its sediments are shallow, its walls bare. Trenches were never rivers. A function for them was hard to discover until the tectonic plate theory of continental drift went shopping for its mechanism. Then it occurred that the ocean floor being made at the ridges had to be disposed of somewhere else, if the world was not expanding. For lack of better, the trenches became locations into which the sea floor plate crept upon encountering another plate, thus disposing of itself tidily. The next chapter will handle this theory, but we cannot leave the trenches without an explanation.

Trench walls are igneous for the most part, straight, and nearly vertical, like fault scarps, say Heezen and Hollister about the Puerto Rican Trench. They belong to the period of great disruption. Their oceanic sides abut continental walls that are much taller and deeper; the connection between the two may not be binding in many or any trenches. The continental wall is of varying chemical composition; the oceanic wall is purer basalt of the mantle. They heat and expand, cool and contract at different rates. The gap or trench may occur as a pull-back of the oceanic basalt or the continent, a drop fault where nothing drops. "The crustal block which forms the floor of the Puerto Rico Trench resembles the dropped keystone of a rising ramp, which once bridged the transition from the thin oceanic crust to the thick foundation of the island arc."[13]

---

[13] *Op. cit.,* 490, 467-9.

Sediments of the trenches are scanty. The same writers say: "It is a general lack of sediment accumulation which is the most notable feature of all the deep-sea trenches. This lack... demands a recent origin of trench topography."[14] Recent must mean holocene or Pleistocene, it appears. But now, the plate tectonicists chase in full cry after the trenches as fulfilments of the need of convection cells and subduction of continental and oceanic material. Are trenches barren because they appeared lately or are they barren because they have just digested hearty meals of sial?

---

[14] *Ibid.*, 483-4.

CHAPTER TWENTY-FOUR

CONTINENTAL TROPISM & RAFTING

A Texas association proclaims the slogan "Stop Continental Drift," in its attempts to foil the trend to believe that the Earth's crust has been, and is, in motion. The crust is thin below the ocean bottoms and thick beneath the continents. It is broken up into a dozen major plates whose boundaries are defined by faulting, heat, and turbulence. The plates show signs of having moved great distances over time. Most scientists have been converted to this mobile perspective from a static one during the past generation. "We now have a new, mobilist orthodoxy, as definite and uncompromising as the staticism it replaced." So writes Stephen Jay Gould in *Natural History Magazine*.[1]

Now and then, goes the theory of continental drift – that is, every couple hundreds of million years - the plates renew themselves. The continents do not. They may be split asunder, or bash their fenders, but they move on and on, majestically riding upon the same material, their own mantle magma.

They carry a two-billion-year record of life, while the ocean bottoms have deposited their sediments periodically beneath the sea shores of continents, or beneath other plates which they may be jostling, so that now they carry no more than the last 160 million years of sediments. The plates are pushed around, it is said, by convection currents. These are hot rock

[1] "The Continental Drift Affair," 17.

moving up to the cooler surface areas and pushing them aside, until these bump into other plates which are being also pushed; then they are forced to descend (or they force the other plates to descend); they melt once more and become deep mantle material.

Much of this theory is incredible, we shall argue. We accept gladly the facts that the continents were once together and then moved long distances and still exhibit minute motion. We accept also the facts showing the ocean bottoms to be very young. Several other facts are grist to our mill, and some minor theories are also credible; we have mentioned several of these and will mention others. But we shall concentrate upon the quantavolutionary theory that an exoterrestrial catastrophe brought about the movements of the Earth's crust recently, and we begin with the most obvious fact in topography, namely that the continents of the Earth are concentrated opposite the oceans. "This means that 82.6 percent of the total continental area is antipodal to oceanic area."[2]

So saying, C.G.A. Harrison goes on to describe how, on a computer, he rotated randomly coordinates representing the continents on a sphere, in order to discover how often the actual antipodal percentage would appear. He simplified the continental areas into circles and fed their numbered forms into a computer which then randomly placed them to see how much land would be antipodal to the oceanic area. He repeated the random placement 2000 times. "The median percentage of continent opposite ocean to be expected from a random distribution of circular continents is 68.0 percent of the continental area. The observed figure of 82.6 percent is exceeded in... 9.6 percent of all cases. Thus from this evidence alone, it would appear that there is a probability of only 0.096 that the present distribution of continents is random over the surface of the earth." He repeated the test with triangular instead of circular simplifications of the proper areas and the results were similar. He concluded that "there

---

[2] "Antipodal Location of Continents and Oceans," 153 *Science* (Sept. 9, 1966), 1246-8.

is less than 1 chance in 14 that the present antipodal distribution of continents and oceans is the result of a random process."

This is hardly surprising. If the Arctic-Atlantic ocean were closed up, if Australia and Antarctica were fitted to their apparent points of departure from southern South America and Africa – that is, if the continents were rendered into a single mass as they appear to have aggregated before the present age of "drift" began, then all of the existing continental land would be antipodal to oceans.

The implication is strong that before the drift began, the ocean areas were in fact land-covered. The major differences between the Pacific Basin and the other oceanic basins, we have noted, indicate that continental material was blasted out of the former and pushed aside from the latter. The movements of the continents since this time can be interpreted upon the premise of a sudden removal of over half the Earth's crust in what is mostly now southern hemispheric ocean. The land of both the eastern and western hemispheres has travelled towards this vacated area. So have Australia and Antarctica.

The so-called plate movements have not been random, nor can they be interpreted in any other way. The continents all exhibit "lunagenic tropism" and nothing much else. They have moved in the particular direction of the lunar-vacated now south-central Pacific Basin under the special stimuli of global fracturing, electrogravity slide, earth expansion, hydrostatic equilibration, and isostasy. We define isostasy here (and elsewhere) as the process by which all mutually affected elements in a system, consequent upon any change in one of them from within or without, share the effects of the change by changing themselves in closest accord with their peculiar sites and natures. When a change is introduced to Earth from the outside, all possible responses of the Earth's motion and masses are drawn upon to incorporate its effects and to do so in accord with their ranked most possible behaviors. Isostasy has to have a function; in the great post-lunar diastrophism, isostasy functions as a tropism. It moves the continents not randomly, nor to the poles or the equator, nor to gather surviving animals for the Ark, but to repair and redress the lunagenic basin.

Propelled by three rifts in all and with a blasted out area east of it, the exception to lunagenic tropism would appear to be the Indian subcontinent. India moved east faster than Africa. But since the continental world was moving generally south as well as east, why did India move north? Relatively Eurasia was moving south, and this is part of the suggested answer. Also India and Australia were simultaneously and together disconnected with a large land mass from Africa and Antarctica by the Atlantic-Indian and Mid-Indian Ocean Ridges; India was simply at the northern end of a plate and could not pivot southwards; northwards lay the old Tethyan Sea region, which was now being compressed and closed up. India was pushed by the largest expansive fracture complex per land unit: this "lava grease" worked upon it like the currents of a powerful river moving a raft downstream. When it arrived at the southern shores of Asia, it encountered the Tethyan shear with weakened rocks and islands, all of which it overran, thrusting and folding its edge over them until the Himalayas were produced, meanwhile elevating, with an assist from the swelling mantle, the great Iranian plateau area. (Two surviving races, one African and the other Indo-European or Tethyan, found themselves on opposite sides of the great mountain mass. They encountered one another thousands of years later, when the Proto-Indian civilization was battered by natural disaster and the Indo-Europeans came down from the Plateau.)

The low continental Pangean mass to the south and east ultimately was partly flooded by the waters of the sky, never to reappear again. Higher elevations constituted the south seas islands of today. This can be called Australasia. It is a land that has the Tethyan shear and moon basin through its northern belt, the northern trans-Asiatic rift to its west, and the South Pacific fork moving eastwards and finally up to mark its southern and eastern limits.

The northern extremity of Pangea was depressed originally by the ice cap and is still rising, although at a decelerated rate. The fjords mark sheared continental mass, sharp, clear, new; the low-lying lands that compose the great flat watery islands and the Artic Sea (which is mostly continental) signal the former land mass under the ice cap load. Some of it was additionally compressed by the new ice cap formed in the Age of Jovea.

The Antarctic Sea was opened up at the south polar forking fracture, and the Antarctic continent, denuded of ice, was pushed southward to center upon the new polar axis. It, too, received a new ice cap beginning in the later "Age of Jupiter," but, to follow Hapgood's "Maps of the Ancient Sea Kings," possibly not until exploration, after a period of civilization, when maps of the coastline were drawn to a considerable degree of accuracy.

Some 450 specimens were recovered at Coalsack Bluff, Central Transantarctic Mountains. Found there were terrestrial amphibians and reptiles of Lower Triassic, typical of the same age in Africa, India and China, especially genus *Lystrosaurus*. These creatures had no "long way around." "The interchange of Lower Triassic tetrapods between Africa and Antarctica could have been only by a direct ligation of the two land masses," probably at Southeast Africa.[3] The Labyrinthadont, the first land vertebrate to be found in Antarctica, has also been found in Africa, South America, and Australia.[4] Marsupials are now placed in South America, Antarctica, South Asia, and Australia.

The infinitely complex faulting of the Earth's surface rocks, aside from the major morphological transformations, is an expected phenomenon of the multiplex pressures of rafting land masses. Additionally, the phenomenon expresses surficially what were more profound upward pressures during the Uranian periods. The "latest" evidence supports Alfred Wegener's view that the continents moved only once, this is in the Mesozoic, and that there were two continental masses, one to the North and the

---

[3] D. H. Elliott, E.H. Colbert, W.J. Breed, J.A. Jensen, J.S. Powell, "Triassic Tetrapods from Antarctica: Evidence for Continental Drift," 169 *Science* (13 Sept. 1970), 1197-1200.
[4] "Continental Drift," V *Ency. Britannica* (1974), 112.

other to the South.[5] These "findings" are expected in our theory. The Tethyan equatorial waters of Pangea probably are the source of the belief that there were two land masses. As for the Mesozoic, 65 to 225 million years ago by conventional reckoning, the period is being rapidly invaded by similar species from both directions, but there may have been a period of terrestrial isolation when the Tethyan waters intervened.

If the continents split asunder or were "born separate" and moved several times, there should be abundant evidence of the events. There is little fresh data on this score. The most prominent basin, the Pacific, hardly gives evidence of having been traversed by continents, though they all drift towards it. Mostly, old theories of the lifting and dropping of land during ancient orogeny have been dusted off and varnished to claim the several periods of movement. Fossil ice ages have been claimed, too, as proof of former dislocations of the continents, but these have been embarrassments to ice age theory from the earliest discovery of pertinent evidence; the presence of pebble drift and till, and of glaciers or high mountain freezing may be referred to dense material fall-outs such as were discussed earlier.

"In the whole of geophysics," Defant once wrote, "there is no other law of such clarity and certainty as that there exist two preferred levels in the Earth's crust."[6] Continents are like ships in a frozen sea. "The continents, steep-sided massive blocks surrounded by an enormous world-encircling sea, have deep roots which project 30,000 or 40,000 meters into the earth's mantle while the ocean crust is but a thin 5,000-meter-thick film frozen over the Earth's massive mantle." Thus report Heezen and Hollister.[7] These continental blocks, I have maintained, are splittable only by a great external force and a responsive expanding

---

[5] *Ibid.*, 108.

[6] Quoted by Jordan, *op.cit.*, and see the chart there (and in *Chaos and Creation*) of the frequency distribution of altitudes of land and sea bottoms.

[7] Heezen and Hollister, *op.cit.*, 521.

force, while the ocean basins are easily producible by fissure volcanism.

The post-catastrophic process is followed by a rapid relocation of the continents and reencrustment of the globe. The continents were not "just drifting;" they "were going somewhere." A. L. du Toit was veering toward reality when he offered in his early (1937) book, *Our Wandering Continents,* the idea of a "gravity slide," the creeping of continental masses toward rimming geosynclinals depressions. He gave at the same time perhaps too much encouragement to the idea of thermally driven currents in the mantle. These were, as we may establish, an accessory after the fact.

When F. Tuzo Wilson, reviving du Toit, and the spirit of Plato's ancient words for that matter, exclaimed, "the earth, instead of appearing as an inert statue, is a living, mobile thing. The vision is exciting. It is a major scientific revolution in our own time...,"[8] he was thinking of continental drift but could better have been speaking for continental trot, or rafting, or lunagenic tropism of the continent.

Continental drift theory has invented convection currents to move the Earth's plates with whatever continental land may be aboard on long journeys over the Earth. The convection currents cycle vertically between the mantle below and the crust above; the currents push the plates about the surface like the uplifted trays of waiters in a crowded café, except that waiters as they weave and duck are known to descry paths between the bar and the tables, while no one can even guess why the convection currents go one way or another, if indeed they exist. For example, Harrison, after proving that a non-random process had to account for the counter-oceanic distribution of land, mentioned large-scale convection currents in the Earth's mantle as a possible cause. This seems to put too much directiveness into convection currents, which are already overloaded with the task of pushing huge

---

[8] "Static or Mobile Earth: The Current Scientific Revolution," in *Gondwanaland Revisited...,* 112 *Proc. Am. Philos. Soc.* (Phila., 1968), 309.

tectonic plates around the globe. A better hypothesis would have been "lunagenic tropism," the tendency of the continental land to move toward the crater of the Moon and to fabricate new crust in compensation for the excisions. Still, the convection current hypothesis is worth considering, if only because at the moment it is the height of fashion in geophysics.

Tall mountains, a trillion geological faults, islands, bays, and most other morphological irregularities denote that the continents were not peaceful bystanders to the creation of the oceans. The Earth is slightly flatter at the poles than its present rotational velocity would explain. The difference is 1%, which is a disappearing relic of the period of rotational deceleration. The great depressions found in the Hudson Bay, Greenland, and North Eurasian areas, which when refitted, fit the shape of the destroyed ice cap, are also relics of the lunarian crisis.

The trans-Atlantic coastlines, above and below the Tethyan transverse fracture area, fit together well. The probability that the jagged and curved pieces would fit as they do by random development is negligible. South America and Africa, North America and Europe, have many points of topographic, lithospheric, and biospheric diversity. The blocks of the continents on both sides of the Atlantic Basin are steep and sharply outlined at the edges of the continental shelves.

The west coast of the Americas is also steep and sharply marked. The western mountains seem to be a unit from Alaska to Chile; this in itself must have great significance: the nearly 180-degree belt of rock underwent a single, simultaneous experience; how can geology, geography, and geophysics ignore the simple meaning of so magnificent a display? The Andes are so continuous with the Rocky Mountains, and the Mid-Atlantic so parallel to the two Americas, and South America so congruent with the Albatross Cordillera that they must all have been engaged by approximately the same vector forces during lunagenesis.

The force of lunagenesis affected profoundly the now western terrain of the Americas even though its epicenter was probably emplaced on the old equatorial Tethyan belt and thousands of kilometers west of Central America. The western shelves of the Americas are of the same age as the East and West Atlantic shelves. But they poorly match the continental shelf morphology across the Pacific Basin. If the steep shelves of the Americas represent one side of a fracture, where is the western fracture to match? The Americas have moved westward; they have risen greatly; they have probably climbed over and rest upon their once opposing land as upon the fracture itself in the north. Part of the sunken continent of Mu, to pre-empt some wag, is below California and explains why it is so.

The East Pacific Rise pushed up and out rapidly; its transverse fractures stuck out far to the West. It met little resistance. The crust had been vaporized, and the upper mantle was boiling, just as it was along the great fractures. To the west of the craters was the larger land mass of Asia, to the southwest the morphological disorganization produced by the fracture system and to the southeast, that affected by the ellipsoid Moon Basin.

The new coasts of the south and west of the Basin were attracted toward the basin; fragments separated and rafted faster than the larger mass to become the offshore islands of South and East Asia. Oceania was born, much of it in arched array, moving inward upon the swatch cut by the main explosions.

The wide Tethyan tropical belt of Pangea was generally trampled upon by the shifting continents, disappearing beneath the Middle East, South Asia, and the Moon Basin. The "Mediterranean" seas were swept north and south and the area was practically fractured and closed as Europe moved down. The Pyrénées, Alps, and possibly the Balkans were then created by shearing forces as Africa rotated southeast beyond Europe.[9] In the withdrawal of Africa, the Mediterranean Basin opened up and waters from northeast, east, south and finally the Atlantic filled it. Later movements may have sunk the Tyrrenian plateau. Later, too, the Triton Sea of the Sahara was emptied, leaving a great desert.

---

[9] "Mother Earth…," *op.cit.*, 12.

The complex Mediterranean morphology reveals deep bowls and large shelves. Hsü has reported investigations of its western bottom, evidencing a dry, desert terrain at one time. The deceleration of the old globe would have promptly dispatched its waters north and south to higher latitudes, supposing it had been tropical before. Huge submarine canyons depict a scene of inpouring waters afterwards. The mouth of the Nile River discloses a narrow, deep gorge cut 700 feet below the sea level of today, which may at that have been several thousands of feet deeper, according to Chumakov. Libyan off-shore canyons are also impressive, as reported by F. Barr and associates.

Geophysics, not having yet considered our hypothesis, has not clearly expounded the original torque of the continents. Earlier I wrote that the Mid-Atlantic ridge veers sharply east at the Equator and explained it as a result of the Earth's sudden deceleration of axial rotation. To this I may now add that five major occurrences signify the same slowdown of rotational velocity. At the same time, they explain the location of several land masses.

The first is that the larger of the two branchings of the Mid-Altantic fracture, arriving at about 50° South Latitude, swings in the direction of the Earth's rotation, east, that is. The fracture is pursuing its original route and continues while Earth as a whole is slowing. It is following the direction of greatest stress, too, where the greatest need to fracture exists.

Second, this bifurcation of the fracture may have happened where the old South Pole may have been; but, more likely, the fracture had not achieved its 'objective,' the old South Pole, before it was forced to split into two. This giant forking might, as M. Cook has suggested, be the normally expected effect of the decelerating explosive fracture of a globe; it might also be an effect of the resumption of mondial rotation, within an hour or so following the halting that may have produced the aforesaid Mid-Atlantic transverse movement of the fracture. The split of east and west now cut off Antarctica, which because it was in a low latitude and neither east nor west, was rendered safe from lateral movement and became a polar continent.

But, third, the fracture, continuing, divided again. This was logical, too, one fork continuing east, but the original torque reversing on the sphere and heading back north. The east fork severed Australia from Antarctica and the north fork cut between Australia and Africa.

Fourth, Australia proceeded swiftly eastward propelled by the crustal slowdown and the attractiveness of the lunagenic basin to the east. Fifth, India, cut off from Africa, rotated clockwise, as expected, and headed, also as expected, north by east.

Since the world-girdling fracture system will be encountered sooner or later no matter in what direction one goes, any area enclosed by fracture boundaries can be called a plate. Ignoring most fractures or rifts that traverse the continents, one may conclude that some ten (Gould) or twelve (Toksöz) such areas or plates exist. Curiously, they are of greatly different size; the Pacific Plate, for instance, covers most of the Pacific Basin, while the Cocos Plate encompasses a smallish region between Central America and the East Pacific Ridge. Inasmuch as a very slight annual movement of several centimeters seems to be occurring at the edges of most plates, and the boundaries of most plates include some portion of the volcanically and seismically active oceanic ridges, it would appear that the whole of the Earth's surface is somehow in motion, and therefore, a science of "plate tectonics" must be devised to account for the "drift" of the combined, inseparable continental-oceanic lithosphere.

That the Earth may have expanded or be expanding in volume along its fracture lines is a theory not to be dismissed, but geologists for the most part prefer to portray crustal, lithospheric drift (carrying the continents) as a perpetual steady-rate movement, which disgorges molten rocks from deep in the mantle, along one and another plate boundary, while engorging rocks at the boundaries of the plate, thus maintaining a constant global surface area. To account for this upwelling and subduction is no small task; "an area equal to the entire surface of the earth would be consumed by the mantle in about 160 million years" at the presently calculated rates of movement up and down.[10] That would be about 510 million square kilometers, of which 310 would be true ocean basin of about 8 kilometers in depth of rock. Granted a uniform rate of exchange and the time allowed for it (which is roughly based upon the age of the oldest portions of the oceanic rocks), something like 1.55 cubic kilometers of crustal rock has to be subducted annually.

The preferred instrument for subduction is the once altogether mysterious but impressive submarine canyons. These line up along the coasts of western South and Central America, also along the western Pacific Basin arc from the Aleutians down to New Zealand, and then too stretch westward off the southern boundaries of Indonesia. Lesser lengths can be discovered in the Caribbean, and in the extreme South Atlantic ocean. Altogether over 10,000 kilometers of submarine canyons are notable.

The time allowed for subduction is conveniently long, so that very little work is required at any given time and place.

To the average kilometer of these canyons is assigned the task of ingesting .00015 kilometer or 1.5 cubic meters of the Earth's surface per year. This is a modest undertaking, but also one can call for help from the old standby, orogeny. India is still smashing into Asia, hence the Himalayan range is piling up debris from the plate edges. South America is being pushed away from the welling-up Mid-Atlantic Ridge and also away from the East Pacific Ridge; perhaps it too is rising from the east while the oceanic crust of the west is being subducted into the long western trench.

The convection cell is a natural heat machine. Hot material deep in the Earth's mantle rises to the cooler regions of the surface, breaks through as a plume or fissure and pushes aside the colder rock; the colder rock is moved along to a subduction zone where it is mechanically forced downwards into the deep

---

[10] M. Nafi Toksöz, "The Subduction of the Lithosphere," *Sci.American.*, 89.

mantle. There it assimilates to the hot surrounding material, and may even return to an area where it will rise once again to repeat the process. The scheme is almost entirely theoretical, although one may, by watching a stew pot, see a similar occurrence, the heated mixture arising from the bottom of the pot to displace the cooler surface mixture which then sinks to the bottom, is heated, and then rises once more.

To observe any part of the convection process, even indirectly, is difficult, but bits of data can be made to fit. Thus, the fact that submarine canyons are coincidental with earthquake and volcanic and mountainous zones implies a turbulent function, such as subduction would be. One cannot deny the evidence of upwelling magma along the great oceanic ridges; there is an output, and there is a movement away from the output.

But is there a subduction? The submarine trenches appear to be cleared for action tomorrow, but not the scene of yesterday's action. They have scanty sediments, whereas they ought perhaps to be full of oceanic sediments, not to mention continental sial that would happen to be subducted. In what was the first attempt at observing an actual subduction of sea floor, Heezen and Rawson made four dives to the floor of the Middle-America Trench in a U.S. Navy submersible, *DVS Turtle*, at around 1600 meters of depth. They observed a set of escarpments moving steplike down to the bottom floor, then an "apron," which shortly encountered the abrupt landward wall of the trench. The apron was bisected parallel to the wall by a "line of contemporary deformation;" this "is interpreted as the sea floor trace of subduction."

But the scene is peaceful. "We observed no features which could be attributed to turbidite erosion or deposition." Further, "at the present time no movement is occurring at the base of the landward wall and... probably no significant deformation has occurred there for decades or centuries... Perhaps the most surprising observation was that most of the steep wall is covered by smooth, undisturbed ooze."[11] The trench here is obviously long defunct or inadequate for the task assigned it. The stepdown escarpment into the trench seems to be a normal faulting occurrence, like much of the African Rift, denoting dropped blocks, in connection with a pull-back motion of the landwards wall of the trench. Still, even if this were an "average" point of a trench, the activity of subduction might be too minuscule to observe.

Several years later, the *Glomar Challenger* was drilling into oceanic sediments north of Barbados at an apparent plate boundary and discovered older Miocene sediments overlaying younger Pliocene deposits.[12] The phenomenon was explained by plate tectonic theory as a product of an underthrusting (subducting) sediment-loaded oceanic plate. As the plate went down, its older sediments were sheared off and ended up overlying its younger sediments. All of these formed now part of a mass that culminated in sub-aerial volcanic mountains. The volcanism would be an effect of the descending slab, which generates heat by friction, shear stresses, and rock faulting. Channels for explosive heat escape would be provided in the course of structural adjustments between unlike rock masses. Earthquakes occur until the masses become thermally indistinct, never below 700 kilometers; there the rock can no longer behave in a brittle manner.[13] The occurrence of earthquakes up to this point is taken to indicate the correctness of subduction, convection cells, and plate tectonic theory.

Here is Toksöz' summary of the current theory of the subduction of the lithosphere.

---

[11] Bruce C. Heezen and Michael Rawson, "Visual Observations of the Sea Floor Subduction Line in the Middle-America Trench, » 19- *Science* (22 Apr. 1977), 423.

[12] Roger N. Anderson, "Surprises from the Glomar Challenger," 293 *Nature* (1981), 261-2.

[13] Toksöz, *op.cit.*, 97.

> The lithosphere, or outer shell, of the earth is made up of about a dozen rigid plates that move with respect to one another. New lithosphere is created at mid-ocean ridges by the upwelling and cooling of magma from the earth's interior. Since new lithosphere is continuously being created and the earth is not expanding to any appreciable extent, the question arises: What happens to the 'old' lithosphere?

> The answer came in the late 1960's... The old lithosphere is subducted, or pushed down, into the earth's mantle. As the formerly rigid plate descends it slowly heats up, and... is absorbed into the general circulation of the earth's mantle.[14]

Interestingly, "in certain areas convection currents in the asthenosphere may drive the plates, and... in other regions the plate motions may drive the convection currents." Lest the reader hoot at the picture of a driver driving the car but sometimes the car driving the driver, it should be interposed that this latter possibility is a broad hint of what may be the truth of the matter, namely, that so far as the evidence goes, the paving and expansion that went on in the past, and their faint stirrings today, would have to, and do, generate currents, even cyclical currents, in the mantle. How could they not do so?

Where plates collide, to resume the theory, a trench is forced open and one or the other plate descends the trench into the mantle, thus letting the ridge, perhaps thousands of kilometers away, continue to churn up lava and pass it along, so efficiently indeed that folds or thrusts are hardly to be found in the vast expanses of the abyss, nor alongside the ridges. The trenches "accumulate large deposits of sediment, primarily from the adjacent continent." (This contradicts another view, Heezen and Hollister's, that the trenches are scarcely sedimented.) "As the sediments get caught between the subducting oceanic crust and either the island arc or the continental crust, they are subjected to strong deformation, shearing, heating and metamorphism...

---

[14] *Ibid.*, 89.

Some of the sediments may even be dragged to great depths, where they may eventually melt and contribute to volcanism. In this case they would return rapidly to the surface, and the total mass of low density crustal rocks would be preserved."

Skillfull drawings enhance the text by showing some sediments being scraped off on the opposite side and other sediment being miscilated and conveyed below. Without wishing to burden this one article with problems universal to its genre, one cannot allude to additional contradictions. There should be enormous masses of plate-served detritus on the inward side of a receiving trench. Indeed, the whole of a previous world of sediments should be dumped in such heaps or carried down into the mantle reluctant, because of its higher density, to receive it.

There are, of course, no such masses. The sediments by the trenches, if they exist at all, are mostly igneous masses, which foregather there in as ordered or disordered a condition as anywhere else. As for the metamorphosed rocks, they show no preference for trenches and can hardly amount to the quantity under consideration, unless this is to be the origin of new granites. The balance of new and old sediments, it appears, is impossibly askew.

The trenches, according to the prevailing notion, have good appetites, but are slow feeders and neat eaters. Perhaps that is why they have never been observed while at dinner. Toksöz refers to low and high subduction rates and draws several diagrams of the subduction process, but offers no proof of subduction other than gravity anomalies (whose findings, he grants, are belabored by uncertainties) and seismology. "The most compelling evidence of the subduction of the lithosphere comes from seismology." Seismic wave behavior in the vicinity of the trenches, where earthquakes are common, has exhibited differences, as might be expected, showing different depths of activity and these have not been interpreted satisfactorily. Now these seismological differences have been assumed to be measures of different depths of the mantle's alimentary canal, so to speak, where different stages of rock digestion are occurring. The argument is almost totally deductive.

If the oceanic plates and basins have been completely renewed every 160 my, then they will have been renewed about 35 times since the Earth originated. Each time these plates would have scraped off some of their sediments upon each other. By now the continents of the Earth should be presented in heaps of sialic rock randomly disturbed as islands around the globe. Such sediments scarcely exist. Or they are unrecognizable as such. The sediments of the ocean are less than a kilometer deep on the average. Call them a kilometer; double this to match the disproportion of sea to land; and multiply this volume 35 times. The result, a column over all the land of 70 kilometers, far exceeds the present continental sediments (if the only source of these is oceanic sediments) nor does it appear in any large sedimentary masses distinct from the indigenous continental mass.

The present mass of sedimentary rocks is about 32,000 x 10^{20} grams. It is about 5% of the crust. From the deepest trench to the highest mountain of the Earth is about 20 km. Some 44% of this is pre-Cambrian, 56% of it is of later origins. Most has been recycled several times, but not all, else we should not possess fossils indicative of all ages. "The whole sedimentary mass has been turned over five times."[15] The oceans are thought to have been in a steady state throughout all of this time, picking up and delivering sediments.

Most of this conjecture becomes nonsensical if a single fact is considered: consistent stratification of species around the world, such that exceptions are considered anomalies. If oceanic plates repeatedly dumped their "young" sedimentary contents at the base of the onshore sedimentary heaps, the Phanerozoic order would be reversed, as in the *Glomar* discovery just reported; the older the sediment, the higher up it would be stratified. Such not being the normal case, one is compelled to reject the theory of

---

[15] Fred T. Mackenzie, "Development of the Oceans," 13 *Ency. Britannica* (1974), 480.

subduction and perpetual plate renewal. Marine sediments are the majority of all organic facies; they are loaded in temporal order according to the principle of superposition. They did not arrive on the land by plate tectonics; they arrived by tides, floods, land rising, and other quantavolutionary mechanisms.

For subduction, forceful convection cells are required. "All the fountains of the deep must be broken up," in a parody of the unique event in the Bible, not once but continuously and forever, over billions of years, enough to move the furniture of all the Earth's land around the world every 160 million years, inch by inch. The path of upwards and downwards movements cannot be smooth; at the least it is different beneath the thin sima than beneath the thick sial; furthermore, some interception must occur at the two or more levels of the mantle where striking seismic discontinuities are observed; indeed, it should perplex the conductionists that these seismic barriers even exist, for would not eons of convection have effectively erased what, after all, can only be levels of chemical mineral differentiation? Seismic studies show that the Earth below the surface is stratified; what else could seismic discontinuities mean?

The thickness of the Earth's crust, as the physicist P. Jordan once said, is a breath of air blown upon a desk globe. The breath should be unevenly blown, for the continental portion is 40 km and the oceanic crust is only 5 kilometers thick. This is using the Mohorovicić Discontinuity as the boundary between crust and mantle. At this "Moho" boundary, the velocity of a seismic signal increases sharply, indicating a density increase from 3 to 3.3, the mean for the crust being 2.8 g/cm^3 and that for the incomparably more massive mantle 4.5. The increase in velocity (and density) occurs within a band of rocks of under five kilometers thickness. Below the oceanic crust, the Discontinuity zone is less than half a kilometer thick. The Discontinuity seems to be caused by "a difference in chemical composition between crustal rocks and the underlying mantle rocks."[16]

The theory of plate tectonics visualizes the conveyor belts of ocean crust moving along between ridges and trenches just above the Moho Discontinuity. In cases where the plate is oceanic and encounters a plate carrying continental material, whether supposedly built up of primordial granites and sediments or of trench debris folds, the conveyor belt (convection current) dips down, and, of course, the Moho dips too and resumes at about 40 km below the continental rock. In all of this process, the Moho is conceived to be independent of the tectonic process presumed to be taking place.

This is incredible. It is much more likely that the Moho Discontinuity marks the level at which the continents marched around the world after the Moon erupted, and, below the ocean, the level above which new crust had to be created from the uppermost magma of the mantle with atmospheric chemical participation. A new, subaerial, low-pressure, hydrated factory produced the oceanic crustal basalts out of upper mantle material. The continents and ocean bottoms are probably still in motion along the Moho Discontinuity, as they were, but much more rapidly, when the Discontinuity was born as the boundary between crust and mantle.

There are, besides the Moho, two more major discontinuities in the mantle, one at 400 km depth and the other at 650 km. In both cases, density and chemical composition are believed to change markedly. Inasmuch as both of these discontinuities, as well as the Moho, pervade the globe as "shells," they must be continuously penetrated by rising, falling, and lateral convection currents. It is perplexing to consider how the currents could be maintained throughout Earth history without erasing the discontinuities. Since there is evidence of the Discontinuities but

---

[16] A. E. Ringwood, "Structure and Composition of Earth," 6 *Ency. Britannica* (1974), 51.

not of the convection, the existence of the convection cells must be doubted. Moreover, as with the Moho, these other discontinuities may represent secondary and tertiary torsion levels, as the Earth, more than once, suffered deceleration of its rotation.

The fact of the general uniformity of depth of the Moho Discontinuity around the world is also an indication that it was formed at the same time as part of an epochal event whose negatively exponential tailing-off was temporally brief. The fact that the continental blocks move at a distinctively different, lower depth in the mantle has less to do with their "greater weight" (relative to the oceanic crust) than with the historical fact of their quite different genesis.

Quantavolutionary theory explains the occurrence of earthquakes along the global fault system, even where no trenches are subducting. And the convection cell theory is susceptible to challenge simply on the basis of insufficient energy, while the theory of plate tectonics as a whole does not pass a number of tests.

Regarding the first point, earthquakes have long been associated causally with faults, even before the oceanic ridge system was known. The submarine trench can be construed as a magnificent type of fault, almost always near an earthquake zone. But a great many earthquakes occur away from trenches. If they occur because material is being stuffed into the bowels of the Earth by a plate, there is yet no evidence of it, and one may as well maintain that the seismism denotes the relative motion of rocks, as was said earlier. The movement is often vertical so that, relatively speaking, some rock is often moving down, but that is not the point. It would be more in order to demonstrate that all the earthquakes occurring landwards of the trenches (and this is mostly the case, except in the Java-Sumatra region) bring about increased elevations as the debris is refused by the depths beyond the trenches.

Moreover, if the continents shifted and the ocean bottoms were repaved by an exoterrestrial and hence surficial force, then the disturbance of the Earth's crust and mantle would form a large area of surface directed as a narrowing cone into the mantle until it reached a point below which seismism could not be energized. Such may well be the case. The points would lie along the global fracture system and also where meteoroidal impacts have occurred.

The fact that an overwhelming majority of earthquakes is registered on the sial of the continents rather than upon the sima of the oceanic crust has surely to do with the greater depth of the continents as contrasted with the oceanic crust, but it also has to do with the greater age and rigidity, hence recent disturbance, of the plutonic land rocks. One may surmise that the sima is "better adapted" to movement because it was "born of movement." The very planar, uniform, featureless character of the sea bottom evidences that it has not participated in terrestrial diastrophism, but has, like the water itself, filled in with molten and flexible rock wherever the land has been removed.

The heat required within the deep mantle to expel excessively heated rock up to many thousands of linear kilometers on the surface is, of course, great; and, at the other end of the conveyor belt, or surface convection current, the rock must be dense and cold enough to sink, with a mechanical force assisting. Elaborate calculations have been made to demonstrate the possibility. None are convincing. It must be in many thousands of degrees Celsius, enough to burn the bottom of the pot, if the favored analogy of the boiling cauldron is pursued.

However, although the presence of radioactive minerals deep within the Earth is only a postulate, rising radioactivity-produced heat is given as the source, rather than some internal fire. Metaphysical figures are not difficult to come by, my critics will have been observing; so I can assert the same. There must be an irregular distribution of giant kettles and small kettles (because the surface areas of the convection process are vastly different) and hence some zones of radioactivity must be chemically different than others.

It is well known that volcanism gives off heat into the atmosphere and beyond. Why, with this naturally effective heat venting apparatus, would the cumbersome convection cell be required? There is no limit but the universe itself to the heat ejected sub-aerially; the bubbling stew is without a lid. Why should there be vast surfaces (between plate boundaries) bereft of volcanic outlets while the enormous mass of molten rock is pushed so delicately sideways as to not break the surface? Repeatedly the convectionists and subductionists use the quantavolutionary words "collision" and "plunge" to denote operations occurring at a scarcely observable rate out of "collisions" between bodies which are already impacted and therefore scarcely able to collide, though capable of jostling perhaps (wherefrom we might receive the submarine trenches). Still we read often that plates "collide;" one plate "plunges" beneath another.

They are in a desperate theoretical fix: their instruments tell them that they have only about 160 million years to sweep around the globe; the energy for this must occur by a relative heat emanating from radioactive decay. Some scholars must long for a young Earth whose interior might still have its "primordial heat" to give away. Their belief in stable astronomical motions of the globe and its solar system neighbors precludes their introducing thermal and inertial forces to abet the heat emerging from radioactivity and pressure.

To speak of flowing rocks as the convectionists do, and to a degree all must, is to employ the word viscosity. "Viscosity is a function of the chemical composition, temperature... and pressure..."[17] A high viscosity marks a slow flow: measured in poises, water flows with a 0.01 poise, honey creeps with 100 poises; and the rocks of the Fennoscandian uplift (where presumably once an ice cap and a polar region had produced Earth-flattening) exhibit by one estimate 2.4×10^{22} poises.[18]

---

[17] M. A. Cook, "Viscosity-Depth Profiles...," 68 *J. Geophys. Res.* (June 1963), 3515.
[18] G. Robert Morton, « Creationism and Continental Drift, » 18 *Creation Res. Sci. Q.* (1981), 42.

Summarizing and developing several studies, Cook publishes figures of about 10^{22} poises as the average viscosity of the crust and upper mantle, a viscosity of 10^{13} to 10^{14} poises at the bases of continents and about 10^{11} poises at a depth of 150 kilometers. A minimum viscosity or maximum fluidity would occur at about the 150 km depth, both geochemical and seismic observations being seemingly in agreement on the matter. But if there are no reasonably short gradients of viscosity thereafter, it is hard to visualize a large-scale convection dynamic in operation, much less a host of a dozen giant cells or a pattern of a thousand smaller convection cells working within the mantle. Not unexpectedly, then, Cook and Eardley calculated that to move the continents even in 200 million years would require forces "a billion to a trillion times greater than those that should be generated by the postulated mantle convection currents."[19]

Some scientific creationists, as exemplified by G.R. Morton, cannot accept continental drift, much less rafting as here described, because by their calculations, "neither convection cells nor any other [lateral] forces could have separated the continents within a few thousand years, if the viscous forces were involved in that movement."[20] The heat generated would have to be in the millions of degrees and would vaporize the Earth. Morton concludes that "either God separated the continents outside of natural agencies or that the Earth expanded in such a way that the viscous forces were no involved." Creationists generally avoid naturalistic exoterrestrialism, Patten being exceptional. So Morton does not consider the possibilities that led the present author to the model of Solaria Binaria: heat can be exploded and fresh atmosphere brought in from a fuller plenum rather than the thin present air of Earth. However, Morton remarkably adds a final

---

[19] "Analysis of Crustal Deformation by Mantle Convection Currents," 1962, unpubl., *cf.* Cook, "Continental Drift: Is Old Mother Earth Just A Youngster?" *Utah Alumnus* (Sept. 1963), 10-12; also critiques and debate, Nov. 1963, Oct. 1964, Nov. 1964.

[20] "Creationism and Continental Drift," 18 *Creation Res. Soc. Q.* (June 1981), 43.

sentence, irrelevant to all that he has said before: "The expansion of the Earth caused by an expansion of each individual atom due to a change in the permittivity of free space (the electric force) is a possibility which could avoid the viscosity problem." Thus he finally grasps for "the electric force," which, we have seen, is a heat-saver.

We are led back to the only mechanism that can produce low viscosity and provide it where needed, an exoterrestrial and hence surficial force suddenly applied to set the crustal blocks containing the continents – that is, the remaining blocks – into lateral motion. First, an explosion of surface must occur, with heavy electrical attraction and expansion. Then what Cook writes (and he uses the northern ice cap as a self-mover, without exoterrestrial assistance) is *à propos*: "Crustal distortion under a force sufficient to cause continental drift should then have amounted to from hundreds to thousands of times more than witnessed in the recent uplifts. In a catastrophic drift process viscosity breakdown along the shear surfaces would permit relatively easy flow compared with that of a threshold drift process."[21] Once in motion away from the rifts, the blocks (or plates) will have provided their own "grease" for a movement enduring several thousand years and exponentially declining to today's minute rates of drift. Overall, the pattern of movement was lunatropic, directed at resurfacing the Earth.

That nevertheless some collisions would ensue was to be expected, for the fractures around the globe necessarily expanded to move crustal fragments towards one another as well as toward the lunar basin. It is not surprising that modern studies detect contrary motions, as, for example, South America is being pushed westwards from the Atlantic Ridge and eastwards from the East Pacific Rise at the same time. These are not constrictory motions, so far as the theory of lunagenic tropism is concerned.

Most geologists and geophysicists today are satisfied that the heat generated and in part used to move the dozen plates of

---

[21] Cook, *Prehistory and Earth Models, op.cit.*, 271.

the world around is not so great as to make life impossible today or for a billion and more years past. A minority, as here, is not so sure. The issue is complex, technical, and abstract to the edge of pure speculation. This, however, is certain: an exoterrestrial and sub-aerial force can require less continuous heat and dissipate it more quickly; it operates with heat as more of a waste product than the key to the movement of the crust. The greatest portion of the heat given off to set the continents in motion would be explosive and would disappear into cold space with the exploded crust.

The resistance to the movement of the remaining crust would be much less than if the crust of the Earth had remained intact throughout Earth history. The continental blocks would require much less energy to move into the large areas heretofore occupied by continental material but now unoccupied, save by an erupting and boiling mantle material. Only several soft kilometers of depth would need to be ploughed through by the continental blocks heading toward the lunagenic basin.

Further the quantavolutionary theory, as proposed here, would rely upon earth expansion, largely owing to electrical discharge, as a precipitator and facilitator of the crustal movement. Except most rarely, as with Carey, the writers on continental drift ignore an obvious probability and even necessity, that when continents drift around the globe and the whole Earth's surface moves – no matter how slowly – the Earth's surface cannot remain a constant quantity, as if some secret ordinance has determined that the globe must have retained its precise figure of today through hundreds of millions of years, no more, no less, no matter what heats burn, what pressures invest the rock masses, what atmosphere bears upon it, what collides, what escapes.

Lately orogeny has come to be added to the marvels created by plate tectonics; the Alps and Himalayas are thus explained; so too the mountains and islands that stand landwards of some submarine trenches. The aforesaid secret ordinance must decree that extra plate is created for every mountain rise, or else admit some expansion of the crust. But if some expansion, why not much expansion? The material that is rising from the hot mantle must bring with it an expansive pressure; it is less dense; but when it cools upon erupting at the ridges does it become more dense? Not if it is like the famous stew pot or porridge in the analogy of convection cells. What remarkable chemical properties the magma must have: having had its backside scraped of sediments by the razor-bladed trench, it returns to the deep mantle millions of years later and hundreds of kilometers away and resumes its former thermo-chemical state.

Scientific advance of an important kind occurs when an acceptable interplay of theory and fact occurs. On the issue of the movement of continents, tropism towards the lunagenic basin was suggested as long as a century ago, by Osmond Fisher. But little was known of the ocean basins and the time scheduled for the event was in the dim beginnings of the Earth. W.H. Pickering of the Harvard College Observatory argued the case in 1907,[22] and it was well publicized. Meanwhile, H. Baker was evolving his theory. "The separation of the continents by fission," wrote Pickering again in 1923, has for 18 years "been attributed to the great convulsion that occurred at the time of the birth of the Moon, from the side of the Earth. This explanation of the origin of our Moon is at the present time almost universally accepted by astronomers. We see the same phenomenon occurring in many close double-stars."[23] He replaced the "center of origin" of lunagenesis off the southern tip of New Zealand.

However, Pickering held to the view that, although terrestrial lunagenesis and the Atlantic fission must have occurred late enough so that the continents possessed their modern forms, the time had to be early: "that a catastrophe involving the sudden removal of three-quarters of the Earth's surface could occur without destroying all life, both vegetal and animal, appears impossible." Therefore, he disputed Alfred Wegener's contention

---

[22] *Am. J. Geol.* (1970), 23, *cf.* CXV *Harper's Monthly* (1907), 120; *Scot. Geog. Mag.* (1907), 523.
[23] 61 *Geol. Mag.* (1924), 31.

that the Atlantic Basin opened up at the end of the Cretaceous period or in the early Tertiary.

Thus an impasse occurred until the present day. Wegener's continental drift theory is accepted but not its cause. Instead geologists cling to their terrestrial ideology and posit convection currents. The effects of ripping some 50 kilometers in depth off of most of the Earth's surface were conjectured to be utterly destructive of the biosphere. Today much new geological and geophysical evidence can be adduced from an examination of the Earth and Moon, tending to support the terrestrial origin of the Moon and the connection between lunagenesis and continental break-up and movement.

Moreover, the Cretaceous-Tertiary boundary is increasingly understood to mark the extermination of most species. Whether this boundary happened at sixty million years or twelve thousand years ago (which I construe to be the case) does not much matter on the issue of biosphere survival. I have pointed out elsewhere that biosphere annihilation was not necessarily predicated during lunagenesis. The immense typhoon that conveyed the crust of the Earth into space would have carried away with it most of the heat generated in the transaction, while at the antipodes of the event, downwards draughts of the then much more voluminous atmosphere would have cooled and regassed the land.

This is only one instance of the physical arguments that can be brought into play to establish that the biosphere would survive. To be borne in mind, also, is the prolific regenerative capacity of all species, no matter what their method of reproduction. The proper question to ask regarding biosphere survival is: what chance did one or more reproductive units of each of a million species have of surviving the conditions of lunagenesis?

To conclude, the tectonic plates are with declining force moving to restore global holospheric symmetry lost in lunagenesis. They are constrained and directed by the global cleavage system. The subduction theory is demonstrably incorrect.

The convection theory, which aside from its weak force and its dependence on subduction theory, depends upon a place to go, is impossible. Quantavolution theory, on the other hand, copes well with continental drift theory, assimilates it, simplifies it, and gives it a strong foundation in cosmogony.

CHAPTER TWENTY-FIVE

SEDIMENTS

We have entertained the possibility that till might have originated from the tail of a comet or cyclonically (tempestites). Using the typical approach of an intruder with an unwelcome hypothesis, I introduced statements of anomaly and bafflement. Thus, where is the till of the seas? Why is the correlation between till fields and glaciated areas not strong? If tektites can be exoterrestrial, why not till -remember that the feather and cannonball of Galileo fall at the same speed? And so on.

Dreimanis could be quoted: "Most of North America, particularly Canada, the entire northern part of Europe and considerable portions of other continents have been glaciated several times during the last two million years, and covered by various thicknesses of till and other glacigenic deposits... It sounds like a paradox, but till appears to have become more complicated with time, in spite of detailed and extensive investigation... "[1] And Kujansuu showed that "the flow directions of the ice sheet in Central Lapland," as indicated by five beds of till, followed five largely different directions.[2] And G. W. White: "In almost any excavation in the glaciated northwestern Allegheny Plateau, a till different from the surface till will be encountered, and in an excavation of 15 feet or more, several till sheets of different ages are to be expected... The tills vary in texture, composition, compactness, permeability and in joint spacing... the till sheets

[1] "Tills: Their Origins and Properties," in Legget, ed., *op. cit.*, 11.
[2] 2. "Glaciological Surveys for Ore-Prospecting Purposes in Northern Finland," in Legget, ed., *op. cit.*, 225.

454   THE LATELY TORTURED EARTH – PART V.

may be separated by a sand layer or a silt layer of varying
thickness... unweathered till may lie upon weathered till, or a
paleosol, or another unweathered till."[3]

As with till, so with all sediments: none is perfectly simple,
or, if so, can be proved to be. Sand occurs as 10% or less of deep
ocean sediments. Basalt does not give up sand; sand is continental.
Is this fall-out, turbidity currents of unobserved ferocity coming
off the slopes, early winds over empty beds? Shelton ends his
book on geology much as I end this chapter, musing about
hypothetical studies, "and finally, before we can do any of these
things, we must be able to tell one rock from another -which is
just about where we started."[4]

About 5% of all crustal rock is composed of sediments
that remain in something approaching their state after deposition.
They veneer about three-quarters of the continental surface to
thicknesses ranging from the merely visible to a dozen kilometers
in height, with the average for the globe at over two kilometers.
Sediments have been classified by priority of deposition and
anywhere from ten to hundreds of major and minor strata have
been allocated positions, sometimes only after prolonged
controversy, some only among certain believers. Besides
containing chemical and mineral traces and distinguishable fossil
remains, an estimated 80% of sedimentary rock are shales
composed of mud or clay, 10% are of sandstone and 10% are of
limestone.

The old problem of sediments missing from the
geological column became more worrisome with the discovery
that the ocean bottoms do not carry their proportionate burden
of sediments, much less the extra quantity to fill the gap in the
geological column. The continental slopes are formed of shaken

---

[3] "Thickness of Wisconsian Tills in Grand River and Killbuck
Lobes...," in R. P. Goldthwait, *Till: A Symposium* (Columbus, Ohio State
Univ., 1977), 160.
[4] *Op. cit.*, 424.

down, wasted down, and blown down debris from the shelves of the continents; but they would constitute only a small part of the supposed accumulation. The composition of the slope deposits is unknown. Perhaps half was carried off the shelves in the continental movements and orogeny following lunagenesis. A fifth may have descended from the sky preceding and accompanying the event. A tenth might have been washed in during the Noachian deluge. The small balance may be divided between river run-off into the oceans and cosmic and volcanic fall-out. Ager reports that "... chaotic deposits and slump topography have now been found at the foot of many present-day continental slopes."[5] The continental shelves and the abysses carry clay. The polar regions and half the remainder of the basins carry ooze. Sand and boulder are confined largely to occasional polar sediments. However, sand composes 10% of the ocean bottoms, too much for long-term sedimentation to have occurred. Carbonates, suggesting organic detritus, are common in the shelf and ooze sediments. Little suggests the continental rock in the oceanic sediments; it is a different world of unconsolidated material.

Perhaps the granite that forms the massive substructure of the continents down to about ten miles is composed of melted sediments, making the original crust out to be a thin basalt covering where the upper mantle has cooled. The chemical composition of granite would deny this idea, however. Nor does the location of the granites or sediments suggest that granitization has consumed sediments. Granite is found below, and intrusively, among sediments, not apparently where it might have been transforming them by conveying some special electrical or thermal force.

Old sediments do not appear to be far less common than new sediments, which they would be if they had been formed and consumed in a special earlier time on Earth. Granites can be formed by subjecting a mixture of albite, orthoclase and quartz

---

[5] *Op. cit.,* 38.

minerals to high pressure (30,000 lbs/in^2) and melting temperatures, and then allowing cooling. Hence it was surmised by O. F. Tuttle that the origin of granites was in hot magma of the mantle.[6] This idea may be the best of the three considered by him (the others being the metamorphosis of mostly sedimentary rock through hot chemical solutions, as above, and metamorphosis of proto-granites from ion exchanges causing crystal changes even while in a solid state); but he does not consider, nor do others, the possibilities of an accumulation of granite from atmospheric (plenum) deposits in an earlier state of the solar system, or of a massive electrical discharge between Earth and external bodies, or of a melt of an earlier crust by an exoterrestrial encounter.

In this book, granite is presumed to be the creation of a period during which the Earth gained dust, charge, water, and heat from the gaseous tube extending between the Sun and its binary partner. We suppose that granite is an exoterrestrial electric welding of a crustal covering for the Earth. It lay under such sediments as have formed out of largely 'cool' fall-out and heavy erosion.

That granite and basalt, both with the hardness of steel, can be quickly reduced to debris is attested by the well-defined Washington scablands; there closely-spaced rushes of water cut many channels of many meters of depth through hundreds of kilometers of basalt plains, before dumping some of their debris in hills, and more debris into the Pacific Ocean basin, where perhaps it was overrun by the continent. It may be added that most of the granite once possessed by Earth was ripped off and exists in a reconsolidated state on the Moon.

With the granite went half of the sedimentary rock as well. Still, much sedimentary rock is found in a largely disarranged condition on Earth, in some places being miles thick, in other places scanty or even nil. And the geological ages of the Earth, largely founded upon the layerings of sedimentary rock of the

---

[6] O. F. Tuttle, "The Origin of Granite," 192 *Sci Amer.* (Apr. 1955), 81.

continents, have long been suspect simply because of the disarrangement and, indeed, chaos of the sediments.

Geologists customarily still speak of erosion as the source of all sedimentary rock,[7] following a process of weathering of source material, transportation, deposition, and lithification which compacts and cements the material into a coherent rock. But to address such rocks with the fixed idea of gradual erosion is inappropriate.

Geologists, writes Ager, generally act on the belief that "the stratigraphical column in any one place is a long record of sedimentation with occasional gaps... But I maintain that a far more accurate picture of the stratigraphical record is of one long gap with only very occasional sedimentation... The gaps predominate .... the lithologies are all diachronous and the fossils migrate into the area from elsewhere and then migrate out again."[8] Ager does not presume to measure gaps of time, perhaps because if nothing happens, there can be no measure of it. Therefore, the gap may be long or short. Here we prefer the brief gap to the long. Indeed, often it can be argued that no gap exists.

In a remarkable survey, Woodmorappe has denoted the presence or absence of the ten conventional geological periods on a sample of 967 equal square areas of 406 square kilometers of the continental lands.[9] Ideally, every square on Earth should exhibit some rocks of all ten periods. Natural history assumes that all areas have undergone similar weathering experiences during any given long period of time; if, as is known, rocks of all ten periods are not found, it is because field surveys have not been competent or complete, or because the weathered debris of given age has been transported as such or as rock later on to somewhere outside the 406 square kilometer area (a journey of a maximum of a dozen

---

[7] As for example, W. G. Ernst, *Earth Materials* (Englewood Cliffs, N. J.: Prentice Hall, 1969), 111.

[8] Ager, 34.

[9] "The Essential Nonexistence of the Evolutionary-Uniformitarian Geologic Column," 18 *Creation Res. Soc. Q.* (June 1981), 46-67.

kilometers), or because the rock did actually form but was eroded and carried off, or because the rock once formed was later subjected to metamorphosis. Some credence can be given to all these explanations, but, too, it is noteworthy that the "presence" of period rocks in Woodmorappe's study often refers to a minor outcropping within the area and not to full coverage of the area.

The departure of reality from the myth is impressive. In no more than one per cent of this sample of the areas of the world are all ten periods of natural history represented. Some of these widely scattered areas are doubtfully complete (in the Himalayas, Bolivian Andes, Indonesia, South Central Asia, and Cuba). Rarely does one find even three of the ten geological periods in their expected consecutive order. Moreover, "42% of earth's land surface has 3 or less geologic periods present at all; 66% has 5 or less of the 10 present; and only 14% has 8 or more geologic periods represented..."

> Individual geologic periods' coverage of the earth's land surface range from a high of just over 51% for Cretaceous ... to a low of only 33% for Triassic. Only 21% of the Lower Paleozoic is represented in 3 or more of its periods; the complete Upper Paleozoic is found in 17% of the areas; the Mesozoic is complete in 16% of the areas. A complete Paleozoic record is found in 5.7% of the areas, and a complete Upper Paleozoic plus Mesozoic in 4.0%. Some percentage of *every* geologic period rests directly upon Precambrian 'basement', especially high percentages of Ordovician (23.2%) and Devonian (18.6%) doing so.

The data confirm the belief of those who argue, with Ager, that there are more gaps than record. Too, the chances are painfully high that one stands upon a seriously incomplete geological column wherever one may be on Earth.

Although the statistics will not suffice to show causation, they support the line of thought here: the Earth's surface has been reconstituted; the reconstitution has camouflaged the earlier surface and the earlier surface has disguised the reconstruction.

Many of the "gaps" in the record are illusions. Fossils are probably as often the perpetrators of unconformities as the indicators of them; they must often have gathered where "they didn't belong" in the course of catastrophes.

The strata of all periods prefer to rest directly upon their prior strata, showing a tendency towards a time-consistency in superpositioning, as conventionally believed; that is, each era tends to be more on its preceding era than on any other era. There is one important exception: all have a greater chance of resting on pre-Cambrian than on the last post-Cambrian eras. Except for the two periods just prior to it, a period has a better chance of resting directly on pre-Cambrian than on any other stratum; the correlation except for two directly preceding periods must be nil. This indicates a pre-Cambrian basement preference of all strata. It also suggests a simultaneity for deposits that have previously been assigned as successions.[10]

So what Price once called the "onion skin theory" of sedimentation is untenable, if it is indeed still retained by many. The essential principle of sedimentation should probably be called "quantavolution." Actually the idea has many antecedents and precedents: this we now well understand. Specifically applied to sedimentation, it means that the rocks of the phanerozoic era convey by their composition, strata, geography, quantities, and geological columns a patterning that suggests intensive, large-scale sudden and brief events, that is, a lately tortured Earth.

Derek Ager takes the position of a macrochronic quantavolutionist. "Changes, cyclic or otherwise, within the solar system or within our galaxy, would seem to be the easy and incontrovertible solution for everything that I have found remarkable in the stratigraphical record."[11] The secondary mechanism, which he employs repeatedly but without criticism of

---

[10] *Ibid.*, Table II.
[11] *Op. cit.*, 83.

its fundamental origins, is plate tectonics. "The theory of plate tectonics now provides us with a *modus operandi*."[12]

He sees a distinction between the exoterrestrial cause and the drifting continents as cause; thus, "we come to one of the great anomalies of the stratigraphical record, with the widespread extinctions of the Frasnian/Fammenian junction" of the Devonian. There is no evident explanation to be found in drifting continents or colliding plates. It seems that here, at least, we must appeal to an exoterrestrial cause. He has several additional preferred temporal locations for exoterrestrial interventions in geology.

He can use plate tectonics to discover and discuss numerous "periodic" and "episodic" catastrophes around the world. This enables him to be macrochronic: "the history of any one part of the Earth, like the life of a soldier, consists of long periods of boredom and short periods of terror."

He offers a wide range of examples, from numerous eras, of the worldwide distribution of various rock-types and fossils; this leads us to the supposition not only of a Pangea in which sediments and life forms might readily become worldwide but also, and perhaps more important, of species that never reached their potential limits, suggesting forceful interruptions of their spreading. Further, it implies worldwide equal conditions for even very special kinds of sedimentation and rocks to form.

He illustrates the bizarre differences in depth of the deposits of the same age in separate regions both near and distant, pointing out, for example, the one foot of Jurassic sediment in Sicily in contrast to the 15,000 feet of one Jurassic zone's sediment in Oregon.[13] Since they do not form on mountains, sediments, which can fill basins to a depth of up to 20,000 meters, would have been below sea level if the oceans existed when they grew.

He alludes to numerous wide differences in rates of sedimentation: a 38-foot fossil tree stands amidst the late

---

[12] *Ibid.*, 100.

[13] *Ibid.*, 40.

Carboniferous Coal Measures of Lancaster; but for the flow of sediments from rivers into the seas he quotes Holmes' measure of only one centimeter per millennium. He estimates the Grand Canyon at under 10 million years; the gorge, that is, provides a case of rapid erosion. "The periodic catastrophic event may have more effect than vast periods of gradual evolution:" this he calls "the phenomenon of quantum sedimentation."[14]

As there are more gaps than record, it is also true that there are more rapid deposits than slow ones, and the two facts may be connected in quantavolution. Rapid rates are easy to discover; Vita-Finzi cites a mid-Atlantic rate of clay deposit that increased suddenly from 0.22 to 0.82 grams/cm^2/year about 11,000 years ago (conventional dating), along with a drop in total carbonate deposition from 2.80 to 1.34 g/cm^2/y.[15] Nearly a 400% increase over an immense area; was it a type of Worzel ash fall-out? Or another case of rapid sedimentation?

At Nampa, Idaho, a well-carved human image in soft stone was recovered at 300 feet depth during well-boring.[16] The drill had penetrated 60 feet of alluvium, 15-20 feet of lava, and 200 feet of quicksand beds and clay, coming upon the sculpture in coarse sand, just below which was vegetable soil, followed by sandstone. One recognizes here a probable catastrophic sequence; the statue's presence, if admitted, wreaks havoc upon anthropology or geology or both.

Doeko Goosen has developed a wealth of related material, yet unpublished:[17]

> Two of my students collected undisturbed samples of a transition zone between a soil of less than 1 m thick and the underlying shale. My hunch was that the soil had not developed from the shale, and mineralogical analysis proved me right.

---

[14] *Ibid.*, 41, 44, 50.

[15] *Op. cit.*, 73.

[16] G. Frederick Wright, 11 *Amer. Antiquarian* (1889), 379-81.

[17] From letter to author, 15 Oct. 1982, Enschede (The Netherlands).

Within cracks of the shale multi-layer cutans were found. Traditionally such is explained by the one-layer per season theory, but when I looked through the microscope I saw oddities not compatible with that theory. [An expert on micromorphology confirmed his conclusions.] The phenomenon must have been caused by very strong tectonic vibrations, causing cracking of the slate and a sudden influx of clay and lime. At the same time fragments of the slate must have been projected upwards violently, passing through the soil, and now found on the surface.

Such tectonic miscibilation must be worldwide and visible under examination according to the quantavolution hypothesis in ground not believed to have experienced tectonism historically. Furthermore, a probable catastrophic cause may be assignable to soil processes that are considered ordinary and gradual. Goosen writes:

The formation of a laminated deposit via the season after season theory occurs only in highly exceptional circumstances. Wherever flooding occurs, there is also biological activity. The Rhine in the Netherlands each year floods pastures within the zone between the dikes, and leaves a thin deposit of clay. In the thus accumulated soil there is absolutely no lamination. The growing grass plus organisms like worms lead to homogenization. Indeed, it will be difficult to find on earth an environment where the season after season theory could be demonstrated. And then, upon seeing a laminated sediment, the inevitable conclusion must be that it is a catastrophic sediment, including the famous Scandinavian varves.

"Sedimentation goes on all the time, forever moving from place to place, forever cannibalizing itself."[18] It accumulates also from erosion of igneous and metamorphic rock. All sedimentary bodies, other than deep sea oozes and volcanic ash deposits, are likely to be diachronous. They stretch and spread out from a node

---

[18] Ager, 58, 52.

over a small or large region, so that the elapsed time from the center outwards may be considerable. Two contrasting illusions, we note, can be created if the same sediment is thinly spread over a large area, first that the sediment is all of the same time, whereas it is not, second that the time itself must be long because of ambient indicators applying to some central segment. That is, dating the indicator, one applies it to the whole, which brings about an illusory dating of adjacent rocks, too.

Rejecting the "layer cake" and "gentle rain from heaven" images as explanations of sedimentations, Ager introduces a rolled carpet that is gradually unrolled with time. We can extend the analogy. A producer of carpets lays down his roll and rolls it out before a salesman; the salesman rolls it up and carries it away to sell to buyers. Sometimes the producer has no carpets; at other times he brings only part of his collection; sometimes he brings in many rolls. The salesman sometimes rejects carpets and they are not sold; sometimes he buys one, or several, or all. A pile of rugs accumulates in the producer's showroom. Piles grow elsewhere. The salesman may even return his defective carpets. He may decide to deal with several producers, even as the producers may deal with different salesmen. Some buyers save carpets as a form of money; others wear them out quickly. In critical times for the economy, heaps of unsold carpets are laid out and accumulate, or are desperately sold in heaps; in inflationary periods, carpets become quickly and widely distributed. These last time periods would quantavolutionize the rug business.

When he is not imagining rugs, Ager's picture of the stratigraphical record is "of one long gap with only occasional sedimentation."[19] But his "occasional" sometimes is rare and sometimes frequent. I have noted this earlier in his view of tsunamis. Also now avalanches: "the frequency of landslides is quite enough to account for a major part of the wearing down of new mountain chains." Three cubic miles dropped in one slide at Flims, Switzerland; 40 million cubic meters of mountains fell into

---

[19] *Ibid.*, 34.

Lituya Bay, Alaska, in 1958. Still, a single earthquake of 10 or more on the Richter scale (and what was the number of the rising of the Sierra Nevadas?) would shake a new mountain range into well-worn shapes with garlands of debris all about below, and enough detritus to provide many moraines; the rise of a mountain range, indeed, may be its own heaviest eroder, then and there. The more rapid its rise, the more eroded it will be when it ceases to rise.

Ager argues convincingly the origin of deep sediments. The production of sediments is independent of subsidence. "It is only when sedimentation and subsidence coincide that the conditions will be right for the preservation of the vast thicknesses that constitute the stratigraphic record."[20] Again, we encounter a falling back upon the old notions of subsidence and uplifting. The phenomena are not mistaken; they are only insufficiently explanatory.

Ager partly realizes this, and sets up a very busy plate welding shop operating episodically over vast periods of time. A number of plates (and he seems to accept many major fractures everywhere as plate boundaries) spend history in roughly their original geographical locations, jostling heavily against one another periodically, episodically, spasmodically. "The continental plates, rather than sailing about the earth until they met in catastrophic collisions, separated and came together again repeatedly along the same general lines. In other words, there were many catastrophes and certain parts of each plate were particularly accident prone."[21] He would better have taken up the simple concept of ocean basins being created *before* the oceans and filled by debris washed down and fallen out of the catastrophic deluges.

We should not diminish one whit or alienate so expert and staunch an ally. We may, as mildly as we can, offer a suggestion. Let us give one more turn to the screws on the lately tortured Earth by computerizing its morphology. Suppose only one index to be composed for a sample of, say, 5 100 sedimentary

---

[20] *Ibid.*, 20.

[21] *Ibid.*, 86.

sequences chosen at random from the 510 million square kilometers of the Earth's surface (one in 100,000; this ratio and size of sample is typical for discovering the political opinions and predicting the voting behavior of the American population).

Call it an Index of Quantavolution "Q/a" (actually this could be a composite of a set of indices). It should contain and combine the number of distinguishable strata; an index of conformity to the ideal sequence of geological ages; the number of discontinuities that might be of diastrophic origin; the proportion of igneous and metamorphic intrusions; the proportion of the square kilometers (as judged by a hexagonal reading from drilling or otherwise) occupied by the central sequence of strata; and a total of the estimate of the lowest possible elapsed time for the deposit of each stratum to the column. Determine the usual statistical parameters of the sample, the sums, means, modes, quartiles, standard deviations etc. of the 5100 sedimentary sequences, and perform the obvious analysis and comparisons.

Some will say that the general information sought here is already known and taken into account, others that it is largely unknown and impossible to achieve, and many (rightly) that it is a caricature of a carefully drawn index. Many will comment that if the MOHOLE could not be financed to drill into the under seas mantle at an especially flushed period of American government finances, this project could never be funded. Many would want "add-ons": for example, "why not get samples of all strata in every sequence while we are at it?" We would eagerly agree. However, plausible conjectures and semi-data might be developed for all aspects of the index by library research and questionnaires addressed to many experts. Substitute sampling could be extensively employed.

Ultimately I would suppose refined summations to emerge such as the following: that the number of strata increase with recency; that superposition is 90% or better, but less than 50% of the recognized sequence is present; that 95% of the discontinuities might conceivably indicate diastrophism; that possible intrusions occupy over 50% of 80% of the sequences; that few sequences preserve their integrity over a square kilometer; that 80% of all sequences might conceivably have been laid down in their totality within 1000 (sic) years and that individual sequences would never exceed 10,000 years using conceivable assumptions.

The report would be entitled, *Reductio ad absurdum*, Part II. Perhaps one of the more entertaining aspects of such a study would be the objections that it is too literally empirical, and that the "total picture" is needed to disprove it and set it aright; the "total picture" is, however, what hitherto has given rise to the cosmogonies and science fiction that have commonly caused distress among geologists.

It would be possible to elaborate the hypothetical findings of such a study and to explain their heuristic and substantial utility, but not here. Thus, if the data is rotated topographically, significant summaries of continental and regional data would be generated. Moreover, as characterizes discussion of empirical data, no matter how crude, the air would be cleansed of some of the purely terminological pockets and gusts that cause turbulence and mental cloudiness. I see in such a project, also, a confrontation of the facts and their consequences that even a most learned and iconoclastic scientist does not consistently afford himself. He may come to realize that microchronism must be employed as a hypothetical model if a catastrophist is ever to integrate his facts.

PART SIX

BIOSPHERICS

Two sets of questions are addressed in these next chapters. What explains the distribution of fossil remains of life, particularly the large number of fossil clusters involving different species? A fossil generally connotes an individual, a herd, or a general disaster. The greater the confusion of species, the more likely an exoterrestrial catastrophe.

Given the increase in studies demonstrating an exoterrestrial connection with general extinctions, can natural history be reordered according to the occurrence, frequency, and type of exoterrestrial disaster? And is large-scale extinction possibly or invariably accompanied by large-scale biological innovation?

Sounds and sights are ordinarily excluded from natural historiography, because they do not linger and one can no longer find their remains. However, we ask two kinds of questions whose pertinence cannot be denied. If exoterrestrial forces were operative on Earth and induced terrestrial forces, what sounds would have been heard and what sights seen? Further, have ancient peoples left us with stories of sounds and sights that indicate certain natural events which they were experiencing? Were these on a quantavolutionary scale?

CHAPTER TWENTY-SIX

FOSSIL DEPOSITS

In coarse quartzose sandstones of stream channels of Antarctica's Transantarctic Mountains, fossil bones of the definitive reptilian genus, Lystrosaurus, were found. Deemed typical of Lower Triassic forms, it has been uncovered also in South Africa, India and China. In the sandstone, mudstone and white quartz pebbles are intruded along with the bone fragments. Logs and coal are at the same depth. Volcanic material is above and below. Remains of between 40 and 50 specimens are among the more than 400 specimens of other species in the same deposit. Numerous fossil relations have been shown between South America and Southern Africa, though not yet the Lystrosaurus. The China parallel introduces properly the Pangean connection.

Pangean world distributions of many species of flora and fauna, both fossil and living, can be traced. Living species that have no way of traversing present-day barriers are discovered to exist on both sides of the barriers, as the tigers of Africa, India and Siberia. Extinct species of one area are alive in another area, impassibly separated by modern geography, as the elephants and camels of North America, probably miscegenable with those of Africa. Specimens of the same extinct species are found in areas separated by modern geography.

A collapsed time schedule for the creation of the ocean basins demands a reconstruction of how aquatic species developed. Pangea was a world of small waters. Small and shallow lakes and swamps are conducive to the generation of individual variations within species and the prolongation of their careers. Whales and sharks travel great distances, but do not need to do so; they can flourish in a Tethyan sea; so with every other aquatic species. The great deeps are a last resort.

The eels from everywhere descend to breed from their rivers into the salt ocean and there find the Sargasso Sea, the great belt of weed-bearing waters on both sides of the Mid-Atlantic Ridge. They die there and their young swim for thousands of miles and years of time to find the rivers of Europe and America. The American eels have 104 to 111 vertebrae, the European 114 or 115, and n'er the twain shall meet.

Igor Akimushkin conjectures that eels originated or dwelt in the intercontinental fissure when it opened up an asserted 130 million years ago, not far from their fresh waters. Then they expanded their mobility to follow the drifting continents.[1]

Fitting the case to the quantavolutionary theory, it would appear that the Sargasso Sea is a part of the old Tethyan world-girdling shallow freshwater sea; that for breeding the eels found the gulfweed more necessary than the saltwater noxious; that there has been too little time to cast off the habit of traversing great distances, or of adapting to seawater for the long adult life; and that the small differences between American and European eels are an additional indication of a recent common ancestry. The Sargasso Sea seems to be growing, which, since it must precede the eels, indicates that it may not have been in existence long. In sum, eel migrations are as much a proof of continental rafting as continental drift is proof of the reason why eels must be astonishing long-distance travelers.

So also with aquaticized birds: if they migrate today intercontinentally, it is a stretching of their original habits; the irregular geometry, followed by birds that fly away from the arctic directly south and then veer at sharp angles to find their winter grounds, and vice versa to return, may be a function of land-mass migrations; the birds seem to be pursuing their original routes. If so, there may have been little time in which to evolve more efficient habits.

---

[1] *Animal Travellers, loc. cit.,* 126-46.

The Pangean shallow waters life centers were mostly wiped out, but survivors could readily adapt to the continental shelves and slopes, and the shallow and middle depths of the new ocean basins. (Wegener once alluded to the exclusive presence of shallow-water fossils in marine paleontology.) A typical succession pattern for the survival of an aquatic species would be to migrate or be turbulently transported from a Pangean center in a flooding action that settled into a temporary pond on the way across the land and towards what was to be the ocean. By the time of arrival at the finalized ocean shelf, where almost all aquatic species concentrate, the ocean waters bordering the land were quiet and cooled enough to permit proliferation. The exponential arithmetic for the growth of the population of the species at this stage would produce numbers sufficient to choke the oceans in a thousand years.

In the oceanic abyss, few species are found, and the same species are more commonly found on the continental shelf with few mutations. There is no exclusively abyssal flora or fauna, nor any "living fossil ancestors." The fact, however, that species do inhabit the abyss signifies that the abyss, were it old enough and the conventional processes of evolution occurring, would be teeming with adapted and mutated species.

The same logic would explain the scarcity of life forms in the high mountains, in the atmospheric bands, deep, below the land surface, and in the deserts. The inhospitability of these environments is only relative to dubious premises. Conventional long-time uniformitarian evolution and adaptation would have permitted all niches to become life-niches. Recent catastrophes provide of extinct niches such as would support a 50-foot winged dinosaur. If the oceanic salt seas carry few analogous niches for today's species, the reason may be limits imposed by the recency of drastic change, rather than limitations of nature.

Rocks dredged from the beveled tops of a number of seamounts carry imbedded fossils of current species that give 8 to 12,000 years readings on C14 dating (probably 4000 years old, then). The abyssal floors contain many bones, remarkably preserved. Large shark teeth of unknown species abound. Elephant teeth are found far down the continental slopes of North America. Their preservation for more than several thousand years is unlikely.

The mountain is a new life-niche for mankind. A swamp is preferred. The altitude of the ruined city of Tiahuanaco is too high for the natives to reproduce themselves readily; they used to descend to the plains for the purpose. Either they were correct or had been living too brief a time up high to be sure. The mountains rose after the city was flourishing. Generally, if mountains were old, they should support many more life forms than is the case. The erect posture of humans is well-adapted to sky-watching and life in the swampland; wading and carrying were greatly facilitated (as probably with certain dinosaur species). The food supply of swamps is lush and the fish and game of swamps easier to catch than the animals of the plain and mountain. It is a common error to portray hominids as living in the African climates of today and exerting themselves in the pursuit of large animals. Findings of bones pounded and scraped by hand-axes relating to hominoids might only signify omnivorous scavengers.

Large attached organisms are rare on the most recent oceanic ridges. The proliferation of such species on such ridges, that are rich in flora and fauna, is to be expected after a brief passage of time. The intense activity of the ridges several thousand years ago blocked their prompt development.

A distinctive southern flora, Glossopteris, found nowhere in the northern regions, is found as a fossil in India, Australia, South Africa, South America, and Antarctica. The case of India is doubly significant because a northern, adaptable, counterpart to Glossopteris exists but has never been found in India which is attached to Asia. This fact not only indicates continental rafting, but also recent continental rafting; there has been too little time for overland diffusion to have occurred.

Identical genera of late Permian fauna are found in Northern Russia and South Africa. A fossil dinosaur of five continents (North America, South America, South Africa,

Europe, and Asia) is known. Pangean distribution is generally confirmed. South America and South Africa, however, do not share mammalian identity today; cats are the only common genera. Many mammals common to both areas existed in Pangean times, before the catastrophes. Flora and invertebrates present a different picture today: there are numerous identities.

Evidences of Paleozoic faunal commonalty between North America and Europe are common. Many extinct Bohemian forms are replicated in extinct Texas forms, for example. During the Paleozoic and Mesozoic, some identical flora were to be found in East Asia and Western North America, and others in Eastern North America and Western Europe.

The age-breaking catastrophes, since they came from the skies, handicapped severely large land animals. Most of the dinosaurs were wiped out at once; the larger mammals were mostly exterminated in one brief period. Elephant remains have been found in South America in Chile, Venezuela, and Brazil, as well as alive in Africa and India. Mastodon remains were discovered in Ecuador and Colombia. Elephant fossil bones were found in a Brazilian bed, or nearer to the sea than that same bed, which contained hundreds of modern human skeletons mixed among numerous marine shells and nodules of carbonaceous matter; these were discovered about 1827; the bed was referred to as of limestone and of tufa (volcanic lava).

Piles of torn and mashed mammalian remains (mastodons, mammoths, bison, etc.) along with remains of many types of contemporary flora and other fauna, are discoverable in Alaska and Siberia. They are found in muck pits. They portray instant disaster by tidal and atmospheric forces. Large deposits of bones are found in Baja California (Mexico) cast up by the same kind of forces, uniting elephants and sharks in death.

Most species of large mammals suffered extinction in undeniably modern times. (In 1975 a radiocarbon dating of a mammoth find placed it at only 400 B.C.) The species that could betake themselves to high ground or fly quickly from one place to another survived in larger numbers. Humans were among the survivors. Maybe it will be also shown that humans were present when the continents split apart. The implication of such proof is that an ecumenical culture must have existed prior to the Lunarian *diaspora*.

The references to the catastrophic extinctions at "the end of the Pleistocene" mark the end of the ice age, which should, according to conventional theory, have been a blessing to most species, but was a universal disaster; life was first threatened by advancing ice and water, and then practically destroyed by the forces that broke up the ice and by ice break-up as well.

Many voluminous deposits of destroyed life occur in areas far beyond the tropical or temperate climate where the same or related species exist today. Injections of space gas at very low temperatures, associated once or several times with the tilting of the Earth's axis, may be evidenced in well-preserved, suddenly frozen life forms found in various places. Moreover, in every area of the globe where collective disaster is manifested among the plant and animal species, the geology of the areas usually confirms the biology: ooze and clay boundaries shift in the deposits of the ocean beds; organic layers are sandwiched between inorganic; ash is generally distributed on several levels of many marine and terrestrial sediments. Each level represents a general disaster; some stand for world disasters. Conflagration, tides, atmospheric violence, and other disastrous forces can probably be discovered wherever the mind is directed. Or so it seems.

Nature lends her occasional favors of fossils in a cruel way -by disasters. Human cult practices provide on occasion fossil cemeteries; otherwise human paleontology, too, would be dependent on the rare, unplanned event of a Pompeii. It is a euphemism, and misleading, to speak of "fossil cemeteries," or even of "' fossil assemblages," but, too, "dump," 'heap," "deposit," 'collection," 'aggregate" and other words are also questionable. Perhaps "fossil deposit" would be best, signifying many life forms concreted with clay, pebbles, and sand.

Fossil deposits may include on the one hand mineralized or petrified remains, or on the other hand preserved organic remains. The basic principle of fossil analysis requires every fossil occurrence to be approached as a catastrophic event. Quick burial of a potential fossil is essential. Then, occasionally, one or more of several chemical processes will preserve some of the organic structure itself, or an image of it, for posterity. R. Redfern summarizes fossilization for us, letting disaster pop out of a fully uniformitarian ideology in an analogy of the "fossil food" in a supermarket.

> Paleontologists sometimes find fossilized animals preserved in an almost complete state: sloths in arid caves, mammoths packed in ice, and men in peat bogs. Such effective preservation was the result of rapid reduction of moisture content or temperature, impregnation with chemicals, exclusion of air, or of a mixture of all four. Although we would hardly call preserved food 'fossil food' when we buy it from a supermarket, there is really nothing new about desiccation, deep freezing, chemical additives, vacuum packaging, and various combinations of all four.[2]

If all the remains of all that has ever lived had been preserved, might they exceed in mass the Earth itself? Termites and many insect species are considered geologically ancient. There is said to be a half-ton of live termites for every living human being. Considering that entire islands and hills have been found composed of mammoth and large mammal bones, and considering the huge fossil beds of vegetation, we can be sure that recent catastrophes have laid down the organic soils of today and a great deal more that has been eroded or quantavoluted since then.

What dies is thus quickly recycled biotically, unless some geological intervention occurs. And this intervention that

---

[2] *Corridors of Time: 1,700,000 Years of Earth at Grand Canyon* (N. Y.: Times Books, 1980). *Cf.* G. M. Price, *Evolutionary Geology and the New Catastrophism* (Mountain View, Calif: Pacific Press, 1926), 234-9.

fossilizes is almost always connected to the cause of death. The fossil record therefore is distorted as to populations of the species and to a lesser degree to the kinds and numbers of species.

Not all is known about fossilization, and less is realized. Ardrey mentions that the waters of Lake Victoria (Africa) were once fossilizing animals quickly and well because of some unknown quality probably not now present. E.R. Milton describes his examination of a petrified tree trunk in Alberta (Canada):[3]

> The piece... was pure clear silica inside, it was coated with a rougher opaque crust of partially fused sand. The tree whose stump was petrified was alive five years ago! After the tree was cut down to accommodate the right of way for a new power transmission line, an accidental break allowed the live high-voltage wire to contact several tree stumps still in the ground. The power was cut off within hours of the break. All of the tree roots which contacted the broken wire were fossilized... Obviously, electricity can metamorphose matter quickly.

One's mind reverts to earlier passages of this book where the presence of heavy electric fields and poisonous gases are given credence; perhaps these may have helped in the fossilizing process.

A fossil is typically an accident, a disaster, an anomaly. We should not find in Ecuador a mixture of mastodon bones, pottery, and coal. Nor reptiles with full stomachs, pterosaurs swallowing food, a mammoth with buttercups in his teeth, or an ichthyosaur mother in the throes of birthing her infants. The very existence of fossils reflects, says C. B. Hanson, "inefficiency in the natural systems for recycling organic material." He experimented with sending mammal bones down a flume in a laboratory in attempts to replicate natural conditions. M. Coe studied the decomposition of elephants in a Kenyan drought, and concluded that only rapid

---

[3] V *S. I. S. Rev.* 1 (1980-81), 10-1.

burial would allow any chance for fossilization.[4] There was no question here of the elephants being assembled to die and then deeply buried away from water and doused with petrifying chemicals so as to produce one of the fossil assemblages so commonly found in natural history. In fact, the best case of a fossil assemblage that geology can afford from historical times is the resort population of Pompeii and Herculaneum smothered and buried by the gases and ashes of Vesuvius in 79 A.D.

The following exchanges concerning a fossil conglomerate of prehistoric Nebraska clarifies the issues, as perceived by uniformitarians and catastrophists.[5] We quote the catastrophist:

> "In the American Museum of Natural History (New York) there is on display in the Late Mammals room (Room 3, 4th floor) a rectangular fragment (about 1.7x2.5 m, and 15 to 50 cm thick) of a bone breccia from a 'fossil quarry' near Agate, Sioux Co., Nebraska. Most of the bones are from a small, two-horned rhinoceros, *Dicera-therium*, with minor amounts from *Moropus* (6%), a clawed mammal related to horses, and from *Dinohyus* (1%), a giant pig like mammal. Extrapolating the quantity of individuals that make up this fragment over the total volume of the breccia layer (360 sq. m 15 to 50 cm thick), one arrives at 8200 *Diceratheria*, 500 *Moropi* and 100 *Dinohyi*. This breccia is believed (by Museum officials) to have formed in quicksand. The accompanying text reads:

> The accumulation of bones is believed to have been formed in an eddy in the old river channel at a time when the valley was not so deeply cut out as it is now, and the river flowed at the higher level. A pool would be formed at this eddy, with quicksands at its bottom, and many of the animals which came to drink at the pool in the dry seasons would be trapped and buried by the quicksand. The covering of sand would serve to

---

[4] *Fossils in the Making*

[5] 2 *Catas. Geol.* 1 (1977), 1-2; *Ibid.* n° 2, inside cover.

protect the bones from decay and prevent them from being rolled or water-worn by the current, or from being crushed and broken up by the trampling of animals that came to drink. But the sand of a quicksand is always moving and shifting around (whence its name of *quick*-sand), and with it the buried bones would be shifted around, disarticulated and displaced, so that when finally buried deeper by later sediments of the river valley they would be preserved as they are seen here, complete and almost undamaged, yet all the bones separate and disarticulated.

"I wonder whether the inventor of this mechanism has done his best to find an *actualistic* example of quicksand sucking up animals (with a lesser density than itself) in such a selective manner. Or is this another example of a *gradualistic* mechanism being preferred at all costs, even if it violates actualistic principles and physical laws? Has the possibility of a herd suddenly buried by a landslide or a liquefied sediment been considered? Are the properties of the overlying sediment compatible with this hypothesis? If so, it would be interesting to investigate this possibility also for other bone breccias, and to find out whether such breccias are more common from certain periods of Earth history than from others."

The story and comments are those of Hans Kloosterman, Editor of the magazine, *Catastrophist Geologist.*

Kloosterman's note receives a reply from Richard H. Tedford, Department of Vertebrate Paleontology, the American Museum of Natural History:

The hypothesis you object to also bothers me. The hall displaying the block of bones is to be revised and that will give us the opportunity to revise the captions for the exhibits. I think the critical evidence here is the extent of disarticulation of the remains which implies dismemberment of the carcasses and transport in a fluid and I see nothing improbable in the ordinary hydraulic agencies in a fluviatile regime. The concentration of remains can also be attributed to irregularities on the floor of the channel (observed during excavation) and the development of local eddies over the larger bones first deposited that trap further remains being swept downstream. The catastrophic factor may be the cause of death of a large group of animals and there are ways to assess this (unfortunately not tried with reference to the deposits in question), but normal stream transportation and deposition seems to me to be sufficient to explain the resulting deposit.

Richard H. Tedford
The American Museum of Natural History
Dept. of Vertebrate Paleontology
New York, USA

And, in rebuttal, Kloosterman writes the following:

If we first of all keep separate the two possibilities: death and deposition by the same or by different causes, the disarticulation of the remains certainly suggests that death has occurred previous to deposition, but the high bone-to-sediment ratio of the layer and the paucity of species suggests rapid burial after death, pointing to a connection between the causes of death and burial. Museum specimens will provide no answer to these problems and we will have to go back to the field, and also compare the characteristics of many different bone layers. Are layers when consisting of only a few species always composed of herbivores? Are their sedimentological characteristics different from other bone layers? Doesn't there exist any classification of bone layers, or have I just been unable to find it?

The issue is attacked by a hydrologist:

The quotation from the American Museum of Natural History implies that a pool, formed at an eddy in a river would have a quicksand bottom. There is only one way such quicksand could form, and that is by upward movement of groundwater through the bottom of the pool (see reference on Ink Pots springs). While this is not uncommon, there is no evidence (e. g. sorting of the sandy matrix of the bone breccia) presented for this.

Again referring to my Ink Pots paper, it is clear that density differences between quicksand and "trapped" animals do

present a problem. The animals may have died from exhaustion, but they would not have been "sucked in". Lacking further evidence for the quicksand hypothesis, I think the mud flow (liquefied sediment slide) solution is more likely.
The only way to solve this question is to collect all the evidence, including grain size distribution throughout the deposit, and detailed description of all "foreign matter" in the sediment.

Robert O. van Everdingen
Hydrology Research Div., Environment Canada
Calgary, Canada

Ref.:
Van Everdingen R. O., 1969: The Ink Pots--a group of karst springs in the Rocky Mountains near Banff, Alberta. *Can. J. Earth Sci.* 6/ 4: 545-554.

And Kloosterman concludes the case:

The problem here is that an equally strong and pervasive uniformitarian influence exists in sedimentology as in paleontology, with, in the interpretation of sediments, an aversion to even such common and minor catastrophes as rapid mass movements. Even if we are willing to consider catastrophist hypotheses, some basic data may be lacking, and thus the "cooperation" of the two specialities may lead to a typical case of "cross sterilization," so common between two different disciplines or even branches of the same discipline.

Enlightening as these comments may be, it is noteworthy that what to this author seems to be the more likely solution of the problem is not mentioned. The animals are of distinct species and were killed together, their bones disarticulated, and their bodies concurrently buried, in a (probably presumed) "eddy" of a now extinct river. No indication of water-wear or scavenging affects the bones. Probably a large cyclone was involved; the animals were picked up, torn apart, dumped, at some distance, and buried in a matrix of debris that was also being transported. If the conglomerate contained more species, further study might reveal a possibility of a water tide as the prime factor.

K. E. Chave's tumbling barrel experiments, in which shells and skeletons of marine animals were subjected to water, chert pebbles, and sand abrasion at 30 revolutions per minute, saw a reduction to under 4 mm grains of most of the structures within 183 hours, with perhaps 40 hours representing a half-life figure for average structures.[6] Complementary reduction occurs biochemically and by the action of other animals.

Clearly, then, given 200 hours of rolling about, little identifiable fossil life would remain. Supposing that the rolling were stretched out in a tide or current, about 300 kilometers of movement at one kilometer per hour would reduce practically all life forms to grain size in a bio-mineral soup, which, when motion ceased, would be deposited and in a matter of days form a strong deposit, partly mineral and partly biological. The tide would be moving much faster in any disastrous scenario. The rate of destruction would increase with the speed. Therefore, a fast tide in a few hours over a stretch of a few kilometers would render the fossil record something readable, if at all, by electron microscope and paleobiochemistry.

If tides had totally overrun the globe, the fossil record would be much less - all the less because tides dig up old deposits as they move, too. On the other hand, is the fossil record so generally rich as to imply large expanses of peaceful, tideless time when shells could find a quiet home, preserved, until pushed into visibility, there to encounter aeolian forces? Looking at the question in another way, where in the world would a fossil go to rest undisturbed by currents, electricity, and chemicals for a million years, or a hundred million, or a billion? "Hitler's *Festung Europa* (Fortress Europe) has no ceiling," we used to say in 1944. Has any fossil anywhere an anti-electro-chemical fortress, a Festung Fossilia with a ceiling? If we had available to us a thorough paleontological survey and map of the Earth above its

---

[6] K. E. Chave, "Skeletal Durability and Preservation," in J. Imbrie and N. Newell, *Approaches to Paleoecology* (N. Y.: Wiley, 1964), 377-82.

granites, we should be able to answer the question of the age of the surface since its last scourings. We do not have it.

Discovered fossil assemblages number in the hundreds, although they are not nicely inventoried. They occur on every continent, in many countries, in high and low latitudes, whenever land animals, plants and marine life have thrived. A large number remain to be discovered. A list of over fifty is before me as these lines are written, and I realize that they are almost all either late Cretaceous (reptiles) or late Pleistocene (large mammals), and that one must take into account many times this number for the aforesaid periods and then every "rich fossil bed" that graces the boundaries of the total Phanerozoic calendar.

An item from *Chemical and Engineering News* comes to mind.[7] Workers "found the fossil skeleton of a baleen whale some 10-12 million years old in... diatomaceous earth quarries in Lompoc, Calif. .... The whale is standing on end in the quarry and is being exposed as the diatomite is mined... The fossil may be close to 80 feet long." A sarcastic reader wrote in (March 21, 1977) that "Everybody knows that diatomaceous earthbeds are built up slowly over millions of years as diatom skeletons slowly settle out on the ocean floor. The baleen whale simply stood on its tail for 100,000 years, its skeleton decomposing, while the diatomaceous snow covered its frame millimeter by millimeter." That is, catastrophes affect the minute as well as the great life forms.

We do not know what proportion of fossils contributing to paleontology was derived from conglomerates as against individual finds. As expected, no one has sorted the assemblages into those involving collective catastrophe and those accumulated by normal individual disasters. A committee of experts would probably find few if any of the latter category, some of doubtful origins, and the majority to be collective disasters.

It would not take long today to conclude, for example, that the famous La Brea (Los Angeles) tar pit and similar pits,

---

[7] *Chem. and Engin. News*, Oct. 11, 1976, quoted in III *Kronos* (Eall 1977), 68-9.

discovered many kilometers away, portray catastrophes. The conglomerates of smashed and disarticulated bones of discordant species (saber-toothed tigers, peacocks, etc.), gravel, and asphalt point to a paradise of wild life suddenly devastated and revived only as the dry, thinly populated land, poor in fauna, of recent historical times. The time of the La Brea incident has had to be lowered drastically; for one thing, human bones have been found there; but also, the assemblage has been connected with other major events, such as the drying of lakes, placed at about 3500 years ago.

On the principle of "the Great Contrary" as the ancient Chinese called it, it would seem that the uniformitarians have received their chief input to the reconstruction of ancient species from the catastrophes that they would deny, just as the omnipresence of strata upon which they depend for their geology carries the heaviest implication of repeated disturbances of the Earth's surface.

Fossil conglomerates are not partial to genera or to epochs. Many recent studies have been based upon material dredged from marine sediments, and concern minute organisms or creatures. These, too, usually mark boundaries ordinarily termed epochal, or climatic, or even catastrophic, for they involve abrupt terminations of some certain composite of species. Thus, when suddenly a thick band of coccoliths is dredged up from the bottom of the Black Sea, aged perhaps three to five thousand years, a sudden end to a regime becomes apparent: a deluge of strange waters, an abrupt climate change, an electric shock transmitted throughout the body of water, or a sudden break in the food chain occasioned by similar events.[8] We speak more of this when we come to discuss extinctions.

At Bearsden, near Glasgow, a fossil conglomerate termed Carboniferous by age is found. Marine and freshwater strata are

---

[8] Egon T. Degens and D. A. Ross, "Chronology of the Black Sea over the Last 25,000 years," in *Chemical Geol.* (Elsevier: Amsterdam, 1972), 4; also, with J. Mac Ilvaine, 170 *Science* (9 Oct. 1970), 163-5.

interlaced; marine and non-marine life-forms are present, not necessarily tied to their "appropriate" rock strata (land plants and marine animals are mixed); crustacean and shark fossils (rapidly decomposable) are found in high degree of preservation[9]. Though often the material of coal beds, they are not carbonized. A series of tidal thrusts is to be assumed; further, coalification does not occur, it appears, unless an independent heated element is added before or after dumping. The evidence is consistent with the catastrophic theory of coal formation.

Coal deposits are fossil conglomerates of a most impressive kind, and call upon the winds, the tides, and the giant bulldozers of ice and rock.

Quotations from botanist Heribert Nilsson are pertinent:[10]

> Even if our peat-moors grew to a thickness of 2,000 meters, *nothing* would be similar to the Ruhr Carbon or any other coal district... If the possibility of an autochthonous formation of the seams is judged from the point of view of the amount of material available, the results must be considered as highly improbable. A forest of full-grown beeches gives material only for a seam 2 cm. It is not unusual that they are 10 meters thick, and such a seam would require 500 full-grown beech forests. *Whence this immense material?* How was it deposited all at once? Why did these masses of living organic material *escape decay*, why was it not completely decomposed?"

To what degree sediments are "rock fossil assemblages" is unknown. They too, with or without fossils, can be transported by high-energy vehicles. If a tree stands vertically in a sediment

---

[9] 5 S. I. S. Workshop 1 (1982) 28-9 citing Nature (17 June 1982), 574.

[10] Quoted by Bennison Gray "Alternatives in Science," VII *Kronos* 4 (1982), 15, from Nilssen's "Summary of the facts and leading principles concerning the non-evolutionary phenomena in the world of biota and the theory of emication," based upon his *Synthetische Artbildung: Grundlinien einer exakten Biologie*, 2 v., Lund: Gleerup, 1953).

does it not demand that its whole depth of burial should be carried throughout its stratum wherever it leads and the whole be considered instantaneous? Should not the vertical great whale referred to above be a measure of a whole stratum's instantaneity? A stratum can only be as thin as its tallest fossil will allow. A poly-strata fossil wipes out practically all the temporal pretensions of the blankets of its bed. Ideally, it should wipe out all identical blankets everywhere.

A famous instance of ancient catastrophic fossilization was introduced by Hugh Miller in 1841 in regard to the Old Red Sandstone:[11]

> The River Bullhead, when attacked by an enemy, or immediately as it feels the hook in its jaws, erects its two spines at nearly right angles with the plates of the head, as if to render itself as difficult of being swallowed as possible. The attitude is one of danger and alarm; and it is a curious fact... *that in this attitude nine tenths of the Pterichthes of the Lower Old Red Sandstone are to be found...*

> At this period of our history, some terrible catastrophe involved in sudden destruction the Fish of an area at least a hundred miles from boundary to boundary, perhaps much more. The same platform in Orkney as at Cromarty is strewed thick with remains, which exhibit unequivocally the marks of violent death. The figures are contorted, contracted, curved, the tail in many instances is bent round to the head; the spines stick out; the fins are spread to the full, as in Fish that die in convulsions... The record is one of destruction at once widely spread and total, so far as it extended... By what quiet but potent agency of destruction were the innumerable existences of an area perhaps ten thousand square miles in extent annihilated at once, and yet the medium in which they had lived left undisturbed in its operations?

---

[11] *The Old Red Sandstones* (Edinburgh, 1941), 48.

The depth of the fossil bed was immediately determined. Miller gives it at over 8000 feet. Hence all sandstones of this type everywhere in the world must be treated hypothetically as quantavolutionary. This promptly casts suspicion upon all rocks in the 360° global ambiance of the sandstones.

It seems that this episode, which fascinated the scientific public over a century ago, is due for a reassessment in the light of current knowledge especially since a new element is found at the well-known scene, radioactivity. "Anomalous high radioactivity has been detected in *Homosteus,* a fish from the same Old Red Sandstone beds in which Pterichthyodes occur," writes Hans Kloosterman.[12] We have mentioned similar cases earlier. Kloosterman continues:

> Latter-day uniformitarians tend to explain the radioactive anomalies by differential absorption of radioactive elements posterior to deposition. Conceivably this will bear out to be correct, but it could be only a partial explanation. Has any study been undertaken to find out whether high radioactivity in fossil bones correlates with the great faunal breaks of the Earth's history?

Radioactivity does not kill and assemble fauna quickly. It is associable with forces that do so and it implies exoterrestrialism: cosmic lightning and electrical discharges; freezing, gassing, and smothering fall-out, and incoming tides that have been radiated elsewhere.

Many microchronic catastrophists, hot on the scent of fossil absurdities, believe in the contemporary existence of species that are conventionally placed in superposition and assigned sequential periods of existence. The number of individual

---

[12] Kloosterman, *et al.* (supra, fn 5) citing S. H. U. Bowie, D. Atkin, "An Unusually Radioactive Fossil Fish from Thurso, Scotland," 177 *Nature* (1956), 487-488; W. R. Diggle, J. Saxon, "An Unusually Radioactive Fossil Fish from Thurso, Scotland," 208 *Nature* (1965), 400.

anomalies - a cold-water clam in a hot-water clam bed or a dinosaur among mammoths - is too small. Indeed, I have read of no incontrovertible case of major consequence for the reconstruction of time and evolution. The most sophisticated of their concepts seems to be fossil zoning, by which, if I understand rightly, is meant the simultaneous growth of ecological sets of a greatly different order. These sets are shuffled about as the scene changes, under castastrophic duress. One ecology is piled upon another and a long temporal sequence is assigned to the whole and its parts.

I can conceive how, let us say, continental tides of translation might sweep in and deposit a life zone upon one area; also I can conceive of another wave, reverse or oblique to the first, carrying upon the same area a second layer of fossilized sediments, and, in the end, of the second being given incorrectly a much younger age that the first. I cannot conceive, on the other hand, of nature being so neat, so orderly, or so given over to long range thrusting. One bears in mind that the longer the transport, the worse the conditions for fossilizing. Also, the chances that a tide or bulldozer will pick up inter-zonal species are excellent and therefore will place not only 'A' upon 'B' but 'B' upon 'A'. But such occurrences are quite rare, and almost always distinguishable. The inconsistency would be noticeable.

One cannot but feel at times that paleontologists have a lore that is locked out of the literature and that would emerge upon systematic questioning. Thus, what are the statistical parameters of fossil deposits *in situ:* how often, for instance, are fossil beds pure and how often apparently heterogeneous and to what degree? Are fossil deposits of ancient ages more likely to be heterogeneous than late fossil beds? If fossils usually travel, as Ager says, do they travel with their own age group? Does the age-pure rich fossil bed indicate, not a long, but a short chronology, because the fossils have not had time to be mixed or destroyed?

No part of the world is without fossil deposits. This would indicate that no part of the world has escaped catastrophic experiences. Marine fossils are of shallow seas: the oceans may be too young to have spawned new species, much less to cast them over the continents.

A great many fossil deposits are assigned old ages. The horrified fish of the Old Red Sandstone referred to earlier are Devonian and given hundreds of millions of years. The theory of this book has been tending toward confining biosphere catastrophes to the nearby ages and to an early period of "radiant genesis," defined in *Solaria Binaria,* with a stable intervening period. Either the ancient assignments will have to be re-timed or we shall have to give up this notion of a long period of Pangean stability during which quantavolutions were in abeyance. (See, e. g. the time charts following the text.)

We cannot conclude here from the study of fossil deposits that all major disturbances have been recent. But these conglomerations lend direct and substantial support to the quantavolutionary theory that Earth changes have been sudden, large-scale, and intense, and that most, if not all, have been very recent.

CHAPTER TWENTY-SEVEN

GENESIS AND EXTINCTION

Man is an exceptional creature, creative and destructive. He is a walking catastrophe for other kinds of life. Rashmi Mayur, in agitating for a "Kalotic World Order," projects that mankind will extirpate most species of life within this generation in exchange for 1.5 billion more people. J. W. Carpenter has cited estimates that 25% of all existing species may become extinct by the year 2000;[1] this is not the work of man alone perhaps, for the Earth itself may be enduring a longer-term decline stemming from its ancient cosmic bouts. But man is failing to protect the Earth.

If our approach is believable, nature requires high-energy forces to extinguish species and must need an equally great force to create them. The forces at the same time maybe subtle and powerful, as with invisible radiation, or flagrant and powerful, as with the crash of a large body into the Earth. Such is quantavolution.

It appears to be easier to discover death than new life. The literature on biological extinctions is getting heavier all the time, but little is forthcoming on genesis. We wonder why. Could it be a taboo against one kind of creation? Perhaps. Might it be this, that eighteenth century economics picked up an idea that common people have always had -and some great ones like Machiavelli and Hobbes, too -that life is a struggle among men; there are few places at the top; one must eliminate competitors to

---

[1] Despatch, UP Int'l,19 April 1982. He is Res. Director, Patuxent Wildlife Research Center.

get one's place; survival is a power struggle. Early modern economists went along with the notion. Thomas Malthus (1766-1834) pushed the line of thought into a world-wide view: goods are scarce, and men will compete for them; who is most effective gets the most; human populations are checked in their growth only by nature's instruments of famine, plague, and war. Later man was excused from the struggle, if he would develop "moral restraint" against excessive breeding. He has not done so.

But the biological world, as young Charles Darwin saw, had no moral restraints and was operating continuously under the pressures of the environment. Nor was there any *rentier* psychology in nature: "Give me my little niche and I will give you yours." Pressure to expand was infinite and this aggressiveness led to the most marvelous adaptations (to other's niches) and actual physical evolutions. So went the line of thought.

The underlying amoral (but moral in its own way) view here found the idea of catastrophism disturbing, first because a moral agent called God was customarily employed to command the disasters and reconstitute the world afterwards, and second because catastrophism without divine controls appeared to be quite disorderly and not progressive, lacking the capacity to create species (Elohim promises Noah *this* Deluge would be the last; nor did the survivors anywhere talk of new species, these all having been created once before.)

Nor when Mendel appeared on the scene with proof of mutations, was it appreciated that a mutation was a micro-disaster, perhaps tied into catastrophe somehow. It was for a later generation of scientists and theorists, impelled by the logic of the atomic bomb, to bridge the gap between an invisible particle and a visible awesome destruction. It was (and is) still too early to say how catastrophe creates as well as destroys; a third line of theory has to be developed to explain the paths of genesis, which despite repeated extinctions, have led to new and different forms of life.

Nonetheless, biological quantavolutions appear to have a large creative element. One of the rare early geologists to perceive this was Clarence King, and, in his attempt to assail evolution on its firmest historical ground, he penned several passages of beauty and importance:[2]

> Greek art was fond of decorating the friezes of its sacred edifices with the spirited form of the horse. Times change: around the new temple of evolution the proudest ornament is that strange procession of fossil horse skeletons, among whose captivating splint-bones and general anatomy may be descried the profiles of Huxley and Marsh. Those two authorities, whose knowledge we may not dispute, assert that the American genealogy of the horse is the most perfect demonstrative proof of derivative genesis ever presented. Descent they consider proved, but the fossil jaws are utterly silent as to what the cause of the evolution may have been.
>
> I have studied the country from which these bones came, and am able to make this suggestive geological commentary. Between each two successive forms of the horse there was a catastrophe which seriously altered the climate and configuration of the whole region in which these animals lived. Huxley and Marsh assert that the bones prove descent. My own work proves that each new modification succeeded a catastrophe. And the almost universality of such coincidence is to my mind warrant for the anticipation that not very far in the future it may be seen that the evolution of environment has been the major cause of the evolution of life; that a mere Malthusian struggle was not the author and finisher of evolution; but that He who brought to bear that mysterious energy we call life upon primeval matter bestowed at the same time a power of development by change, arranging that the interaction of energy and matter which make up the environment should, from time to time, burst in upon the current of life and sweep it onward and upward to ever higher and better manifestations. Moments of great catastrophe, thus translated into the language of life, become moments of creation, when out of plastic organisms something newer and nobler is called into being.

---

[2] *Op. cit.*

The breaking of an age is the occasion for instant creation and instant destruction. The quantavoluting high-energy forces concentrate upon reducing and at the same time increasing the variety of species. Otto Schindewolf, from 1950 on, was tracking what he called faunal discontinuities, for which task D. L. Stepanov called him the "most important and consistent spokesman of the idea of neocatastrophism in contemporary paleontology." In him one finds a more stringent scientific tongue than King's but the same view. "Faunal discontinuities... involve not just the dying out of old, but also the more or less sudden emergence of new phyla. This phenomenon can no longer be successfully accommodated under the term catastrophe in the true meaning of the word: it should rather be described as *anastrophe*."[3] (that is, 'upturn, ' not 'downturn'). It was partly for this same reason that the term *quantavolution* was chosen. Probably most species are born or die out at the disastrous junctures of natural history whence the rocks and fossil seas, too, provide evidence of commotion.

Pietro Passerini cites estimates of 1.5 million extant species and between 3 million and 8.5 million species as existing but still unidentified.[4]

G.G. Simpson estimated the number of existing species at two millions, and the all-time average since the beginning of life at between 500,000 and 5 millions. He guessed that the average species endured from 500,000 to 5 million years. He put the time since life began at from one to two billion years. When he performed his arithmetic he emerged with a high total estimate of all species of four billions, a medium estimate of 341 millions and a minimal estimate of 50 millions.[5]

Sometime later, Teichert estimated the number of discoverable or fossilizable species at ten millions, lower than

---

[3] *Op. cit.*

[4] "Knowledge and Entropy," 3 *Catas. Geol.* (June 1978), 17-9.

[5] '*How* Many Species?" 6 Evolution (1952), 342; Teichert, "How many Species?" 30 *J. Paleont.* (1973), 967-9

Simpson by a factor of five. The vertebrates among them were guessed at a round million. Cook used many less, accepting 1,105,000 for the living species, and then proposed that a figure of 130,000 for fossil forms discovered be considered a fairly complete sum of all past species. He asserts grounds for believing that most fossilizable species have already been discovered, implying that most or all species were created in short order and that a tenth or so have been eliminated. If algae and worms can be traced in the sediments, what would not have been traceable?

Schindewolf comments that "good conditions of preservation existed even for the most delicate, soft-bodied organisms in the Precambrian;" furthermore, it is incorrect that the rock strata of quantavolutionary times are missing or totally destroyed along with their hypothetical fossils.[6] Cook's view accords with his microchronic view of Earth history, which would permit one or several catastrophes and a natural dissembling of the fossil record to tempt exaggerations of the expanses of time and the progress of evolution.

Between Cook's one million and Simpson's two million for living species, reconciliation is conceivable. Between his 130,000 (say 200,000), and Simpson's maximum of four billions, there is no hope of ultimate agreement. Even Simpson's minimal figure of fossil species, 50 millions, is 250 times larger than Cook's. Altogether we are in a state of ignorance on what nature has afforded as candidates for extinctions. For that matter, no one is so bold as to define absolutely a species, much less to maintain nowadays that the conditions for speciation have always been the same. There may indeed be one or more dubious premises in all reasoning on the subject.

We may be confident that at least all major forms of life and many manifestations of each have been recovered from the past. In this sense, for the philosopher anyhow, there are no important gaps in the record. Yet, evolution demands ancestors,

---

[6] *Op. cit.*, 12

and its theory becomes dubious if the extinct are not sufficient in numbers to provide ancestors. Or at least the same few ancestors would have had simultaneously to branch in numerous directions; this is not impossible to argue; and a shortness of time would be no handicap to the argument. For the moment, to hold in abeyance an opinion on stasis and evolution, I shall accommodate my thinking to a million or more living species and over a million fossil species.

For several additional issues beg introduction. With the painful realization of gaps in the record that refuse closure, the reality of quantavolutions, and the improbability of point-by-point evolution no matter how much time is allowed, some scientists have spoken forthrightly for a new look at the record. They find that the path of evolution has been irregular, that there are times to evolve and times for quiescence. (Nor is this an artifact of time estimates.) Writes Brough, concluding an extensive review:[7]

> Evolution seems to have worked in a series of more and more restricted fields with large-scale effects steadily decreasing. Evolution at the present time is a slower and much more restricted phenomenon than it was earlier, and seems to be concerned with speciation in a pattern of larger systematic units which was laid down in the more or less remote past, and seems to have been standardized for a long time.

Genesis may not work at the will of God, but it does not work uniformly either. "Given a more or less even mutation-rate, and Natural Selection as a cause in evolution, there is difficulty both in accounting for the early and relatively rapid phases of evolution giving rise to major groups and also for the great decline in this phenomenon in later geological time."

Brough holds to spontaneous mutation as the source of genesis and speciation, and "Natural Selection merely works on

---

[7] James Brough," Time and Evolution," in *Studies in Fossil Vertebrates,* Atlone, London, 1958, 38, 34, 36.

these;" furthermore, "changes in organic forms have nothing to do with external factors." So he gets into a tight corner.

> There seem to have been evolutionary surges in the past when large changes of organic form took place, and produced the larger systematic units... There is plenty of evidence suggesting that during these evolutionary surges changes produced by mutations were not random, but were directional; this is well seen in such groups as the mammal-like reptiles, and in the higher bony fishes where several independent phyletic lines undergo the same sort of changes at about the same time.

Natural Selection may have assumed more importance when this process slowed down. An example of the evolutionary surge would be the "sudden appearance of a highly-developed fauna in the Cambrian," after diligent search of undisturbed sandstone, shales and limestones of the pre-Cambrian for hints of what was to come.

We speak here of simultaneous physical changes in a collectivity of species that may be unrelated. Within a species a saltation of individual changes must be also occurring. Hence there should suddenly occur a heavy branching out of types, some to survive, some soon to die. But then we encounter two additional phenomena of the fossil record -a lack of transitional types and an absence of short-lived sports.

In the case of all the thirty-two orders of mammals, Simpson tells us, the ancestral record is very poor. "The earliest and most primitive known members of every order already have the basic ordinal characters, and in no case is an approximately continuous sequence from one to another known. In most cases the break is so sharp and the gap so large that the origin of the order is speculative and much disputed.[8] ' E.C. Olson, reviewing the literature lately, reports: "under the very best circumstances...

---

[8] *Tempo and Mode in Evolution,* (NY: Columbia U. Press, 1944) 106; see review of his *Splendid Isolation* (New Haven, Conn: Yale U. Press, 19800 by Jill Abery, *S. I. S. 4 Workshop* (1981), 25-332.

morphological and stratigraphically graded transitions between classes and subclasses have been found. At the level of phyla and higher categories, any information on transitions as far as the fossil record is concerned is essentially non-existent."[9]

T.H. van Andel surmises that missing links "may have been expunged from the record."[10] The *Glomar Challenger* found one-half of the assigned 125 million-year record missing from deep cores drilled in the South Atlantic Ocean: he implies a catastrophic removal of the layers.

Other paleontologists, specialists in other evolutionary fields, agree: as with the rocks, so with the life forms, there are more gaps than record. In treating of this important point, discussion has focused upon "transitional types." It can be said that for no phylum, class, order, family, or species is there an indisputable succession of types that is predicted under the neo-Darwinian theory of evolution.

If, as Rodabaugh points out, micromutations must account for all observable variations between species, then the number of transitional species must be exceedingly large. "Furthermore, each species must be exceedingly viable in order to survive long enough to give rise to some 'evolved' descendent."[11] He then proceeds mathematically to demonstrate, with a probability approaching certainty, that transitional forms have not in fact existed. A "transitional form" is the species of life that is both intermediate and ancestral in relation to any two discovered fossil or living forms. "Missing link" would be a synonym for it. Where, for instance (if birds are indeed descended from reptiles), is the reptile who is just starting to sprout the wings of the bird? And the ancestor of the horse is nowhere to be found. D. M. Raup

---

[9] "The Problem of Missing Links: Today and Yesterday," 56 *Q. R. Biol.* (1981), 405-40; *cf* Mark Ridley, "Evolution and Gaps in the Fossil Record," 286 *Nature* (31 July, 1980), 444-5.

[10] *Nature* (3 Dec. 1981).

[11] David A. Rodabaugh, "Probability and the Missing Transitional Forms," 13 *Creation Res. Soc. Q.* (Sep 1976), 116-9.

and S. M. Stanley[12] are quoted: "Unfortunately, the origins of most higher categories are shrouded in mystery; commonly new higher categories appear abruptly in the fossil record without evidence of transitional forms."

Until lately, the ape Ramapithecus was in favor as the possible ancestor of the hominids. In 1982, it was reported that close study of a skull of Sivapithecus dated at 8.5 million years, and regarded as practically identical to Ramapithecus, showed definite relationship to the orangutan and hence was deemed not to be a transitional form to man.[13]

Nevertheless, although it is already admitted that transitional forms are absent, Rodabaugh computes, from the number of fossil birds estimated to have been found, the probability that a transitional form will exist. He finds the possibility so tiny as to be absent, quoting Emil Dorel: "Events whose probability is extremely small never occur." Rodabaugh concludes that, either the present biological world got here by macromutations ('hopeful monsters') or by special creation.

The "hopeful monster" is the new species, containing many changes, thrown out by a general mutation, and hopefully satisfying the conditions of survival. Rodabaugh declares that the concept "is rejected by nearly all evolutionists." Still, it has been reported, "within certain of the dying families [of Upper Cretaceous ammonites], an increase in size and the presence of bizarre-looking forms may be noted. This is a common accompaniment of extinctions of many groups."[14] It suggests catastrophe, accompanied by radionic mutating storms that both alter and destroy species. At the end of an "age" (defined as a "more settled" period), the species-mix and distribution of the biosphere suffer revolutionary change. Whereupon the struggle for life niches renews under more and more uniform conditions,

---

[12] *Principles of Paleontology*, 1971 (San Francisco: Freeman, 1971), 306.

[13] 121 *Sci. News*, (Feb. 6 1982), 84.

[14] V *Ency. Britannica* (1974), 576.

which may, however, not be the uniform conditions of the past age.

Charles Hapgood, another catastrophist, whose work has already been cited, confronts the same problem and although admitting that the major proponent of macromutation or "systematic mutation," Richard Goldschmidt, is opposed by the majority of writers, believes that a sudden shift of the Earth's poles and crust could produce the requisite shortening of the tempo of evolution. I am treading upon uncertain ground. In what has been said of the sacred and divine elsewhere (in *Chaos and Creation* and *The Divine Succession*, both works of the Quantavolution Series), I maintain that the historical gods are scientifically explainable within the framework of natural causes and human nature, but merge into a philosophy of religion that is not germane here. Hence enlightenment on the scientific level has to come through a uniform explanation of the fossil record or through macromutation in a catastrophic setting.

It is possible that a trillion "sports" have been disposed of by quick extinction, and that the few fossils that come down to us represent trillions of individuals of the standardized species. In this case, the absence of transitional "missing links" is not so improbable as some make it out to be. That is, if during a billion years, the average number of individuals per "long-lived" species has amounted to, say, a trillion and the average aborted and transitional form had to "make" or "break" on no more than one thousand specimens, then the chances of finding and recognizing such a necessarily handicapped form in the fossil record are negligible. That is, the transitional species would be a small population. If successful, it would spread with exponential rapidity.

If, for every significantly mutated species which survived there were 10,000 that did not, then even $10^4 \times 10^3$ would give only 10^7. By contrast, the surviving species averaged 10^{12} specimens that might enter the hall of fame of the fossil record. The relative chance is then 100,000 to 1. Consequently, if even a single showcase of transitional freaks has entered into the fossil record, there is enough to satisfy mathematical expectation.

It is more likely that a form of quantavolution operates (it is discussed in *Homo Schizo I* and *Solaria Binaria*). The absence of transitional types, if it proves anything, probably goes to prove that something like quantavolution must exist in genetics; there is then no expectation of transitional types. A mutated reptile has wings and it flies, without a long time of flight-prone ancestors.

However, transitional forms are not the most bothersome problem. Nor is it the continual relapse into Lamarckian environmentalism that characterizes the literature of many professed Mendelian-Darwinists. It is the nagging intuition of purposefulness that afflicts both the religious and atheistic observers alike. The species, from the virus up to the human, appear to be put together meaningfully. The species function in the weirdest, meanest, most wonderful ways to exist -not to progress, adapt, change, or intelligize, but simply to carry on an existence as best they can. Every species appears probabilistic to the point of impossibility.

A species may be "fantastically" constructed; but it is functional. A billion cases of an animal or plant cannot be denied. Its every trait relates to every other trait, just as in a culture every culture trait relates to every other culture trait somehow, no matter how "senselessly." The species is a whole, just as a culture is a whole. How can it be that, amidst the millions of chemico-physico transactions always occurring in the human body, a shot of adrenal hormone, prompted by a scare, is practically simultaneously counteracted by a hormone to prevent over-reaction to the scare, as the classical work of Cannon on homeostasis, or *The Wisdom of the Body*, first elaborated?

Stanley's calculations show that species of European mammals of today have on the average survived for one to two million years by conventional calculation (middle Pliocene mollusks had a mean duration of 7 my). Very few species of short duration (less than 0.4 my) occurred in the record. No ephemeral species appeared and disappeared. He concludes that "much more than 50 percent of evolution occurs through sudden events in which polymorphs and species are proliferated."[15] So here we find no sports, no transitionals, and a suggestion of macroevolution or quantavolution or "punctuated equilibrium." Also Stanley and Harper have noted a lack of correlation between rate of evolution and generation time.[16]

Life forms have widely varying generation-lengths. The human, who lives relatively long, reproduces from dozens to millions of times more slowly that most animal species. The human, therefore, should have had less evolutionary change in his past than a great many 'lower' and 'simpler' forms. Too, if the capacity to mutate is considered a positive feature of a species in "natural selection," then the human and many another 'advanced' species should be regarded as handicapped in the struggle to survive and adapt.

The biologist will probably agree with this and go in search of other advantages afforded these handicapped species in natural selection. When his search fails, he must grant that biology has always had an in-grained prejudice for the complex 'higher' animals, especially man. Man, like other advanced mammals, and indeed like all specialized as opposed to primitive, general, and simple organisms, is poorly designed for survival.

Nevertheless, this dismal picture includes a seed of hope, indeed a new hypothesis of quantavolution. If generation rate and evolution rate do not correlate, it may be that evolution occurs, whether in simple short-lived forms or complex long-lived forms, at an instant time that is absolutely short and therefore, reversing the history of the Colt revolver, "makes a big man equal to a little man."

More importantly, a long-lived form may inherit a genotype which all life forms share, no matter their generation time. This would be the ability in a mutation to change

[15] S. M. Stanley, "Stability of Species in Geologic Time," 192 *Science* (16 April, 1976), 267-8.

[16] C. W. Harper, Jr., comment, 192 *Science* (16 Apr. 1976), 269-70.

instructions for the largest and most *complicated* cell assemblage as readily as for a single-celled animal. One result would be equalization of evolution effects; the concomitant would be quantavolution or macroevolution, that is, the instant all-around change when a mutation occurs.

We have already noted the conspicuous absence of flora and fauna of the ocean bottoms and high mountains. The matter is relevant here again. The charts of extinction of species are also charts of genesis of new species. When species are exterminated in large numbers, new forms follow. Paleontologists question whether the new species are alterations of the old, or descended from earlier forms that failed to appear in the old fossil record, or evacuees from other zones of life. The first would seem logical but we are given to believe that first the old die out and then the new appear. This is an aspect of the problem of missing transitional forms. Yet it seems inexplicable.

Should not the dying dinosaurs and mutated mammals appear in the same strata? If heavy radiation is killing off one form but creating another, the stratigraphic gap should be inconsiderable, or the old and new forms should grade continuously into one another. It should not require more than several centuries to prove the fitness of a new form and to find it in numbers upon the next catastrophic occurrence. Perhaps this is what did happen; however, we are used to placing a million years between any two highly visible events in the record. Or at least one should be able to locate first a catastrophized conglomerate of fossils and then in succeeding uncatastrophized strata the new forms appearing as individual fossils. Else we should have to double the number of catastrophes, one for extinction and a second for genesis.

But is it "flesh or fowl?" Or, as Velikovsky asks: "Were all dinosaurs reptiles?"[17] Live birth among dinosaurs seems now fairly certain and not rare and there may have been a large mingling of important features hitherto believed distinctive

---

[17] II *Kronos* 2 (1976), 91-100.

between dinosaurs and mammals. Western USA rocks (Hava Supai Canyon, Colo.) produces drawings of dinosaurs, elephants, ibex, and human figures, as well as pictographs. If this ensemble is of the same time, a shocking reconstruction of the holocene period must ensue, absorbing time all the way back into the Cretaceous and up into the Neolithic. But all those creatures exhibited may be pre-selenian, and were extincted, even the particular human race of the artist, around 12,000 B.P.

Leaving this perplexing issue, we return to the problem of the ecological niches. These should be quickly occupied upon the demise of old species. Cameroun and Benoit found algae, fungi and bacteria thriving in volcanic lava laid down by volcanic eruption on Deception Island in Antarctica. Elapsed time was one year.[18] Krakatoa's little island received new life, too, within several years of being exploded and completely burned out, not only microscopic life but amphibia, reptiles and birds.

Yet, to repeat an earlier fact, large attached organisms are rare on the most recent oceanic ridges, according to Heezen and Hollister.[19] At 1000 to 4000 meters of depth, the ridges should be rich in flora and fauna, of established species. This signifies either an extremely young age for the ridge system as a whole, or for the most recent millennia a very heavy general eruptive activity.

In the end, so far as concerns genesis, we hold to quantavolution in biology and geology. The holospheric principle continues to be productive; the lithosphere, atmosphere, hydrosphere transact continually with the biosphere: all are affected by high-energy forces ultimately originating exoterrestrially. Genesis or the new in life occurs hand-in-hand with the destruction of the old life forms. This is nothing more than Schindewolf's "anastrophism."

No more revolutionary times than the present have struck geology and biology since the victory of gradualism and evolution

---

[18] NASA, news release 69/ 80, 27 May 1969.

[19] *Op. cit.*, 550-7.

over a century ago. The most striking signals of the change are emitted from the new studies of the extinction of species.

In 1961 Schindewolf prepared for the 113th General Assembly of the German Geological Society a status report on neocatastrophism.[20] He claimed major faunal discontinuities on the boundaries of the Precambrian-Cambrian, Permian-Triassic, and the Cretaceous-Tertiary eras. "On the divide between the Precambrian and the Cambrian there was a relatively sudden and thorough-going transformation of the animal kingdom, in which durable hard parts were deposited for the first time." There is a partial species overlap of short duration as the Permian moves into the Triassic as he notes in 4 groups of fauna, but he names 24 that expired and 24 that newly appeared. At the Cretaceous-Tertiary boundary, "the dinosaurs represent only one aspect of the much wider extinction process and the profound change in the composition of the faunas..." The larger mammals then came into being.

P. S. Martin and others trace the extinctions over most of the world.[21] D. A. Russell draws a picture of losses of 50 out of 250 terrestrial genera, a third of floating marine genera, half of the bottom-dwelling genera, and least of all in losses, about a fifth of the swimming marine genera.[22] He estimates that 75% of all species died alongside the dinosaurs, and in a period of only 1000 years, in conjunction with magnetic field reversals instigated perhaps by blasts of supernova radiation from a nearby star. He argues that "it is beyond the capacity of forces within the crust of the Earth to produce global catastrophe on this time scale;" conjectures of glaciation are inadequate, especially since no evidence is to be had of a general temperature change. Nor does Russell grant that the Sun could expel such high bursts of

---

[20] *Op. cit.*

[21] S. Martin, and H. E. Wright, Jr., ed., *Pleistocene Extinctions: The Search for a Cause* (New Haven, Conn.: Yale U. Press, 1967).

[22] 3 *Catas. Geol.* 1 (June 1978), 8; additional data in 10 *Geos* 3 (Summer 1981), 8.

radiation. Schindewolf here denominates 16 faunal groups as exterminated, 3 as overlapping briefly, and 24 as newly arising. As young Darwin wrote in his *Journals* (Jan. 9, 1834), "certainly, no fact in the long history of the world is so startling as the wide and repeated extermination of its inhabitants." (How could such observations end up in uniformitarianism?)

Schindewolf also dismisses explanations offered for these quantavolutions, none of which he deemed valid, such as gaps in the rock and fossil record, epidemic diseases, climatic changes, ice ages, differing depositional characteristics of species, reduced salinity of seawater, competition and natural selection, mammals eating dinosaur eggs, and changing sea levels.

Then he reaches into the skies. "Since faunal discontinuities are universal phenomena, they must arise from *universally active causes*. This has compelled me to look for agencies that would (1) have worldwide effects and (2) could extend to the totality of biotopes in the sea, on dry lands, in freshwater and in the air, as well as to stocks of most varied habitats and ways of life."[23]

His explanation lies in radiation storms:

> Since 1950 I have favored the hypothesis that sharp fluctuations in the high-energy cosmic radiation reaching the Earth should be considered among possible causes... I proposed that, on the one hand, the direct impact-effects of ionizing radiation should be considered, and, on the other, especially the increased generation of radioactive isotopes, which would become incorporated in the living organic matter and the molecular compounds of the chromosomes. Here they would unleash a twofold mutagenic activity, through ionizing radiation, on the one hand, and by the liberation of electrons in the decay of the isotopes on the other.

He cites theories of supernovas as the source and media for the transmission of the anastrophic material, and credits E. A.

---

[23] *Op. cit.*, 18-9.

Ivanova with "a connection between the faunal discontinuities and the migration of radioactive elements." Schindewolf points out that the exceptional survival rate of insects compared to other fauna may be due to the fact that "the resistance of insects to radioactive radiation is about ten times greater than that of human beings and other organisms."

Schindewolf's conclusions, including his exoterrestrialism, have been supported by later studies. In a summary report of 1982,[24] W. Sullivan added the Devonian-Carboniferous and the Pleistocene-Holocene boundary periods. In the former some 30% of the animal families disappeared. In the Pleistocene climax, 70% of the large mammals extincted. In both eras, marine life suffered greatly as well. He separates the Permian-Triassic into two extinction periods, 50 million years apart. Raup estimates that 96% of all marine species may have died out in the late Permian. Valentine and others before him (1974) have noted the petering out of highly innovative evolution.[25] The origination of phyla, classes, and orders came successively to a halt; families declined, but diversified in the Mesozoic-Cenozoic. A macrochronometrical paleontologist would say that there has been no major innovation in life for 40 million years (present company excepted).

Species, as we have indicated earlier, are an unknown quantity, with gross discrepancies in estimates of their historical numbers. Species are also more susceptible to genesis than the statistically concocted general groups with their assigned, more basic features; this is in accord with theory, whether microevolutionary or macroevolutionary. Probably species have been extincted and ramified on disastrous occasions that did not affect the existence of the basic forms to which they pertain.

Mankind may be one case in point; small differences are all that can be observed between man and ape, but as with the absence of major differences between men and women, in the

---

[24] NY Times, Jan 19, 1982, C1.

[25] 48 J Paleontology (May 1974) 549-52.

words of the French deputy: "Vive la petite difference!" In the two volumes on *Homo Schizo*, the origins of the differences between hominid and homo are discussed.

We uneasily recognize the need to consider together at the same time a new chronology, a new theory of mutations, better data on numbers and extinctions of species, and the observed quantavolutions of the Earth. For only by such means will we be enabled to answer a question such as the suddenness of extinctions. Somewhere in the space between a day and twenty million years, a line has to be drawn to distinguish catastrophe and gradualism.

The studies and critiques of the work of Alvarez and associates on the Cretaceous-Tertiary extinctions illustrate the point. The superseding of dinosaurs by large mammals is known, with their accompanying less dramatic extinctions and creations. Also now a chemical boundary is known. By one count, "Iridium-rich layers marking the end of the Cretaceous Period have now been found at more than two dozen locations around the world." Freshwaters and sea bottoms were affected along with dry land. Iridium is much rarer in the Earth's crust than in presumably exploded and space-affected meteorites. Hence a cosmic event is predicated, the Alvarez group holding to a middle-sized meteoroid explosion as the source, and a several months darkness accompanying the explosion as the killer of the dinosaurs.

Critics argue that the dinosaurs did not extinct with the end of the Cretaceous and took much longer to die out anyhow. Others say that the iridium is a product of heavy deep volcanism and slow sedimentation. Another maintains that the dinosaurs died from a drying up of their swamps. Still another claims that a mere several degrees of temperature rise or fall would halt the incubation of reptilian eggs and in a short time destroy the species.

After the Cretaceous comes a "nine-million-year" period of the Tertiary known as the Eocene. Geologists (Ganapathy, W. Alvarez *et al.*, and O'Keefe) now speak of a "terminal Eocene event", a catastrophe marked, as in the case of the end of the Cretaceous, by high iridium concentrations and microtektite fields. Do tektites and iridium always occur in exoterrestrial crashes? Or does this suggest that the two events, post-Cretaceous and post-Eocene, were one and the same, the "Eocene" and other eras having been concocted for differing fossils and strata of the same time.

An impatience and frustration seizes a person who is imbued with the perspectives of quantavolution and recency in biology and geology. Ordinary accounts of animals, plants, volcanos, winds, rocks, etc. come to look lame and foolish. The author, riding a KLM plane across the Atlantic in 1982, puts aside this chapter and glances through the *KLM News* magazine. There a puff is given KLM for flying seven small lemon sharks from Florida to Holland. The sharks needed "tender, loving care," "had to be massaged constantly," "sprayed continuously," "given extra oxygen," - these being beasts "having inhabited the oceans since some 50 million years before man made his first appearance." How, one wonders, could the sharks have prospered through one catastrophe after another: either the extinction would not be complete and exponential reproduction would quickly make up the difference, or else sharks are young species and much of their ecology must be young as well, including, say, manganese nodules that form around shark teeth in the abyss.

There are then the meteoroids and the supernovas as sources of anastrophic radiation. Could "cosmic" radiation come from volcanism? If deep, heavy, and worldwide, radiation closely akin could fall out from volcanism. But such volcanism, as we explained earlier, must look for high-energy excitation from the skies. As for the supernovas, in *Solaria Binaria* Milton and I attribute heavy radiation to at least three novas -a preliminary outburst of the Sun creating its binary, and two explosions of its binary in subsequent millennia. We also designate several other possibilities of radiation, that would be heavy enough to account for periods of intervening radiation, not from novas but from impact-explosions and crustal removal in passing encounters.

Here and there now reports are issuing of excessive radiation levels in rocks and fossils. Kloosterman was earlier quoted on the subject. Salop speaks of a primary enrichment of uranium in dinosaur bones. Numerous similar findings have been reported since 1956 in Brazil and Argentina. Some bones from an undated red sandstone were radioactive. J.E. Powell summarizes these findings. Fossils from Mongolia also show high levels of radioactivity. Kloosterman located these facts and also discovered that almost none of the world's natural history museums have measured radioactive levels in specimens of their collections.[26]

However, the prevalence of fossil conglomerations around the world implies brief periods of extinction, and forces not alone of radiation, and pre-existing ecologies quite different from those that came after the catastrophic periods.

So many rich fossil deposits occur in circumstances that reveal high-energy processes to be at work. In Baja California, fossils were laid down over hundreds of square miles of the desolate terrain, exposed by surface erosion. Living and extinct species mingled in broad confusion. Flint and obsidian artifacts lay also upon the fossil sediments. The bones of mastodons, ancestral horses, a giant tortoise, camels, bison, sharks, whales, sea cows, and fish were plentiful. A shark species found in Mississippi turned up here in the Pacific. Assigned times, prior to investigation on the spot, ranged from 50,000 years (in the case of the artifacts) to 60 million years. Dating aside, the giant, confused, and rich fossil fields signal a catastrophe or a series of catastrophes at short intervals of time, from floodwaters sweeping in from land and sea.

S.J. Gould, who has pursued assiduously the study of extinctions, has had to go well beyond gradualism, uniformitarianism, and natural selection. Luck or chance figures heavily. Random macromutation can substitute for isolation, by creating two species in the same niche without the benefits conferred by travel. Commenting on the Permian-Triassic catastrophes, where an estimated ninety-six percent of the families of marine organisms ceased their existence, he says:

---

[26] 3 *Catas. Geol.* (June 1978), 4-7.

There are few defenses against a catastrophe of such magnitude, and survivors may simply be among the lucky 4 percent. As the Permian extinction set the basic pattern of life's subsequent diversity (no new phyla and few classes have originated since then), our current panoply of major designs may not represent a set of best adaptations, but fortunate survivors.

Would the stripping of half the Earth's crust and an associated expansion and cleavage of the Earth, together with a paving of the ocean basins, all occurring within several thousand years and most of it very quickly in a single action complex, exterminate entirely the biosphere? Even the most determined catastrophists have passed over so frightful a concept. If, as has been conjectured, a meteoroid explosion of a few kilometers' diameter would destroy the dinosaurs, the colossal event portrayed here would annihilate all life.

To counter this universal skepticism, there is the fact that life does flourish today despite the event, so that if the event were proved, then the skepticism would have to vanish. However, taken as a problem in its own right, instead of an inference determined by an external logic, we should stress certain possibilities in the event of lunar fission.

1. The atmosphere at the time might have been enormously greater and so extending far into space to permit a reviving reverse flow to replace the escaping atmosphere, and to act at the same time as a great vacuum cleaner against the heavy dust clouds and heated air.

2. Although an enormous number of species may be extincted, only several survivors of a species may guarantee a replenishment of continental scope within centuries.

3. The possibility must be entertained that hitherto unused intra-species genetic adaptability can permit survivors of modified form under stresses seemingly quite destructive.

4. Holospheric catastrophes by their very complexity can block each other's effects, allowing some life-preserving niches to survive and even fabricating niches where none existed before.

It is no longer rare to hear scientists arguing an intervention from outer space to push evolution along. Objections arise from extreme proposals, whether of intelligent visitors or of lower orders. "Extraterrestrial footsteps on the sands of history," R. E. Dickerson has remarked, "do not seem to be mandatory."[27] They would be superfluous, for that matter, if a quantavolutionary theory has laid down the sands. Further, as detailed in *Solaria Binaria*, if exoterrestrial voyagers had landed on Earth they might well have felt at home. Until quite recently, their former planetary abode would have provided a genetic milieu in the same vast plenum of atmospheric gases that the Earth enjoyed. However, Mars and Mercury have lost practically all of their life-support systems while the Earth has retained a crucial halo of air and a vast supply of water.

In itself this can be made into an argument for a short term of life on Earth. The more one studies the possibilities of natural disasters the more likely it appears that, over long stretches of time, these would have been so frequent as to make a total disaster much more likely to occur. That is, if several disasters are granted, given the same Earth and Universe, why did not many occur and why not worse? Assigning the Earth and its species five billion years of self-development may turn out to have been a frustrating detour in the history of the human mind. By contrast, encapsulating the disasters within a unified theory, quantavolution, may prove enlightening and progressive.

---

[27] Letter, *Sci. Amer.* (Dec. 1978), 10.

CHAPTER TWENTY-EIGHT

PANDEMONIUM

Polite language exclaims, "Pandemonium ensued..." In ruder language, "All hell broke loose..." In either case *all* (pan-) *demons* (daemon-) are in action (ium). "Pan" was a god as well as the word for "all." He was the son of Hermes (Mercury) according to one story. He was a noisy disturber of the peace, a collector of disorderly crowds, an orgiastic god of revelers. He was by no means a symbol of sounds alone but of general tumult. Great noises are all-absorbing and entrancing, as the rock-music discotheques aim to prove. Pandemonium is not only the sounds and their effects in themselves but also the meanings that their auditors place upon them. In the end, the catastrophic pandemonium evolves into music.

A pandemonium is how high-energy sounds to people as it bursts upon the human world. A specter is how high-energy is seen by people as it occurs. Smell and taste are affected also in the processes studied by the earth sciences. A natural catastrophe, especially, is a holistic event: every human sense, and every part of the habitat, is affected.

Pandemonium is the capital of Hell in Milton's *Paradise Lost*. Elsewhere, Plato offhandedly mentions a catastrophe that he does not name and says that the survivors came down from the mountains with their ears ringing. Hesiod, in his *Theogony*, speaks of Mother Earth (Gaea) groaning under the pressures of Ouranos in primordial times. Here are reasons for treating of sounds in earth sciences: their natural origins and their effects on the biosphere. Observers of high energy forces without exception dwell upon their sounds. When we learn more of them, we shall know more about the earth sciences. There will be a place for a few acoustical geologists among volcanologists, seismologists, meteorologists, and paleontologists.

When a bad local flood occurs, as it did at Wilkes-Barre (Pa.) in the spring of 1973, the physical processes are mediated by television through their sights and sounds; there occur physical destruction, economic dislocation and distress - all of them mediated through the eyes and ears. Admissions to nearby mental hospitals went up sharply; also the use of hard drugs, and the suicide rate.

In the great Alaska earthquake of 1964, the destruction, death, terror, sounds and sights all together made their lasting impact on people. Psychiatric symptoms such as depression, withdrawal, guilt feelings, and irrational blaming of people were common reactions. The churches came alive with repenters and worshippers.

Modern cases permit us to empathize with the ancients. Exaggerate them a thousandfold and one gains an impression of the ancient experience. We read in the log of a ship's captain at sea near the exploding Krakatoa: "So violent are the explosions that the eardrums of over half my crew have been shattered... I am convinced that the Day of Judgement has come."[1] The climactic blast was heard 3000 miles away. A crazed survivor ashore insisted that "the Arch Fiend stood everywhere, implacable, unpitying, offering help to none, listening to no imploration."

We are told that when the volcano at Cosequina, Nicaragua, erupted on January 30, 1835, the explosion was heard in Jamaica, 850 miles away. The blast was so terrible that at one village "300 of those who lived in a state of concubinage were married at once."

Tornados have their own repertoire.

A tornado, like thunder, is heard many miles away. As it approaches, there is a peculiar whistling sound that rapidly

---

[1] Furneaux, *Krakatoa*, 188.

changes to an intense roar, reaching a deafening crescendo as it strikes. The screeching of the whirling winds is then so loud that the noises caused by the fall of wrecked buildings, the crashing of trees, and the destruction of other objects is seldom heard. The bellowing of a million mad bulls; the roar of ten thousand freight trains; like that of a million cannons; the buzzing of a million bees (when the tornado is high in the air), and, more recently, the roar of jet airplanes -these are some of the phrases used by those who have experienced a tornado.[2]

And so to meteors: Frank Lane writes of the meteoric shower of February 9, 1913 that was first seen at Saskatchewan, Canada, and last seen off the Brazilian coast, 6000 miles away. "As they passed southeast over Ontario they grew more brilliant and great explosions were heard. Detonations and earth tremors were caused along the path of the procession to distances of 20 to 70 miles on either side." In 1958, L. LaPaz wrote, "To listen to the sound effects produced by a large meteorite fall is a unique and awe-inspiring experience. Neither a hedge-hopping jet nor a keyholing rocket gives rise to the sky-filling reverberations set up by a falling meteorite."[3] Neither a nuclear blast, with its single report, one might add.

The rumbling, grinding, screaming sounds of earthquakes are well-known. Velikovsky quotes the plaint of the Egyptian scribe of Papyrus Ipuwer at the time of *Exodus:*

Years of noise... There is no end to noise... Oh, that the Earth would cease from noise, and tumult (uproar) be no more.

The ancient Greek poet Euripides speaks in *Hippolytis* of tidal waves near Corinth:

An angry sound, slow swelling
Like god-made thunder underground

---

[2] Frank Lane, *The Elements Rage, loc. cit.,* 1958,
[3] *Ibid.,* 179.

A wave unearthly crested in the sky;
Till Sciron's Cape first vanished from my eye,
Then sank the Isthmus hidden, then the rock of Epidaurus.
Then it broke, one shock and roar of gasping sea and spray
flung far, and shoreward swept.

In assessing what such sounds do to humans, it is well to recall that the age of firecrackers, firearms, cannon, dynamite, and nuclear blasts is young. The first detonation of dynamite occurred in 1881.[4] Primeval sounds were entirely of nature, apart from the pathetic imitations of sounds made by humans. If the paradigm of this book is correct about Pangea, the pre-quantavolutionary, pre-human period of late times, the world was peaceful and orderly, with overcast skies and little celestial or terrestrial turbulence. Man's ears were not made for explosions any more than his eyes were made to stare at the sun. A tiger's roar, an elephant's trumpeting, squeaks, whines, growls, yells, the splashing of waters, the snapping of twigs, the slumping of old trees - this in our theory was the pre-holocene acoustical environment.

However, and it is argued so in *Chaos and Creation* and *Homo Schizo I,* awful noise descended upon the first humans, as they were being born. Great noise was from the first heard as a manifestation of the gods, a theophany. When meteoroids broke through the skies, when cataclysms began, then came a pandemonium that terrified humankind, that drove people mad, that deafened them, and that catastrophized human nature and culture, together with their ecology.

Stephens reports that accidents, absenteeism and other factors indicating degradation of human performance can be correlated with infrasonic waves arriving from storms 2000 miles away.[5] Infrasonic waves cause nausea, disequilibrium, disorientation, blurring of vision and lassitude. All of these have

---

[4] IV *Ency. Britannica* (1974), 955.

[5] R. W. B. Stephens, 7 *Ultrasonic* (Jan. 1969), 30-5.

been described as accompanying earthquakes, ball lightning and volcanism.[6]

Some thunderous and strange sounds accompanying the passage of meteorites are attributable to the friction and collapsing vacuum of passage, but others have been theorized as products of the conversion of kinetic energy into electromagnetic radiation. Romig and Lamar have studied this problem. The high velocity of such waves would explain why some meteoric sounds are heard during and even before the visual sighting of meteors.[7] C. S. L. Keay has recently summarized from New South Wales many reliable reports of a large fireball in the atmosphere, tens of kilometers high, whose sounds reached the ears instantly with hisses, hums, swishes and crackling.[8]

Frederic Jueneman has speculated, on the basis of apparent acoustically provoked mutations in a London bomb crater from World War II, that catastrophic acoustics may have been an active mutator in ancient times.[9] The sensitivities of plants and animals to sounds has been widely surveyed by P. Tompkins and C. Bird.[10]

The splendors of auroral displays vary with the behavior of the Sun and the Earth's magnetosphere, among other factors. They stretch from 90 to 400 kilometers high, and on occasion seem to dip down to the very plane of the viewers. They, like all other fascinating phenomena of nature have been held responsible for the allegedly mad legendary accounts of catastrophes. Thus the ancient Teutons might recite their sagas of a world on fire, but uniformitarians, unimpressed, would see in these only the auroras that the northern peoples were lucky enough to view. This is a

---

[6] See Corliss, *op. cit.*, CrSD-045, GI-232 from *Monthly Weather* R (Feb. 1895), 57.

[7] 28 *Sky and Telescope* (Oct. 1964), 215.

[8] C. S. L. Keay, "The 1978 New South Wales Fireball," 285 *Nature* (1980), 464-6.

[9] I *Pensée* 4(1973), 112.

[10] *The Secret Life of Plants* (New York: Harper and Row, 1973).

topic for another time and another author: specters of colors, rays, and lurid skies were plentiful in cosmic disasters, exceeding the auroras. Every disaster has its color scheme and geometric figures.

It has its sounds as well, and the aurora can join other natural forces even today in suggesting the pandemonium of catastrophe. An account by Hans Jelstrup, a Norwegian astronomer, in 1927, exemplifies the auroral visual and auditory experience:[11]

> When, with my assistant, at 19h 15m Greenwich Civil Time, I went out of the observatory to observe the aurora, the latter seemed to be at its maximum: yellow-green and fan-shaped, it undulated above, from zenith downwards - and *at the same time* both of us noticed a very curious faint whistling sound distinctly undulatory, which seemed to follow exactly the vibrations of the aurora.

They later proceeded to record the impulses on an instrument and found "the vertical component was greater than 100 microvolt/meter."

Many years earlier, another Norwegian had polled persons from "all parts of the country" about the sounds of the aurora and received "92 affirmations against 21 negations."[12] Apparently many people provided a surprisingly large set of descriptions. They used words and phrases like: sizzling, creaking, soft whizzing, the sound of tearing silk, "hoy, hoy, hoy," a rustling stream, crackling, rolling din in the air, clashing, like a flapping flag, flapping of sails, hissing of fire, the sound of a flight of birds, the buzzing of a bee, roaring of wind, soft breeze, roaring of the sea, a distant waterfall.

What can be made of this, aside from its entertaining aspect, is that the sounds of nature are legion; that these join the hundreds of electrical sounds; and that a record is to be had of all

---

[11] Reported in Carl Störner, *The Polar Aurora* (Oxford: Clarendon Press, 1955), 137.

[12] S. Tromholt, 32 *Nature* (24 Sept. 1885). 499-500.

these sounds in these mild times of the Earth that can be used to identify ancient and legendary metaphors of sound. So that when dragons hiss and flaming rays dart from their nostrils, one does not simply say here is an especially exciting auroral display, but assigns to the dragon hypothetically the electrical qualities and sounds of the aurora or of bolides whose "sounds are described as hissing, swishing, whirring, buzzing and crackling" when they have the "brightness of the full moon" and reach the observer at the same time as the visual image does.[13] So, too when in *Ezekiel* (XLIII, 2) it is said that the voice of the Lord "was like the sound of many waters." Ancient records and legends are rich mines of electrical allusions from which not only the state of electrical phenomena can be assessed but also the electrical technology of early cultures can be surmised; this field, ignored hitherto, is being researched by J. Ziegler.

The Books of Moses carry testimony of great celestial noise that cannot be rationalized as ordinary thunder. And Noah, it is said in Jewish legend, was spoken to by a voice from the sky amidst a great commotion. This followed the failing of things upon the earth and was followed by the Deluge. The story of Job, later on, reads: "Hear ye attentively the terror of his voice, and the sound that cometh out of his mouth." Again: after the ends of the earth are lit up, "a noise shall roar, he shall thunder with the voice of his majesty, and shall not be found out when his voice shall be heard." A circum-global sound.

As the Jews passed from Egypt in the tumult of Exodus, they paused at Sinai. "I am Yahweh," heard the people during the night at the Mountain of the Lawgiving. "And all the people saw the roars, and the torches, and the noise of the trumpet, and the mountain smoking: and when the people saw it, they trembled and stood far off." Ten blasts of the trumpet sounded out the Decalogue, legend tells us also.

---

[13] Daniel S. Gilmor, *Scientific Study of Unidentified Flying Objects* (NY: Bantam, 1969) as reported in Corliss, *op. cit.*, C1-235, GSH-001, from M. D. Altschuler paper.

Great sounds were reported from around the world: the Babylonian *Gilgamesh* epic: "Loud did the firmament roar, and earth with echo resounded." Hesiod's *Theogony:* the huge Earth groaned when Zeus lashed Typhon with his bolts -"the earth resounded terribly, and the wide heaven above." Velikovsky pursues the name Yahweh elsewhere: he finds Jo, Jove (Jupiter); Yahou, Yao (Chinese emperor of the age); *Ju Ju huwe,* (an Indonesian invocation to heavenly bodies): Yahou, Yo (in the Hebrew Bible); Yao, Yaotl (ancient Mexico); Yahu (ejaculation of the Puget Sound Indians and other Amer-lndians when they performed the ritual of raising up the fallen sky off the earth).[14]

It is perhaps of some significance that Cohane has found Haue, a Middle-English god-name, in the names of gods, sacred places, rivers, salutations, and objects all over the world into the hundreds of instances. "In the landscape of the Old Testament part of the world is still overflowing with Hawa place-names."[15] All sound alike despite spellings such as oa.., ua.., awa.., huwa.., oua.., wa.., and so on. Provisionally, we may entertain the idea that the sound of great natural events were incorporated in the basic vocabulary of new-born humanity. If so, the popularity of the "awah" sound is at least as ancient as the time of Moses (circa 1500 B.C.), and probably several thousands of years older, and would also then be carried on down to modern times. "Yow," "wow," and "ow" are everyday American slang exclamations. The divine voices were also heard later on. A Babylonian hymn to Nergal (Mars) is of the first millennium B.C. and reads:

> His word makes human beings sick,
> It enfeebles them
> His word - when he makes his way above -
> Makes the country sick.[16]

---

[14] Examples here are from Velikovsky, *Worlds in Collision.*

[15] Examples here are from Velikovsky, *Worlds in Collision.*

[16] Velikovsky., *Worlds in Collision,* 263.

The motions, noise, and gases of a heavenly body of large dimensions seem to be indicated here. The god Mars is referred to in Babylonia as the God of Noise. There is an insistent connection of noise with the planet Mars.

The connections between heavenly sounds, sacred events, and the beginnings of music appear to be secure. From Chernikov, in the Ukraine, Soviet scholars reported finding mammoth bones converted into skull drums, shoulder blade kettle drums, and lower-jaw xylophones, at an estimated date of 20,000 years. If the instruments, all of the drum family, are correctly identified, it would mean that the settlement was fully human, with a religion. For nowhere is there any indication of musical instruments or musical sounds that are not connected with the heavenly host.

When the Wonguri tribe of Australia conducts today its holy dream time ceremonies, the assemblage beat sticks together; the dancers keep rhythm; and the stories of earliest times are recounted, of the time the Moon left their land forever and the morning star accompanied her. Ancient Greek myth tells of the infant Zeus; he was being hidden from his father Kronos who would swallow him; his nurses, the Curetes, drowned his cries with drums, cymbals, and dances.

Drumming and whistling may be the oldest emulated sounds. The bull-roarer is an ancient and world-wide instrument, a primitive noise-maker that whips the air into a sound like a falling body. It thunders and whistles. Perhaps whistling also developed with a pipe or fife. The horn, whence the trumpet, might follow; it is a piercing and blasting instrument. The arched string instrument - harp, lyre - must have joined the sacred group quickly. All together they reproduce the music of the spheres and of the gods.

In earliest China the drums were used to communicate with heaven. The drum comes from K'uei, a green ox like creature who came out of the sea shining like the sun and moon and making a noise like thunder. He was captured by Huang-ti who made him into a drum skin. But the same K'uei is also the master of music who alone can bring harmony between the six pipes and the seven modes. Without this harmony heaven and earth would lack their essential music. K'uei was also master of the forge, of dance, and of regulating floods.[17]

The sickle with which Kronos (Saturn) castrated Ouranos (Uranus) was also the harp (lyre) of Demeter who had taught the Titans to reap. The strings of the lyre were ultimately five or seven, corresponding to the number of spheres counted as planets. Vail thought that the arch of the harp and sickle came from the opening of the boreal hole of the north when the regime of canopy skies began first to break down; the arch was the sickle; it was also the arch of the lyre, and the strings to be plucked were the beams of light playing down upon the earth.

Here, as in many other cases, an issue is whether the sacred image came before the invention or the invention was made and compared with a later celestial image. As usual I incline towards the position that the sacred example preceded the profane.

The correspondence between the number of planets and the number of strings on the lyre is an instance in point. It is only one of many. The number of observed planets obviously determines the number of strings on the instrument, not *vice versa*. An old Chinese text says that "the calendar and the pitch pipes have such a close fit, that you could not slip a hair between them."[18] This seems an odd expression until one realizes that the sacred calendar is replete with a synchronous musical calendar - from Easter music to Christmas music, for example. The pipes are pitched to heavenly sounds and numbers; the calendar is an arrangement of heavenly events.

The Pythagorean philosophy of ancient Greek culture generated the theory of music and the theory of numbers out of the behavior of the heavens. The "harmony of the spheres" of which the ancients spoke was probably first the sounds of heaven

---

[17] Santillana and von Dechend, *op. cit.*, 125-8.

[18] *Ibid.*, 4.

of the "better" sort, to which humans might adjust, and which, to them, presaged a tranquil stability, and then later, inferentially, the visual reliable order of the heavenly bodies as noted and welcomed by philosophical astronomers.

Robert Temple has been able to locate a fundamental connection between geodesy in Egypt and Greece. The Greeks and Assyro-Babylonians had the heptatonic or seven-toned diatonic scale of today. The Egyptians possessed a musical octave of seven degrees (that is, an eight-tone scale, such as the West has today). The same seven degrees was the geodetic principle followed in the topographical surveying of Egypt. For the Egyptians, 1° North was at Behdet and 8° was at the southernmost limits, by the Great Cataract of the Nile. Further, Temple, with suggestions by L. Stecchini, established an octave of centers for oracles: running up the lines of latitude and musical scale at equal intervals, thus: Barce, Triton, Paphos, Omphalos (Crete), Kythera (or Thera perhaps), Delos, Delphi and Dodona.[19]

Robert Graves has reported an octaval version of the name of Yahweh, *Jehuovao*. The sacred name can then be pronounced and chanted as a set of vowels running the gamut of a musical scale. We are reminded of the connections between Egyptian and Hebrew culture, when Demetrius wrote: "In Egypt the priests sing hymns to gods by uttering the seven vowels in succession, the sound of which produces as strong a musical impression on their hearers as if flute or lyre were used." The seven vowels were uttered in succession as the divine unspeakable name.[20]

Musical sound, and also noise, can be broken down into pitch, rhythm, timbre and volume. The first instruments specialized in rhythms, for instance, and had variations of pitch, timbre and volume. The pipe or flute specialized in pitching different tones and a whistling timbre. Using such elements in combinations, music could be built up. But it would not have been

---

[19] *Op. cit*, 29. 20. *Ibid.*, 266
[20] *Ibid*, 266

possible without the basic psychological changes that were taking place in people. Control of themselves and the gods was the paramount motivation behind the people who originated music and all other aspects of culture.

The humans had a compulsion to repeat their first experiences, which were naturally terrible; this is explained fully in my work, *Homo Schizo I*. The repetition of rhythms is the repetition of the sounds of the gods at work upon the world. The orgiastic side of music - the furious beatings, poundings, amplitudes, blasts, whirling dances and frenzied lyrics - is an imitation of the behavior of the gods in the days of creation. The orgiasm is the basis of the plot of song and chant; it gives the melody line, the beginning, middle and ending.

Repetition and orgiasm shape the four elements of music, and lend form both to the instrument and to the unique composition prescribed for it. The very design of an instrument is intended to supply a limited span of capabilities to the musical elements. Not only does the music itself follow patterns under strict general rules, but the instrument is a mechanical contrivance to see that the rules are obeyed.

To all of this is added from the start the sublimation that the music affords. Tests of endurance, involving the basic, and destructive, elements of earth, air, water, and especially fire, sometimes are incorporated into the dance and music. Battles of the gods, too, may be emulated. The gods are being controlled at the same time as they are being celebrated and honored; the audience is being controlled as it celebrates and honors the gods.

"Heavenly sounds" are a contradiction; they are actually the suppressed and sublimated sounds from heaven that destroyed the world. In The Holy Dreamtime of the Australian Womburi is a Holy Dreamtime of all other peoples - for all peoples have them. Sacred myth, song, dance, and music provide an escape from horror by saying and doing all that was said and done in those days in a way that remembers in order to forget.

Contemporary music that is avant-garde has the subconscious ambition, certainly doomed to fail, of confronting the terrible days of catastrophe directly. It brushes aside the sublimation, and the compulsive repetitiveness of music. It destroys expectation, and unleashes the gods. It destroys form by atonalism and arhythmism. It randomizes the four elements. Whatever happens in a sound-producing setting - "a happening" - is "music." The computer is used to reduce dependency upon skills, pitch, volume, rhythms, and timbre. It creates the mixture that is the "true reality". All this is often done without full realization. It is nevertheless a largely honest attempt to return to the primeval chaos in which humanity was born.

I have known geologists to taste stones and drippings, to smell in crevices, to feel the texture of rocks, to tap a fracture and listen, and of course to hold up a specimen to view by every angle of light. Hence it is not a radical departure from the earth sciences if we carry our inquiry into broader realms of sound and light. Our intent is not to create a marriage of sciences and humanities: that is good in its own right and if it is a by-product of this interest, so much the better. Our motive is to understand and possibly to reconstruct natural history.

Whereupon it happens that, once the idea of the constancy of natural events through long eras of time is put aside, and another model of inquiry is advanced, we must take advantage of the treasury afforded by human history. The dumb rocks can tell their stories in part through human lips. All the motions that are forbidden in the dead past are resumed through the sights visited upon early human eyes. The sounds and sights of events that witnesses and their descendants describe are clues about an Earth that is less static and more dynamic than the earth sciences have heretofore portrayed.

CHAPTER TWENTY-NINE

SPECTRES

To recount the visual experiences of ancient humans in regard to natural phenomena would be a work of thousands of pages of agonies, joys, and revelations. However, the reader is probably aware of their nature through voluntary and inescapable exposure to fairy tales and horror movies. The earth sciences will profit more from a discussion of some relationships between natural events and the specters that accompany them. I shall avoid speaking of the eyes when used functionally, as, for example, to assess damage or to organize a new life. Rather I shall concentrate upon the visual effect in itself, and what it conveys about natural events.

Uniformitarians usually abandon their position on change when it comes to what ancient voices convey about natural events. That is, in order to hold on to their belief in a natural world that changes by gradual evolution rather than by quantavolution, they say that humans have changed their "exaggeration-rate." They often deny ancient testimony, using pseudo-anthropological arguments that early mankind was superstitious and excitable, hence quite unreliable. What he claimed to see were in fact illusions and delusions; what he passed on as memories were gross exaggerations. In other words, nature behaves in the same way; man has changed.

This theory we find unacceptable, as also we do its accompanying statements, that typically proceed like this:

"Ancient people are not to be believed if they say that a large body was spotted and approached in the sky. Or that bodies of all shapes and sizes rushed high and low through the skies. When mountains and land are seen to rise, and at other times watched as they sink, this must be an illusion and an exaggeration. No one could have seen a wall of towering water. That there should have fallen sheets of flame and weird colored waters or dense substances, including even life forms, that ice and hail should fall in deluges and wind should sweep away forests: these again were delusions. Seeing the landscape dissolve in an earthquake, while even the air is rendered into visible shock waves, and seeing the Earth explode and pour out boiling magma from cracks and cones: again illusions. Telling of the destruction of almost all that was living: people must have been psychotic to make up and pass along stories of such events."

The quantavolutionary position is that they were probably psychotic, but partially because of the nature of such events. Thus, to some extent, we become uniformitarian in respect to human psychology as we become quantavolutionary in regard to nature.

An increasing number of studies of modern mankind in disaster lead us to accord greater reliability to ancient stories. A severe trauma of terror, such as the nuclear blast at Hiroshima, leaves the survivors quite catastrophized. What happens thereafter matters little to the survivor. Subsequent sights are likely to fall upon a numbed and hopeless creature. Where survivors are reduced to hopelessness, few lift their hands to help others. The prognosis of the group is poor. Studies of the aftermath of Hiroshima have shown this to be the case. Each succeeding horrible sight is seen by eyes becoming too jaded to respond. We should bear in mind, too, that Hiroshima was a local event, a minute fraction of what many a fossil agglomeration and extinct volcano chain tells us once happened. When we see millions of trees all felled at once buried in the Fens of England, a blast many times greater than Hiroshima has to be postulated.

After the explosion and tidal wave of Krakatoa, a survivor spoke of scenes "too horrible to remember; incidents that reminded of the animal instinct that enables people to do the impossible."[1]_When a fireball blazed erratically across the Southern States of the U. S. A. on March 24, 1933, people were terrified. "Ninninger (1936) says that seasoned cattlemen, accustomed to facing the vicissitudes of life and who ordinarily knew no such thing as fear, told him they despaired of their lives during these 'terrible moments. ' Yet they were 75 miles from the fireball's nearest approach !"[2]

If, however, people on the periphery of a disaster survive, these will be terrorized but hopeful of themselves. Even this was noted at Hiroshima. If after days, months, years or centuries, a disaster of the same dimensions strikes, and again some survive - some of a new generation, too -then the memory and meaning of catastrophe is reinforced. But again the survivors are active, self-preservative, and hopeful. They still can believe in some surcease and control. They have meanwhile established relations with gods and nature, the very forces of wrath. They can immediately interpret the events, and produce one or more inventions to propitiate and control the gods and, therefore, the events.

Prophets will help them to remember and to react:

> The Lord will smite you with madness and blindness and confusion of mind; and you shall grope at noonday, as the blind grope in darkness... (Deuteronomy)

> And they shall go into the holes of the rocks .... for fear of the Lord and for the glory of his majesty, when he ariseth to shake terribly the earth... (Isaiah)

> The great day of the Lord is near, near and hastening fast... I will bring distress on men, so that they shall walk like the blind, because they have sinned against the Lord; their blood shall be poured out like dust, and their flesh like dung... (Jeremiah)

---

[1] Furneaux, *Krakatoa,* loc. cit., 108.

[2] Lane, *The Elements Rage,* Loc. cit., 179

> I am about to shake the heavens and the earth, and to
> overthrow the throne of kingdoms .... the horses and their
> riders shall go down, everyone by the sword of his fellow...
> (Haggai)

These are the visions of prophets and there are many
more like them, posed as promises, to be sure, but with the full
assuredness that comes from past experiences. We note marks of
genuineness: going into caves during earthquakes (for the sky
brings worse terrors); the kingdoms are overthrown, then the
survivors attack each other; the survivors are stunned, maddened,
functionally blinded. The preventatives are difficult, if not
impossible: that all should worship faithfully and properly, and
obey divine commandments.

Now here is a legend of the Indians of the Badlands of
South Dakota. It tells of how the Badlands came into being, laying
it onto violations of the will of the Great Spirit who had granted
plenty but had decreed peace, and there was no peace. Warriors
prepared for battle:

> At last all were assembled and the day had come for the
> advance. And now the Great Spirit took matters into His own
> hands. Dark clouds hid the sun from the face of the world.
> Lightning streaked across the blackness and thunder rumbled
> high over the hills. From the ground flamed forth fire, and the
> earth shuddered and rocked. A wide gulf opened and into it
> sank the mountain tribe - all their people - all that they
> possessed. With them sank all life - the waving grass - the clear
> springs - the animals.
>
> As suddenly as it came the storm ceased. The earth became
> fixed in waves as it had rolled and shaken. There was only a
> barren waste on which nothing has ever grown or can grow.[3]

---

[3] M. E. Gridley, *Indian Legends of American Scenes* (NY: Donahue,) 101

After a catastrophe, the sights of doom are only partially capable of recall. They are personalized, humanized. Then they tend to fade over time. They are sublimated in many ways. The history seems to us strange; it is literal, detailed, yet surreal, as in the Bible story of Sodom and Gomorrah. I discussed the geology of the story in Chapter 22. When the family of Lot, warned by an angel, was fleeing the doomed Cities of the Plain, it was forbidden to look back. The Cities were utterly destroyed. Lot's wife turned to look and was transformed into a pillar of salt. Thus did subsequent generations, perhaps even the descendants of the family (who violated the taboo against incest to perpetuate themselves), remember the event and tie themselves personally and visually into it.

Of salt in the Great Rift Valley of the Jordan there was plenty or perhaps just then it came to be plenty and is plenty today. It was a convenient "memory tag", to imagine a seen horror encased in a pillar of material produced out of the holocaust itself; then 'this is where Lot's wife was frozen with fear and died' becomes 'this is where Lot's wife became a pillar of salt because she viewed the terrible wrath of the Lord. ' That is, the story is tied into the event all the more closely. That there may have been nothing left of her except a location and new salt would help, if true, to explain the story. The others, who dared not look back, would have no way of knowing.

I am not arguing for literalism but for "spectralism" which I would define as subjective realism: first, a sympathetic and fully possible truth has to be searched for and, then, whatever is left over as "false" has to be explained in the vision of the subjects and of their immediate descendant, and finally in objective psychological and anthropological terms.

A much broader range of cases may advance the argument. There is, for example, the dragon. Everyone knows what a dragon is. All do not know that it is a theophany, a divine manifestation. And that the creature is closely tied to visions of events in the sky, many times repeated.

Chamber's Encyclopedia, defunct now for many years, carried a charming passage on the dragon:

> The dragon appears in the mythical history and legendary poetry of almost every nation, as the emblem of the destructive and anarchical principle; ... as misdirected physical force and untamable animal passions... The dragon proceeds openly to work, running on its feet with expanded wings, and head and tail erect, violently and ruthlessly outraging decency and propriety, spouting fire and fury from both mouth and tail, and wasting and devastating the whole land.

The dragon is regarded as a benevolent creature by the Chinese, however. And no people has been so devoted to the symbol. Its iconography was as intense as that of the crucifixion of Christ in Medieval Europe. Recently, Carl Sutherland found that the dragon made its appearance in Chinese art around 1500 B.C.[4] This date is a well-marked catastrophic boundary, known in radiochronometry, archaeology, geology, legend and history. Eliminating bit by bit "all later accretions," he thinks that he has "attained some understanding of the sight observed by the ancient Chinese: a writhing, bright, elongated thing. It was irregular in outline; it was apparently on fire... This thing, the dragon, seemed to be driving off the terrible flaming globe and so to be benevolent as well as powerful." Later on it was given legs and scales. It is almost always shown in the heavens. Flame symbols show the sky to be on fire. The globe carries lightning and thunder symbols as well as fire symbols. (Probably the lightning generated the moving legs of later representations.) The Chinese Emperor with a "Dragon Face," sat on the "Dragon Throne" wearing robes of state on which dragons were displayed.

Dwardu Cardona has presented first-hand descriptions of comets that compare them with dragons.[5] The accounts range

---

[4] 4 *Pensée* 1 (1973-74), 47-50; see also V *S. I. S. Rev.* 280-1, on the cosmic serpent.

[5] I *Kronos* 2 (1975), 35-47.

from England to China. The comet of 449 A.D. stretched over
England from beyond Gaul to the Irish Sea, "a ball of fire,
spreading forth in the likeness of a dragon, and from the mouth
of the dragon issued forth two rays..." Thus wrote Geoffrey of
Monmouth. Some comets "lash their tails" wildly.

The Chinese "Kung Kung" dragon flung himself in rage
against the heavenly mountain, turning the skies around, and
tilting and flooding the world. He had a son-dragon, "K'au-fu"
who wished to keep pace with the Sun. K'au-fu tried to quench
his thirst *en route* by drinking up the rivers of China but succumbed
finally of thirst. Cardona identifies the myth with the Phaeton
myth and episode. Phaeton, eager to drive the Sun's chariot, did
so incompetently. Legends recite that he came so close to Earth
that the rivers of Asia, Africa and Europe dried up. Strabo's
Geography mentions the terror of the Syrians and Aramaeans at the
sight of Typhon, probably the same as Phaeton.[6]

> That the myth of Phaeton describes a shifting of heavenly
> bodies, we know from Plato. That Phaeton was a comet, or a
> 'blazing star, ' we know from Cicero. That this 'blazing star'
> became a planet, we know from Hesiod. And that this planet
> was the planet Venus, we know from both Nonnos and
> Solinus.[7]

Then Cardona takes up the question of the Chinese "fire
pearls," or "tear drops of the Moon." These we have discussed as
the tektites, which are scattered over the Earth. He concludes that
they splashed upon Earth after great meteoroids or cosmic
lightning discharges had blasted the Moon. Possibly it was the
work of the cometary Venus, for the dragon Lung is pictured
chasing a great pearl across the sky. And the fear that the Moon

---

[6] VII *Geography* (1924 ed.), 3,8.

[7] Cardona, *Supra* fn5, 37.

will be devoured by a comet is part of some legends and modern anthropological reports.

That the ancients may have actually observed such bursts upon the Moon is argued by astronomer Jack B. Hartung.[8] According to the *Chronicles* of Gervase, for June 18, 1178, at Canterbury, England, five persons witnessed with their naked eyes the explosion of a crater. Hartung estimates it as perhaps 13 miles in diameter. In Gervase's words:

> A flaming torch sprang up, spewing out, over a considerable distance, fire, hot coals, and sparks. Meanwhile the body of the Moon which was below writhed, as it were, in anxiety...

Whether this writhing was an illusion created by air waves or an actual rolling seismism of the Moon's surface is not to be known. Bancroft once reported an Aztec legend that the sun and moon emerged equally bright, but to the gods this was not seemly; so one god took a rabbit by the heels and slung it in the face of the moon, dimming its luster with a blotch whose mark is seen to this day.[9] Great events have impacts on human behavior and human behavior can be sometimes used to conjecture upon possible great events. One must reason back and forth, trying all the while to avoid circular argument. A difficult case is the similar duration of the lunar cycle (today) and the menstrual cycle of women (today). The one is 29.5 days; the second can vary from 21 to 35 days, but concentrates upon 29 days. Gestation occupies generally nine moon cycles. Various scholars have mentioned these 'coincidences. ' Recent studies have shown that the Moon cycle is more closely followed when women of varying menstrual periods are shut up in a room where they cannot be aware of

---

[8] Dwardu Cardona, "On the Origin of Tektites," II *Kronos* 1(1976), 42-3.

[9] Related in Donnelly, *loc. cit.,* 169.

moontime and suntime; they unconsciously tend to approach the lunar revolution.

It is ordinarily believed that the Moon was on the present cycle long before the first human evolved. Anthropologists have maintained that the coincidence ultimately reinforced human attention upon the Moon and also provided specious grounds for marking the peculiarity and witchcraftiness of the female sex. Menstruation is often the subject of taboos.[10] In some places, women in menstruation must not be seen. Harsh penalties for violations of menstrual taboos are common.

Under the quantavolutionary theory here, it would be possible to view the "ideal" menstrual cycle as itself determined by the cycle of the Moon. Only the human female behaves on the monthly cycle. A psychosomatic response to the greatly feared and revered goddess and god of the Moon, newly in place and settled into a regularity, could be achieved by disciplining a varying physiological function. People will go to any lengths to harmonize their behavior with that of their gods. (I discuss this subject in the volumes on *Homo Schizo.*) To bind a whole sex and indirectly a whole people by its important reproductive cycle to the Moon god who passed them in daily review would appear to be a principal invention of the human race. There was strong incentive to devise this proof of devotion to the great god: it had ceased to bring ruin on the world and was guarding the new peace.

"Spectralism" might propose another case for consideration. How long have nights and days characterized earthly existence? A legend has persisted down to our times on the high plateau of Bolivia, around the impressive ruins of Tiahuanaco, that the city existed before there were stars in the sky. Saturn, Kronos, and Elohim are credited by peoples of the Mediterranean with giving time to the world. The Hebrew creation story has the Lord on High declare: "Let there be lights in the firmament of the heavens to separate day from night; let

---

[10] Wolfgang Leader, *The Fear of Women* (NY: Harcourt, Brace, Jovanovich, 1968).

them serve as signs and for the fixing of seasons, days and years."
Whereupon the Sun and Moon were placed in the sky. I would
suppose that the Moon, after terrorizing humanity by its
assemblage and irregularity, promptly became the basis for
calendars everywhere, once it began to obey the laws of Kronos
(Chronos or *Time*). Time-factoring in earliest mankind was a way
of following the gods in whatever regularities they might exhibit;
Marshack has reported Paleolithic lunar marking extensively.[11]

Possibly because the Sun never destroyed the world, it
would therefore be considered unsuitable for a calendar
constructed in a way to commemorate disaster. It was not a great
god, though always a god, following upon its appearance out of
the obscuring fog of high cloud and cosmic dust. Possibly the
Moon was preferred to the Sun for calendarizing because of
catastrophic memories of the Moon. Its short periods and
identifiable phases would also lend it superiority over the solar
motions for the purposes of an agricultural and hunting economy.
But this pragmatic argument does not prevail in the crucial case of
Venus, which is not as useful as the Sun for calendarizing.

The Sun was rarely calendarized; yet Venus was. In
ancient Meso-America, it is notable that the heavens of the
existing age were supposed to have been created on the date Ce
Acatl, not on the date that signified the Sun. Ce Acatl was the
Morning Star, Venus, and identified with the great god,
Quetzalcoatl.[12]

Quetzalcoatl was also the name of a bird of gorgeous
plumage. Marcus Varro, the learned Roman author, reported that
once long before his time the planet Venus changed its color, size,
form, and course, a strange prodigy which, he said, had happened
never before or since. That Venus displayed colors more
frequently is suggested in an article on color in the *Reallexikon von
Antike und Christentum (1969)*, speaking of very ancient times: "In

---

[11] Alexander Marshack, *The Roots of Civilization* (NY: McGraw Hill,
1972).

[12] *Codex Telleriano-Remenesis* II, PI. 33.

foretelling the future, it was taken into consideration whether the planet Venus was wearing a black, white, green, or red headpiece."

It is to Velikovsky's credit that he not only uncovered the Venusian approach cycle, which put many peoples in terror of the destruction of the world even well into the modern period (for example, the Aztecs of Mexico), but he also was finally able to demonstrate that the Egyptians stuck to a Venusian calendar down to Roman times.[13] Contrary to pragmatic logic, it was the wicked, destructive, adored, and possibly eccentric Venus whose behavior was calendarized, while the routine sun was taken for granted. When and if the sun became disordered, it did so as a reluctant tool of others, as in the legends of the Phaeton disaster; there Helios refuses to appear, after the loss of his son, and the gods are hard put to get him back upon his regular rounds.

A specter is something seen that is there and not there. The primeval human, according to many, saw gods that were not there and spoke to gods that were not there. The noise and sights were pure hallucinations. Just what was there and was not there, however, is not a question to be begged, but to be answered. No one, today or ever, has seen a personal disaster with the cool eye of a scientist thousands of years from the scene. But the cool eye should not claim that the disaster did not occur -or that it happened in a way to conform to his daily newspaper accounts of earthquake, floods, and meteors. One must grant appropriate credence to the primeval scream; the skillful doctor listens studiously to the patient's complaints.

The popular Revelation of John, Apostle of Jesus, is a magnificent mad vision of the destruction of the world. The Catholic Bible says that "the Apocalypse is a revelation of things that were, are and will be."[14] Revelation aims to picture how most of the world and its people (among whom the wicked outnumber

---

[13] "Astronomy and Chronology," Supplement to *Peoples of the Sea* (NY: Doubleday, 1977).

[14] 14. Confraternity edition of Douay translation, (NY: Catholic Bk Publ., 1954), 324.

the good) were and will be destroyed. The good are imperishable, and will be judged and admitted to heaven.

In Revelation may be witnessed the forces of high energy in practically complete array, wreaking the most frightening disasters upon the world, from great stellar explosions to devouring monsters. The forces are commanded by, indeed are, angels. Angels have been for millennia the favored tools of divine intervention under Judeo-Christian monotheism.

Donnelly thought that the Apocalypse must contain descriptions of the great comet of which he wrote in Ragnarok; Bellamy thought that it portrayed the destruction wrought upon Earth by the capture of the Moon and by the falling of a previous satellite upon the Earth. Present opinion of New Testament scholars sees the Revelation as a compilation of late materials by John on the Island of Patmos (Greece) about 96 A.D. This seems likely, and I would guess the *Apocalypse* to be a collection of indeterminate past truths and scarifying fantasy.

Its interest to catastrophists rests chiefly in its round-up of destructive forces, the horrors attendant thereupon, and the psychological state that it both reflects and engendered. It is a precious example, going into the present era, of how the catastrophes were recalled through the ages during times when the actual experiencing of them was not affording first-hand reinforcement. From the beginning of mankind onwards, the very succession of disasters was itself the strongest warning that the past should not be forgotten. The great popularity of the Bible is probably due to the capacity of many of its passages to re-enact the terrible days of chaos and creation.

The Bible is instructive, too, on experiences of cosmic darkness. In the Genesis story of creation, the record of man begins in a world growing lighter, but still sunless and moonless. Elsewhere, I have discussed the atmospheric developments that coincide with this account, which is by no means the sole account passing down to us. The cherished light was not to be turned on forever, for the Bible itself and every single mythology of the world tells with dismay of various succeeding ages when a darkness fell upon mankind. The Götterdämmerung (or Ragnarok) of the Norse and Teutons is both a twilight of the gods in the sense of a universal darkening and in the sense of an approaching struggle and death of the old gods.

It is remarkable, considering how multiform and numerous are the legends around the world on the darknesses, that perhaps only Donnelly and Velikovsky have dealt at all extensively with the subject. Darkness is very much a part of the Biblical catastrophes. In the story of the Lord's visit to Abram and ordering of sacrifices may be seen the sixth catastrophe mentioned in the Bible (after the Creation, the Garden of Eden expulsion, the Deluge, Job's trials, and the destruction of the Tower of Babel). There Abram fell asleep at twilight and a "great fear and darkness" came upon him. And in the darkness "a smoking furnace and blazing torch passed."[15]

Later on occur the catastrophes of Sodom and Gomorrah, Joseph (Egyptian famine), Exodus, Joshua, David, Elijah, Amos, and Isaiah. The catastrophe of Exodus brought complete darkness for some days: "They saw not one another, neither rose any from his place for three days." (10:22) An Egyptian stone inscription about what was probably the same event states that "during these nine days of upheaval there was such a tempest that neither men nor gods [the royal family] could see the faces of those beside them."[16]

Darkness figures prominently in most accounts of catastrophes whenever the period. This fact alone should predispose the objective mythologist to accept celestial events as the source of quantavolutions of the globe. Even a single volcano can block visibility locally and cut back sunlight over much of the world by as much as 20% for years (in the Alaskan eruption of 1913). However, reading carefully the legendary accounts, one is

[15] *Genesis* 15: 12, 17. D. W. Patten, R. R. Hatch, and L. C. Stinhauer, *The Long Day of Joshua and Six Other Catastrophes* (Seattle: Pacific Meridiam, 1973).
[16] Velikovsky, *Worlds in Collision*, 59; *cf.* 58-62.

compelled to see in them a much more horrendous and prolonged experience. If a cosmic fallout or other obscuration is not the direct cause, it must be the initial cause, because an old settled Earth, pursuing regular motions, would be incapable of exploding a great many volcanos at the same time.

Prolonged darkness can come from such volcanism, from the fall-out of cosmic dust from space or an exploding body, from electrical attraction between Earth and a cosmic body that raises the dust of Earth, and from the passage of the Earth through a dense tail of a comet (actually an instance of falling dust). Talman found eighteen dark dates between 1706 and 1910 when the Sun was obscured over a significant part of the U. S. A. or Canada by forest fires. In only three cases did the darkness endure for as much as five days.[17]

Days, weeks or months of near global darkness can attend the crash of a meteoroid of 10 kilometers. Scholars studying biosphere extinctions now refer regularly to such effects, as in the study by the Alvarez group referred to earlier. Years of darkness have been claimed in rare cases, the Exodus period being one of these. Heavy winds are reported during the days of Exodus; Talman found the dark days of forest fires to be windless. Perhaps volcanism of a rare kind produced the Exodus dark skies, but more likely is the combination of large-scale volcanism and a prolonged fall-out of cometary dust. Yet Velikovsky mentions two legends of a temporary failure of the Sun to set in Middle Asia and China around this date, and wonders whether the Earth's rotation could have slowed for so long before resuming.[18]

The close of the Cretaceous age with its heavy extinctions saw a darkness of only weeks or months, according to one view, which suggested as the cause an exploding meteoroid of middle size. Nevertheless, most species of animals and plants were extincted, and great physical devastation occurred, so we may suppose that various events combined to worsen the darkness and

---

[17] C. F. Talman, 112 *Sci. Amer.* (6 Mar. 1915), 229.

[18] *Worlds in Collision*, 62.

that they operated holospherically. We suspect much more than the meteoroid was active.

The most impressive of all sights, to judge from many accounts from the earliest records and legends to the most modern of writers is that of a comet approaching the Earth. Unlike the strike of a nuclear missile, the comet gives the fullest visual warning, as well as causing a number of electrical effects from afar. It is "the most provocative apparition of all," in Calder's words, referring to Halley's comet, due to approach the Earth once more in 1985.[19] When the Roman Emperor Nero saw the comet of about 60 A. D., he had many leading Romans murdered to avoid the death he saw for himself in the heavenly portent. "The Incas of Peru regarded comets as intimations of wrath from their Sun-god Inti... In twentieth-century Oklahoma, at the apparition of Halley in 1910, the sheriffs arrived just in time to prevent the sacrifice of a virgin by demented Americans calling themselves Followers."[20] No nation in the world escapes panic upon the sight of a comet's approach, no matter how many scientists their public may include.

That the sight of a comet in itself could so impress people, without ever having caused harm, as so many such as Calder declare, is highly doubtful. Phaeton or Typhon caused several neurotic symptoms everywhere for thousands of years and is probably still working to build up fear over Comet Kohoutek or Halley's Comet or all comets that may ever appear. As attested to by the behavior of modern tribes of Amazon jungles, literacy and historiography are not required.

Peoples picture comets in many different forms, none of them impossible. They tie comets into many lessons, symbols, rites, and stories of their religions. Beyond religion, they integrate the comet-complex into sex, work, play, politics, and war, in highly disguised ways. They dread new apparitions and revere substitute portrayals of past comets. Nor could this universal fear be diffused

---

[19] *The Comet is Coming!* (NY: Viking, 1980).

[20] *Ibid.*, 12-3.

from one cultural center to another, like the sweet potato or noodles; the fear must have a basis in historical reality. As we have demonstrated in so many writings, the comet as an apparition that is followed by catastrophe is a substantially true memory retained of mankind.

To conclude, specters and pandemonium accompany catastrophic events of the earth sciences. In themselves they do not leave vestiges. Still, little by little, research will build up rough measures of the intensity and scale of the events from the visual accounts available in legend and reports. In the case of every important god stretching back before the dawn of classical history, we can elicit and reconstruct from legends of sight and sound the workings of high energy forces that connote catastrophes.

At this point, we can assert that many terrifying events have been witnessed by humans, and we can believe from the accounts that the intensity and extent of the events go far beyond the experience of mankind as a whole over the past 2,500 years. Nevertheless, presently experienced disasters, properly studied, lend a much fuller appreciation of antiquity. When the Egyptians suffered terribly from the natural catastrophe of the time of the Hebrew Exodus, a scribe wrote that women became barren and men lost their hair; the Ipuwer papyrus was known and read long before the nuclear bombs of Hiroshima and Nagasaki, but a new sensitized generation was required to perceive in these scarcely intelligible lines the awful news of radiation disease.

PART SEVEN

DIMENSIONS OF QUANTAVOLUTION

High-energy expressions of nature number a round score. Each has its pertinent sphere, in a way, as hurricane to atmosphere, thrusting to the lithosphere, and pandemonium to the biosphere; but all high energy events are effectively polyspheric, as a hurricane mows down forests, sends tidal waves over the land, flushes the air, and erases or builds mounds.

When a high energy even achieves quantavolutionary proportions – that is, is of high intensity, broad scope, and suddenness – it is invariably holospheric. Certainly this fact has made difficult the organization of materials for this book; it is not unusual for all spheres to be highly associated in the event. An examination of the Index will reveal how often a given force expresses itself aside from the events of the chapter in which it figures most prominently.

Overhanging all is the exoterrestrial event, the theme that pervades this book and lends it its unity. It is the cosmic power that supplies the Earth with its quantavolutions. Left to itself – an absurdity – the Earth would pursue an evolutionary and uniformitarian path through the ages. But Earth has not been left to itself, nor is it likely to be.

CHAPTER THIRTY

INTENSITY, SCOPE AND SUDDENNESS

The eye of the poet, quotes Ager from Shakespeare, "in a fine frenzy rolling, doth glance from heaven to earth, from earth to heaven." "So," says Ager, "ultimately must the eye of the geologist, in seeking the nature of the control. One always seems to come back to climate as the primary explanation of the sort of phenomena I have been discussing, but for the ultimate control, sooner or later, we must face the possibility of an extra-terrestrial cause..."[1]

Meanwhile, the Soviet geochemist, Y. P. Trusov, is writing that "the fundamental motive cause of geochemical processes is the contradiction between internal -physico-chemical -and external -macroplanetary, nuclear, and cosmic -factors active in the earth's crust."[2]

We shall see more and more of such intimations of the Earth's exoterrestrial transactions, until the earth sciences will undergo their own theoretical quantavolution. In this process, poorly equipped though we may be to move between geology and history, we shall have to reconcile the two modes of thought and bodies of fact. There ought to be no logical conflict between natural laws and historical events. Either historical occurrences - counting ancient voices, too, as historical events -will contribute to the affirmation or display of natural laws, or they are false or falsely interpreted. Either natural laws conform to validated

---

[1] *Op. cit.*, 83.

[2] *Interaction of the Science in Study of the Earth, loc. cit.*, 252.

historical behavior or the "laws" are not laws and require limitation or correction.

To pin down a quantavolution, even a single one, is like wrestling, "no-holds-barred." One grabs at any possible fact, at any method, hoping to take advantage of it. Tactics that scholars ordinarily spurn are demanded. If the geologist wants to know whether the Earth has long rotated at its speed of today, he asks the astronomer. If the astronomer is conventional, he replies "Of course." If the astronomer is a true empiricist and even a sceptic, he says "We don't know," and asks the paleontologists, the geophysicists, the ancient historian, and the mythologist for help.

M.G. Reade, a confectionary engineer, navigator, and scholar, addressed himself to the evidence of the *Panchasiddhantika*, documents of ancient India. There, at a time suspected of being around the eighth and seventh centuries B.C., he found evidence of "aberrational," slower rotations for the Earth from data given for five planets then known, amounting to a 360-day year. The same Hindu figures suggest "that the whole solar system may have been slightly more compressed than it is at the present day, the Earth and all the planets being rather closer to the Sun than they are at present."[3] This, with other pieces of evidence from wherever they occur, in a dozen fields of study, becomes valuable, once belief in the constancy of the historical skies is held in abeyance. The struggle to know becomes, as was said, "wrestling, no-holds-barred."

It may be argued that the most ancient cosmogonies of the world hold a consensus that amounts to a model of recent natural history. Perhaps scholars would agree that the following thirteen complex experiences are recited in or can be derived from the earliest sources and from the oral accounts provided by existing belief systems that pretend to refer back to the "beginnings." I imply in each case that proofs of fair reliability are accessible to expert ethnologists, linguists, and mythologists from

---

[3] S. I. S. Workshop (1982).

among the many collections now available from all parts of the world.

1.   Earliest man could make out no sharply visible lines between far sky, air and earth; they merged.

2.   Earliest man asserted that the atmosphere cleared somewhat amidst a chaos, and that, here and there, the ceiling of clouds broke.

3.   He claimed to see a great body appear in the "North" that was not the Sun, was more vigorous than the Sun, and remained in the sky for many centuries.

4.   He observed the dense planets (Mercury, Venus, Mars, and others) to be present and close in, while the gaseous planets were part of, or grouped close to, the binary second sun.

5.   He determined that the planets moved with some regularity, with occasional changes of motion and place, in a heavily gaseous space, but the gases were diminishing.

6.   He viewed a series of explosive 'battles, ' during which Earth suffered heavily, and whose outcomes provided a succession of gods of the same family.

7.   Archaeo-history says that the last active binary principal second sun was Jupiter (by many names), the others having retired into farther space as indifferent gods (becoming the *deus otiosus* of theology).

8.   They assert that planets passed close to the Earth and that comets and debris both passed by and struck the Earth.

9.   Early legends reported that the whole Earth was deluged with waters, fires, and other material fall-outs from the skies on at least several occasions.

10.   Earliest man says, too, that the Earth exploded a great deal of material into the sky, including possibly the Moon, which, in any event, he claims to be a late arrival.

11.   He claims that the Earth changed its motions repeatedly and that its surface morphology was drastically modified.

12.   Primeval humans refer to electric discharges of the type of St. Elmo's fire and thunderbolts as much more frequent, even continuous at times, and often of much greater intensity than at present.

13.   Finally, early humans thought that they had observed their own "creation"; that is, immediately upon being humanized, they felt capable of observing their distinctive internal psychic processes and their external relations with others and with nature.

Modern explanations of this primeval cosmogonic consensus, should it be agreed to exist, are various. Perhaps it developed from a single diversifying human race that might be said to have taken off at the time of the Ice Ages, the early Holocene Epoch, or some such baseline. Or perhaps it developed when numerous sub-cultures, already diversified among early mankind, witnessed events independently. Or it may have diffused later on from a single powerful political-religious movement with a highly persuasive ideology.

In assembling this cosmogony, it may be appropriate to make no distinction between gods and nature, taking it for granted that when an ancient legendary voice says 'god' it means a discrete and powerful natural force or body which may ('the known god') or may not ('a new god') be behaving in a characteristic (i. e. predictable) manner.

These thirteen event-complexes of primeval natural history constitute, I think, a consistent, if presently non-authoritative, model of natural history, one which I have adapted to contemporary science in several books. Their numerous anonymous discoverers were fully human observers who imputed the phenomena to animated beings (gods) for compelling reasons, especially in an attempt to control them, so as to assuage terror and get on with the business of survival under most unfavorable conditions.

To put the hypothesis absolutely: nowhere on Earth is a people to be found whose legends contradict this total set of claimed experiences. No ancient people asserted a linear or uniformitarian history. Then the questions arise: could all this have been a universal set of illusions affecting all people? Was a universal genetic archetype of the human mind bound to erect this cosmogony? Was it a consensus of observers?

Scientists are not dealing here with 'anomalies,' but with a universal set of consistent allusions. The detail is so extensive as to rebuff facile explanations; one ought not merely to conjecture 'archetypes, ' or 'grand delusions. ' 'Euhemerism' may provide the answer; it interprets myths as traditional accounts of historical personages and natural events. But euhemerism should not prejudge the case in favor of uniformitarianism by retrojecting current history.

Anthropologists finally established as research doctrine that primitive cultures are to be taken seriously; the statements of informants are to be examined, not ridiculed. And the examination can be conducted and completed without conversion of the anthropologist to the views of the informants. *Pari passu,* the most ancient "fossil voices" are to be audited seriously, even sympathetically. In this case, the voices would have to be translated into a model that would begin to make sense to modern physics and psychology.

The results would be foreseeable. Considering the intellectual revolution that would follow, the ancient cosmogonical consensus would be rejected by most scholars in short order. For the following principles of physics and natural history would be among the most likely to be inferred from the ancient empirical beliefs:

A.   All planets and satellites would have to exhibit evidence of very recent extreme thermal and explosive experiences.

B.   The solar system bodies would have to show a declining but considerable set of electric fields and electromagnetism, and solar system space would be in the process of clearing up its ionized gases and plasmas.

C.   Remanent binary behavior would have to be evidenced by Jupiter or by the outer planets as a group.

D.   Continental "drift" theory would need to permit a negatively exponential rate of movement from a very late breakup of the Pangean crust, and a socket from which the lunar material was wrenched must be shown on Earth.

E.   Astronomical motions would have to be reckoned as short-term, empirically observed behavior until a new mathematical model could be developed.

F.   The biosphere, lithosphere, atmosphere and hydrosphere must be capable of interpretation according to which major elements and features were quantavoluted or saltated, and present constituents and behavior are comprehended as "tailing-off" phenomena.

G.   The human brain (behavior) would have to be compatible with convulsive original experiences that set it upon its present course, hologenetically, in a quantavolution.

H.   Human culture would have been hologenetic, too, arising abruptly as a total response to the requirements of a quantavoluted mind.

I.   Explicitly, as implied in all of the above, a basic error in radiochronometry must be demonstrated, and long-term geology heavily revised to admit numerous occasions of late, large-scale quantavolutionary phenomena.

If some such model is physically impossible, then we should have to discover and explain some other structural-functional mental dynamic, universal among human groups, that made necessary its elaboration as science-fiction.

If the early scientific catastrophists had gone on with their work, we would have learned enough by now to make what I have just stated an epilogue rather than a prologue. "If the catastrophists had gone on..." The German methodologist and sociologist, Max Weber, once wrote at some length about the scientific justification of "if... then..." historiography." If Lincoln had not been assassinated, then the U. S. A. would have become more unified during the Reconstruction period: may such a thesis be posed and dealt with scientifically? The answer Weber gives is "yes." So some legitimacy (the "legitimacy of scientific authority" Weber would have said) is owing us for proposing this line of thought for some future historian of science. The theses just presented and those yet to come are then monuments to a science that might have been and a budget of a future science.

Popular geology believes that the Earth is stable and quiet, and that where it is not so, the explanation has to do with a hot turbulent mantle that continually causes surface disturbances. Geologists were responsible for this belief and mostly share it. What can be labeled as the conventional geological position is summarized by Shelton:[4]

> Most geologists look inside the earth for the ultimate driving force of diastrophism; no known exterior forces are sufficiently versatile to account for the variety of deformation we see... It would seem that plastic creep, perhaps in the upper part of the mantle, is the active element, and the brittle crust on which we live is passively riding on this very slow flow. Of course, discernible forces arise from the rotation of the earth, from the tides, and from gravity acting differentially on irregularities in the crust and its surface topography, but these influences probably can do no more than modify and locally complicate

---

[4] *Op. cit.*, 423.

what is probably the essential mechanism of crustal deformation - very slow plastic movements at about the level of the upper mantle.

Shelton goes on the show why "this concept is attractive," why the presumed "plastic creep" has most of the essential capabilities needed to mold the Earth's surface over great lengths of time. "The combination of gravity with variations in the density of the material" operates so that "circulation in the deep plastic zone probably involves rising and sinking columns as well as horizontal currents... Some kind of very slow thermal convection -the rise of relatively warm columns and sinking of relatively cool ones -is a favored hypothesis for the ultimate cause of diastrophism." This is about as far as the theory of 'land-based geology' has come.

In contrast, we have been offering a space-based geology. Here the "ultimate cause" is exoterrestrial. Quantavolutions - intense and abrupt events of large scope - occur. Without exception these involve exoterrestrial transactions. No intrinsic Earth-force can produce quantavolutions. These events can be given values and measure; they can be comprehended and subjected to at least as much quantitative modeling and manipulation as is afforded by 'land-based geology. ' Unlike evolutionary theory, which deals in bulk low-energy transactions, quantavolution pursues bulk high-energy transactions.

The forms that high energy takes have been discussed heretofore, and will shortly be summarized. In a score of guises, all will have ultimately originated by space transactions of particles and masses. The space transactors are galaxy, planets, Sun, Moon, comets, meteoroids, plasmas, electric charges, and so on, sometimes taken as independent, sometimes dependent, variables.

From the standpoint of the Earth, an expression of high energy denotes an exoterrestrial force when it achieves a specifiable level of intensity, scope, and abruptness. Invariably, it operates to include other forces and to develop, with them, countervalency, as well as extended effects. Here we shall give hypothetical examples of what would be *prima facie* demonstrations of the operation of quantavoluting high energy expressions originating exoterrestrially. Together with the materials assembled earlier in their respective chapters, the examples run the gauntlet of 'land-based' alternatives; the thesis is that they cannot have occurred without a direct or near relationship to an exoterrestrial event. The examples are hypothetical; they are conjectural approximations of what could at a later stage of the earth sciences assume a more qualified and varied quantitative formulation, of what could later on be historically located.

Supposing, in the first instance, we were seeking evidence of a "cosmic hurricane," that is, of a high-energy wind of ultimate unearthly origins. One will be entitled to claim exoterrestrialism with "the discovery of three heterogenous fossil agglomerations of the same age within an area of 1,000 kilometers diameter from which sediments of the same age are patchy, missing, or abnormally continuous." And, we may add "provided that tempestites and other wind indications can be assembled for the area" since aquatic tides will invariably provide associated data, and in fact, by the principle of mutuality of high energy transactions, no effect is single. The proposed discovery is of unknown difficulty; it has not been attempted; it may be simple or practically impossible. Yet how else can we search for "fossil winds."

And this could denote cosmic cyclones as well, the uplift of immense agglomerations of material and their erratic deposition. Of many thousands of geological and atmospheric studies, is there one on the cosmic fossil cyclone? None, though we have mentioned the evidence of single fossil tornadoes. Yet we know the effects and conditions of cyclones, how they occur in multiples, of their transporting power, of their relation to volcanism and explosions, and of other characteristics that make them invariably part of a catastrophic scenario. In *Solaria Binaria*, Milton and I posit thousands of downbursting cyclones as the most logical means for a deluge to bring huge sky waters down to

Earth, shaping itself thus with the help of the also inevitable electric discharges.

Let us posit another example, trying to isolate fossil electrical discharges, while granting their presence in every high-energy expression. We would seek "Metamorphosed rock on one-fourth the prominences of a 100 km diameter mountain range, which is not otherwise metamorphosized." Or we might seek "non-assembled heavy biotic dissemination in contemporaneous sediments taking the form of fusain and calcination and extending over an area of 500 km diameter." This latter may be from a conflagration as well.

The detection of thermal change goes beyond electrolysis and conflagration into non-calcinating fluctuations, and of these is climatology composed in part. The correlation of climate with exoterrestrial phenomena is proceeding apace. When we offer as a suggested criterion, "Cold and warm weather fossil species occupy contiguous strata or are mixed in the same deposits," we are probably opening up many strata of natural history to quantavolutionary exoterrestrialism. Unless it can be shown that the changes are gradual, the exoterrestrial presumption is justified.

The search for fire effects is broader because it admits the provenance of ashes: "Simultaneous fires devastating 3+ areas of 1,000 km^2, each of which is 1,000+ km distant from the others." Without such evidence, the world can scarcely be said to have burned up, even in significant part, as the fossil voices insist. Paleochemical analysis, a field in its infancy, may be the appropriate technique; still, the very material to sample may have been blown or washed away, and there is the high energy of volcanism, to which are generally ascribed the ashes that cover many parts of the world. Somehow, we must go beyond the ancients, who united, in the concept of fire, spontaneous and celestial conflagration, volcanism, and electricity.

Especially for volcanism, there would occur evidence of "Plinian outbursts simultaneously of 20+ volcanoes anywhere on earth." This figure is modest; yet it would indicate exoterrestrialism; few volcanologists would deny the repeated occurrence of such phenomena and some might dwell upon much grander episodes.

Earlier we have sought evidence of fall-out. The archives of anomalistics, as R. W. Wescott has employed the word,[5] and which William James referred to as "the unclassified residuum," are replete with minor cataclysms, many of them traceable back to an origin on Earth, others patently exoterrestrial, and some of questionable origins. One might here venture in search of "Cataclysms of water, minerals, fluids, gases, biotica, and dust 100+ times greater than norms of the twentieth century, happening in a period of less than a year, and often continuing for many years."

And perhaps one should seek "Poisonous chemicals in similar strata at 4+ points at least 300 km from each other." But the mention of poisons could send one in search of "Six or more fossil conglomerates of similar sediments anywhere in the world exhibiting 2+ times the normal background radiation of modern age bones."

As for deluges of water and other space debris, one would raise the factorial on some of the above fall-out, and explore "A type of non-fossiliferous deep sedimentation discoverable over an area of 100 km diameter." Some creationist scholars, using the flood of Noah as a unique all-encompassing event, and pushing the principle of the mutuality of high-energy transactions to its limits, have managed to interpret all diastrophism and catastrophic morphology as effects of flood and tide.

Proving precisely a deluge, as distinct from, even although associated with, floods and tides, is a difficult problem for geophysics. The evidence is of a kind elaborated earlier in this book -the search for the sources of oceanic water, chemistry of seawater, and so on. Still it may be possible to discover a true exoterrestrial deluvial sediment by, if nothing else, the exclusion of all other explanations from related features.

---

[5] V *Kronos* (Spring 1980), 36-50.

Sometimes fossil lake and sea basins are detected and, rarely, a sudden displacement of waters from the bed is the subject of comment. The "outrageous hypothesis" of Bretz governing the sudden emptying of now extinct lakes in a barrier-bursting flood of northwestern U. S. A. - the Channeled Scablands - is a case in point. Where one lake is emptied, exoterrestrialism is doubtful. If "2+ bodies of 100 km 3 of water were abruptly displaced at the same time," exoterrestrialism would be indicated, possibly an axial tilt, or secondary events following the exoterrestrial event, such as a massive thrusting, a deluging and bursting of barriers, an ice surge and melt, a tidal damming and bursting, and a 9+ Richter seismism.

Fossil tides are also difficult to distinguish. One may propose "Tidal waves attaining 100+ meters in amplitude at 10+ land points not less than 400 kilometers apart. This might achieve a satisfactory level of confidence in an associated exoterrestrial event. For cosmic flooding, one would repeat the deluge hypothesis, where uniformly fossil-bearing strata are included.

We appreciate, however, that flooding of this kind may originate in the sinking of land followed by its rising or by the melting of an ice cap, flooding, and then either withdrawals of water for new ice or a rising of the land. Explanations of this kind are common, and usually omit any causal explanation beyond its mere statement, *viz.* "Here we find a marine-fossil stratum of age 'A' probably due to the ending of ice age 'III.' "

It is doubtful that the Earth can contract globally, if only because an exoterrestrial electrical discharge that might compact it would be associated with a thermal force that would expand it. That the margins of the continents can be flooded is probable; the ice caps contain enough potential water for the purpose; and that ice accumulations have melted in times past is fairly obvious. However, it is also now fairly plain that, for ice masses either to accumulate or melt requires a quantavolutionary exoterrestrial transaction.

At the same time, for the land to rise, carrying the biotica of shallow seas with it, also requires an exoterrestrial transaction.

Here the criterion may be "An absolute rise of 300+ meters over an area of 100+ km diameter." Actually this may occur at the sea bottom as well as on continental land. To identify an absolute local, much less a worldwide, expansion is again difficult. It may be that the universal lava venting, circular bubbles, high plateaus, and broken crystal grid of the Earth reveal global expansion.

Convection current theory, unfortunately vulnerable, can seek a merely terrestrial explanation of these phenomena. But if it tries to engage its currents to shape the Earth's surface as just described, as well as to push the continents around, it will logically have to posit a very young and turbulent Earth, which it will refuse to do. Or a large Earth expansion, which it refuses to do. Or an Atlantean concept of great sunken continental areas, which it would hotly reject and furthermore do not exist.

Land might be removed by explosion into space, with some fall-back. It can also accrete, meaning that a mechanism such as a cosmic wind or typhoon has incited local "minor" turbulence. Suppose "A heterogeneous conglomerate of 100+ km 3 filling a basin or composing all or part of an elevated range." Must this accretion be exoterrestrially caused? Probably indirectly by induced winds, tides, and bulldozing.

Bulldozing with a rock, or ice, or aquatic shovel can accrete (filling basins and forming hills); it can produce thrusting, and it can remove land features. Thus if "Five or more blocks composed of similar rocks, of 30+ km^3 each are separated by 100+ km from each other and are 10+ km from kindred strata," we can speak of thrusting by bulldozing provoked by exoterrestrial transactions. The figures used may, in fact, be much reduced. Thus, also, if "Two or more consecutive eras of sedimentary rock are missing over a region of 200 km diameter," the removal will have been accomplished, if not by bulldozing, then by hurricane or tides on a cosmic level of intensity.

The relation of cosmic pressure (electro-mechanical) to expansion and thrust may be explored by the detection of "Expanses of 10,000 km^2 with frequent granitic and metamorphic outcroppings designating a prior period of heavier overhang rocks and a thrust or blast removal of the overhang." The argument here would follow along the lines of argument against "plastic creep" in general.

When speaking of thrusting, the original event will have been a large-body collision or encounter near-in with a great body -Sun sized or greater to the eye, yet seemingly far removed. Inertia comes into play in the atmosphere and lithosphere. "A ten-second deceleration of the Earth on a single day" will produce many local thrusts, expansions, and probably every other high-energy manifestation. Even at this seemingly modest deceleration, one would be able to find later on extensive macro-and microfracturing of the lithosphere. Raise this deceleration to hours and the surface of the Earth would be extensively altered.

As evidence, what would be demonstrable is probably already present and awaiting discovery, that is "Areas of 500 km: exhibiting fractures of 85% of all included grids of 15 kilometer diameters in one or more strata." Perhaps here one should expect tortured sea bottoms and igneous flows that would have been nonexistent or molten during the events. Axial tilting would also be denoted by patterns of inertial change probably by now totally confused in the morphology and petrology of the Earth, except as we have pointed out in earlier sections of this book, where axial tilt can be detected on gross features of the global map.

The most obvious form of exoterrestrial transaction is the meteoroid explosion, which is provable on its face, of course. That a great many of such intrusions are not yet discovered has also been shown. That the Earth should have fewer craters than the Moon only occurs by reason of their quick erasure here. As detecting techniques improve we should be able to speculate reasonably, in the manner used for fracturing above, that meteoroid impact craters arc present over all of the globe save where erased by other quantavolutionary processes.

Several biosphere phenomena may be hypothesized as indicative of exoterrestrialism in quantavolutions. The pandemonium accompanying quantavolutions is not likely to have left a geophysical record. "World-wide sound at 100+ decibels,

approaching human physical limits" can be considered, given, for
instance, the thousands of square kilometers of high audibility of
the Krakatoa volcanic explosion; still such effects would not
fracture rock nor (probably) affect the hearing of species
genetically.

Other acoustical effects might, however, be mutating, and
even chemically effective on the molecular level of the atmosphere
and lithosphere. Legends do describe great sounds that suggest
exoterrestrialism: we have alluded to them. So too, have we
mentioned specters as often the greatest contribution of ancient
voices to the proof of exoterrestrial events affecting earth; thus we
would allow as evidence of an exoterrestrial transaction "Reports
of an observed cosmic intrusion of an apparition the size of the
Moon when at meridian, or larger." We would expect such a
specter to be associated with at least several other high-energy
effects mentioned above.

A final trio of expressions may be advanced. We should
be alerted to magnetic effects. These may be indicators of axial
tilting. But when localized, they can point to meteoroid explosions
and peripheral and subsurface melts, where post-event
magnetization differs from magnetic orientation, as displayed in
the circummagnetic field of central Canada, having south Hudson
Bay as its focus. "Inconsistent and strongly deviating rock
magnetism over 5° grids of latitude-longitude" proves
quantavolution.

In biospherics, "Biosphere extermination over 100 km²
in 1 incident of under 1-day duration" would be proof of
exoterrestrialism. So would "The extinction of 3+ species in less
than 1 year," or "Depopulation by 70% of 1+ species of 10
biological families in less than a year in an area of 1,000 km
diameter." A number of phrasings may be formulated to denote
physical catastrophe in biological terms as well as in terms of
physical science.

At the same time, genesis may be used as an indicator of
quantavolution. Thus. "The simultaneous appearance of 3+ new
species" will suffice to indicate a catastrophic innovation.

"Simultaneous" means genesis within a century, or the smallest frame visible in the fossil record. An "appearance" should not prompt an assumption of missing transitional ages.

Thus we have possibilities of operationally defining quantavolution as a happening of high intensity, everywhere, at the same time, and, as we shall shortly argue, quickly. In a great quantavolution, many things change at once, overlap, transact, follow in quick succession. A quantavolution of one kind, once initiated, has a prelude, a climax, a procession, a recession, a stabilization and finally a uniformity.

The concept of negative exponentialism holds that the initial quantity (intensity, number, frequency, amount. volume, locations, incidence, etc.) of a type of event decreases sharply with the passage of time, but ever less sharply as time is extended. Finally, the rate is indistinguishably uniform, that is, the same, and the activity being observed is constant. One notes that this may be accomplished theoretically (i. e. imaginatively) or by the use of empirical data. In theory one can annihilate change by stretching time: increments of volcanism are spaced so far as to provide a negligible rate of change, or spaced so tightly as to provide catastrophic rates of change.

To denote negative exponentialism realistically (empirically) requires data on the beginning, the end, and at least one point of time in between. Thus, if 500 volcanos were active 11.500 years ago around Auvergne in France and 0 are active today, a decrease is undeniable (from 500 to 0 in 11,500 y) but the decreasing might have occurred at any point and in a number of ways: all 500 might have stopped erupting last year, for all we know. At least one estimate of the number of active volcanos at some point of time between the two given ones is required to permit an elementary idea of the progression.

Let us suppose -which, alas, may be the fact -that all activity ceased before history began; further, that no evidence of relative youth is to be observed by geological examination; worse, that no radiometric test and not even Carbon 14 dating is capable of assigning relative dates. All we can say is that the first local

references were 2000 years ago and no mention of volcanic
activity is to be found. So the curve is flat for the past 2000 years.

What next? One can go searching for records of other
volcanoes in other areas, the Mediterranean say, where history of
a kind goes back another 2000 years. There we would have to
discover some evidence - whether on official tablets, in legend, or
in archaeological excavations of extinct human activity laid upon
or beneath lava or ashes - that an ascertainable level of volcanism
was occurring, whereupon the indicators of this would be
presumed to indicate what was happening in the Auvergne. But
this logic, of course, violates the ordinary supposition of most
volcanologists, that volcanism in one area does not suppose or call
up volcanism elsewhere, a supposition true in these days, it would
seem, but not necessarily true if volcanism were more rampant.

One may resort to widespread ash layers as well. If layers
are thick and far-flung, it may be reasonable to suppose the
Auvergne would be afflicted by the same activity as is producing
the ashes generally, and, with ever better chemical analysis, the
ashes may even be traced to the neighborhood of the volcanoes
in question.

Still, as the chapter on volcanism reflects, at the moment
historical volcanology has to put together bits of evidence from
widely separated localities in order to supply what is largely a
conjectural statistical foundation to the generalization that at
certain historical points in time volcanism leaped to peaks,
subsided quickly, and then evened out, thus lending the
appearance of a uniform activity but, if one wishes to assume our
position, also letting us guess that volcanism is delineating
negatively the exponential principle. .

For each and every type of expression of force involved
in a catastrophe, there would exist a statistical curve delineating its
individual intensity over time. Each expression would possess its
peculiar rate of decline from its initial peak -its own "disturbance
constant" -giving us various exponential or hyperbolic functions.
Then, for instance, as more and more data illuminated the dispute
over the late Cretaceous extinctions, curves might be drawn to

depict the fate of biosphere segments and of inorganic expressions of the catastrophe, answering ultimately the questions: "How intense, what scope, how sudden?" for portions of each sphere, the several spheres, and the holosphere of Earth.

If peak catastrophic and holospheric turbulence has occurred, say, at five points of time in the holocene, there will be a new negative exponential curve to assign to the effects of each set of events. If these curves are merged, one gets a kind of roller-coaster curve, rising and falling in conformity with each set of events, while at the same time maintaining a momentum of generally falling activity until, at the end, like when the roller-coaster ride is ending, the past two thousand years become practically a smooth glide.

Such is the negative exponential curve of quantavolution, taken as a whole or in its subsets. For these phases in any high energy expression are subject to the successive sets of phases of the quantavolution of other kinds, subsequently, yet concurrently, initiated. These may accumulate intensity at any phase, or countervail, that is, diminish intensity.

The whole Earth is acting out the several stages for numerous forces at any given moment in time, and the state of the Earth may as a whole be deemed uniformitarian or disastrous as it is working its way through a low cumulative effect of the forces or a high cumulative effect.

The low effect - the world as it is today is mostly a descendant effect from original high effects. This we have considered as the principle of exponentialism, which results, in the end, as an almost uniform rate, with exceptional cases of high activity. In an article on "Landform Evolution" (geomorphology), interesting in its vagaries and confusion, the *Encyclopedia Britannica* cites many catastrophic conceptions, ascribes erroneously the beginnings of scientific catastrophism to Bishop Usher's Biblical literalism, and summarizes uniformitarianism today as holding, "Although present processes are similar in kind, process rates must have been variable." But it is doubtful that any scientific catastrophist ever believed that processes were dissimilar. It has always been an argument over rates.

It is also of interest, and insufficiently addressed by the many commentators who recognized that C. Darwin took from Malthus the idea behind his theory of the origin of species by means of natural selection, that he did not see the larger consequence of Malthus' idea of exponentialism. This latter idea expressed in the belief that while population rises geometrically, the means for its subsistence increases arithmetically - points to catastrophism but inversely, that is, negatively, implying that the catastrophe is a sudden leap and then an exponential decline from the leap in the direction of increasing gradualism. Ignore the leap and the character of exponential decline, as Darwin did, and natural history is stripped of its salient behavior. What appealed to Darwin and those of like mind, such as Spencer, was the competitive struggle as the means of subsistence grew scarce in relation to population; and the notion, of course, that "fitness" is an objective concept, in nature as in society.

There exists little speculative or empirical literature on the abruptness of catastrophe. Catastrophe by definition connotes an abrupt disintegration of an existing course of natural behavior. How is "abrupt" to be conceived? Suddenly, quickly -but is this seconds or millennia, or something in between? Should we say that, to have a quantavolution, an event or set of them has to occur in less than a million years? This would please some conventional geologists who have given themselves some five thousands of such units to reckon with. Five catastrophes distributed over the period would consume only one-thousandth of the time allowed.

But what kind of catastrophe is it that would take a million years to happen? Suppose some poison slowly entered the atmosphere or suppose the Sun for a million years was hyperactive, and radiated the biosphere beyond the sufferance of many species. Even conventional scientific gradualism would find the postulation of such slow "catastrophic" processes implausible.

Natura facit magnum saltum: that nature, when she leaps, leaps high, is a more believable axiom. This is no place to argue, as we do in another book, *Solaria Binaria,* for a million year history of the solar system. But if we were to cast dice, giving each possible source of catastrophe, whether slow or fast, an equal chance, we should very probably cast forth one of the fast catastrophes. That is, of the half dozen major types of catastrophe that are possible, only a special variety of particle and dust bombardment produces a slow catastrophe. And this, as we have implied, should be measured in hundreds rather than millions of years.

The many scientists who today make dire predictions about the effects of a carbon dioxide pollution of the atmosphere or of the removal of the ozone barrier to exoterrestrial particles, couch their forecasts in hundreds of years; why would the same and other scientists wish to insist retrospectively upon tens or hundreds of thousands of years for the same phenomena to have occurred? If they did, it would be for irrational, that is, ideological, reasons: they would be unconsciously straining to support an evolutionist view of natural history.

Luis Alvarez, and perhaps his associates as well, after suggesting that the sweeping extinction of the biosphere at the Cretaceous boundary came with a solar obscuration by dust raised by a meteoroid crash, elected a period of about three years of dusty atmosphere, then lowered the effect by a factor of ten, to three months.[6] Again, what is abruptness? What is "geologically instantaneous?"

Eicher notes a "huge" recent Chilean ash fall which is never over 10 centimeters deep away from the central volcanic area.[7] Yet in the Upper Cretaceous strata of Colorado, over thousands of square miles, there occurs a bed of bentonite, highly compressed volcanic ash, which is a meter thick. This may have

---

[6] *Op. cit. cf contra* R. Jastrow, "the Dinosaur Massacre," *Sci. Digest* (sep. 1983).

[7] Don L. Eicher, *Geologic Time* (Englewood Cliff, N. J.: Prentice Hall, 1968), 72-3.

been coincidental with the boundary events of which the Alvarez group speaks.

Smit and Hertogen inform us that the great biosphere extinction marking the Cretaceous-Tertiary boundary "was abrupt without any previous warning in the sedimentary record."[8] O'Keefe at the same time accounts for the devastation of fauna at the end of the Eocene (assigned 34 million years ago) by radical climatic change induced by a ring of microtektites and tektites circling the Earth for perhaps a million years and obscuring the Sun.[9]

Ogden discusses abrupt changes in American forestation about 10,000 years ago, also climatically impelled, with the pattern of pollen deposits in lake sediments moving at the rate of a mile a year.[10] Hapgood has compared what arc regarded as 'normal' rates of ice retreat with the results of carbondating, and allows some 60,000 by the one and only 17,000 by the other. He believes that the carbondating must be in error.[11] Cracraft, in expatiating upon the "punctuated equilibrium model" of macroevolution, argues that speciation is a "geologically instantaneous phenomenon."[12]

There is, in sum, a growing body of paleontology and geology that perceives abruptness of change as a feature of natural history. What means "sudden" and "abrupt" is likely to be a much-discussed question in the near future. We can suggest here merely that every feature of the holosphere enjoys its idiosyncratic manner suddenness.

A species, a land mass, a body of water, and an atmosphere all change according to their nature, and measured in human terms, this may be fast or slow. When a quantavolutionist speaks of abrupt change, he can only mean the margin between explosion and extinction on the one hand, and the rate of change

---

[8] 8. 285 *Nature* (1980), 198.

[9] 285 *Nature* (1980), 309.

[10] *Op. cit.*

[11] Hapgood, *Path of the Poles*, 127-8.

[12] *Phylogenetic Analysis and Paleontology* (NY: Columbia U., 1979), 26.

peculiar to a given organism or natural process when the rate is affected by a disaster produced by a specified high-energy expression.

Similarly, when speaking of energy of high intensity, the quantavolutionist is describing known natural forces proceeding at abnormally high rates. A recently discovered ash layer in E1 Salvador covers 1300 square miles and a once flourishing Mayan civilization. (The fall of ash was dated much earlier before the culture was unearthed.) Some 45,000 of such eruptions would be needed to blanket the Earth. The volcanoes, mostly extinct to be sure, are present; how many of these were ever exercised simultaneously?

One more, when speaking of scope, scale, or simultaneity, the quantavolutionist seeks limits appropriate to the effects of a high-energy force, between total immediate transformation and a highly significant change. Isaacs and Schmitt address themselves to oceanic energy sources; they provide global figures on the great energy sinks and low energy manifestations involved in currents, waves, tides, thermal gradients and salinity gradients. The rising and falling of waves is an energetic type of movement. When it occurs as a tsunami, or is pulled up tidally in an exoterrestrial encounter, it multiplies exponentially its force, as was said earlier, so that nothing can withstand it finally except the Earth itself.

The rotational energy of the Earth can he translated into 6×10^{15} Megawatt years. All the electrical needs of the world projected into the 21st century amount to 3×10^{17} MW: if continuously mined from the energy of the Earth's rotation, the length of the day would be increased by five minutes per million years. This is the latest and one of the finest comparative measures by which the forces of nature are converted into everyday terms and may be used to explore the dimensions of catastrophe as well.[13]

---

[13] "Ocean Energy: Forms and Prospects," 207 *Science* (18 Jan. 1980), 265-73.

Until recently, to take another example, only three cubic miles of petroleum have been drawn upon for the useful and often unpleasant industrialism of modern times; if, as we suspect, the origins of petroleum are largely cometary and cataclysmic, many an ungovernable object in the sky may contain that much and many more cubic miles of the substance or its components; awaiting the occasion of manufacture may be an abundance of cosmic electric potential.

Hibben once voyaged the far North with an eye for catastrophic remains. He remarks that the Pleistocene ice sheet (if it truly existed as such) never covered the central regions of Alaska nor parts of the Aleutian Range. He reports, as have others, the several hundred feet of frozen muck deposited in various unglaciated areas. In the muck are volcanic ash layers, peat, animal and vegetable matter in vast quantities, and ice fragments. Below the muck have been found mammoth bones, human artifacts, and tree stumps in their original position as they had grown. The total effect is of several simultaneously interacting high energy forces, whose total rate of burnup of the Earth's rotational energy must have in hours, not in a million years, taken up the equivalent of five minutes of the Earth's rotational energy, and perhaps then, indeed, as a prior condition, the Earth's rotation may have slowed by that much, or more.

All effects of high energy deteriorate exponentially, we repeat. Often, as with a hurricane that expends the energy of many hydrogen bombs, the force is largely employed within and against itself. Forces also act by the principle of countervalency. Bursting into operation, one force generates another, which may not only bring on a third, but may turn against the first and moderate (as well as heighten) its effects. A volcanic wind can halt a lateral hurricane; two sets of rocks can counterthrust. An extinction of one species can promote the survival of another species. Cross-tides may create destructive vortexes but also moderate each other. A deluge can dampen the fire with which it originated from a third force. And so on. The possibilities are very many; if the

Earth exhibits patches of peaceful history here and there, these may be effects of countervalency.

Countervalency may occur on the grandest scale. Repeatedly the theory of the eruption of the Moon from Earth is challenged by the conviction that so large-scale and destructive an event would have destroyed the Earth's crust entirely, or at least its biosphere, or at least all vertebrates and forests, or at the very least mankind. Such is not the case.

The energy of the lunar eruption may or may not have exceeded the energy involved in wiping out the Martian atmosphere and biosphere; the gross energy expended (transformed) is not the issue; the counterrailing operations of the energy forms, the coincidences, are the determining factor in the extent of destruction.

On several occasions, the Earth's atmosphere may have been destroyed and transformed. The presence, according to the theory of *Solaria Binaria,* of a gaseous tube enveloping the solar system, even until a dozen millennia ago, allows for a drawing off of the atmosphere over half the world, for a rush of atmosphere from the opposite hemisphere, and for cataclysms of atmosphere from the plenum, not irreconcilably different from the atmosphere that it displaced.

In other cases, the very motions of the Earth itself will tend to deprive a catastrophic force of complete victory. If 50,000 volcanoes erupt simultaneously, the whole atmosphere will be put to work with electricity and water to bring down the dust, part of which, for that matter, may erupt into space in pursuit of the body that produced the motion changes and eruptions in the first place. In Saint-Pierre, there was a prisoner in his dungeon, sole survivor of the volcanic explosion of Martinique. In Hiroshima there were the unexplainable uninjured survivors of the blast and holocaust. Once again, problems posed by catastrophes find their solution in the behavior of catastrophes.

At the present stage of the earth sciences, there are probably many fewer persons who will insist upon finding the ultimate source of great turbulence inside the Earth alone. Still this conviction - or is it a hope - persists. Geologists tend to believe that nothing grave ever happened in the skies; biologists often look upon the rocks as gift-wrappings for their fossils; astronomers are inclined to believe that nothing serious happened upon Earth; anthropologists and historians usually believe that ancient times were as serene as nature today. This consensus is suspect. Some scholars apparently are still reassuring one another, so that all might eventually come to believe that no event of great importance has happened in any sphere of existence.

I hope to have suggested in this chapter some orderly means of bringing forward and considering exoterrestrially provoked quantavolutions. Most such means are difficult, even impossible. But what else can be done? Most of us, whether from timidity, distaste, or because expertly qualified for other forms of combat, will not engage in "wrestling, no holds barred."

CHAPTER THIRTY-ONE

THE RECENCY OF THE SURFACE

If a fossil whale standing on its tail can disprove "millions of years" of sedimentary accumulation, perhaps a live animal can try to do the same. Igor Akimushkin tells us how "The cepola fish... may sit still for hours on the crooked end of its tail, with its wide-open jaws turned upwards expectantly, waiting patiently for heavenly manna to fall into its mouth."[1] The cepola is a fish of the abyssal ocean, where it lives in perpetual darkness.

If it feeds this way for twelve hours a day and collects one millimeter of material with its mouth, in which enough nourishment is contained, then in a year it might be said that a column of 365x 2 millimeters will be striking the ocean floor. This would amount to a column of 730 meters in 1000 years, assuming that inedible waste and compression cancel each other out. Since the ocean sediments average one kilometer, and our live precipitation meter may be at a typical location, the column will reach the average depth of sediments in about 1350 years, under uniformitarian suppositions. With a negatively exponential fall-out, cepola would have once fed more quickly than he does today. So the ocean bottom cannot be older than 1350 years, and ethology becomes the queen of clockmakers.

Quantavolution should be embarrassed to joke so, if science were not on some occasions a theatre of the absurd. One can reflect upon the history of geology when, blessed by the *nihil obstat* of Lyell, geologists would simply draw upon time without end to do away with complexities and perplexities. When Poulett

---

[1] *Animal Travellers, loc. cit.,* 87.

Scrope prepared his famous studies of the volcanoes of Auvergne (France), his theories might be liberated from temporal restraints, such that a recent commentator on his work, Rudwick, could refer to "unlimited drafts upon antiquity" as his necessary and useful tool.[2]

Continuing until today, the time scales have been even more expanded, much more, so that many a geologist has felt free to mount his facts into any frame of time that can hold them; the duration itself would scarcely be accosted for proof. Owing to recent discoveries such as the youngness of the ocean bottoms, and to late criticism of biostratigraphy, the license to capture time has become more restricted. But radiochronometry, newly developed, reigns supreme over time and is dizzied by success.

Conventional chronology today gives about 15,000 years to the Holocene and latest period, and about two million years to the Pleistocene. Then some 35 my go to the Tertiary, with its Pliocene, Miocene and Eocene; 55 my to the Cretaceous; 27 my to the Jurassic; 23 the Triassic; 33 the Permian; 74 the Carboniferous; 72 the Devonian; 22 the Silurian; 57 the Ordovician; 92 the Cambrian; and some 2000 million years (or much more) to the Precambrian era.

By this point, the reader is well-aware of our scandalous departures from the conventional text. We have been arguing, in the whole of our Quantavolutionary Series (and see page 497 below), that all of the preceding ages probably have occurred within a million years, and especially that major elements of the Holocene, Pleistocene, Tertiary, Cretaceous and Carboniferous have occurred within the time usually allotted to the Holocene, namely some 15,000 years or less. The great disparity has occurred, we maintain, owing to the displacement of time by catastrophe. And to denote these catastrophic intervals, we have used certain disruptive episodes that we have tied into

---

[2] M. J. S. Rudwick, "Poulet Scrope on the Volcanos of Auvergne: Lyellian Time and Political Economy," VII *Brit. J. Hist. Sci.* 3: 27 (Nov. 1974), 205-42.

astronomical events, bringing a sequence of periods that we begin with the Pangean, and then go on the Uranian, Lunarian, Saturnian, Jovean, Mercurian, Venusian, and Martian, each marked by catastrophe, until the present or Solarian period to which only some 1600 years are allotted.

Many salient events are disallowed to quantavolution theory by conventional science not because they take too long to happen, not because they did not happen, but because they happened very long ago. Notable among such cases are the fission of the Moon from the Earth, the transportation of hydrocarbons by comets, the prolonged great heat of Venus, the desolation of Mercury and Mars, and other impliedly catastrophic occurrences, whose number is surprisingly large - even determining - when plucked out of the pages, for example, of the special account of the solar system contained in the *Scientific American* for September, 1975.

Some fifty-nine techniques of determining prehistoric duration and fixing distant events were summarized by the present author (1981) and deemed faulty in one or more regards. This, even when taken together with the sources to which it refers, does not constitute definitive disproof of the validity of long-time chronometry. However, it does permit us to entertain a short-term model of solar system history. The evidence of *Solaria Binaria* is such that all previously existing tests offering macrochronic conclusions are either modified to suit our model, or declared invalid.

With regard to geological and biological tests that assert long duration of processes, evidence is accumulating rapidly that quantavolutionary transformations are physically possible. Independent of historical argumentation, geological and biological time are collapsible in theory and in the laboratory. Astronomers figure time in light-years over vast distances, but this is a convenience, not a measure of history. Empirical tests are, however, also theory-dependent, as, for example, the "thermonuclear" Sun whose dynamics are invisible, and the potassium-argon radioactive decay tests performed upon moon soil that presume a three-billion-years-old Moon, or the radiocarbon test that believes in a practically constant atmosphere.

Every discipline advancing long-time claims would today be in a defensive posture were it not for the heavy investment, both intellectual and material, in radiochronometry, which is believed to be paying rich dividends. The bedrock defense of radiochronometry is that radiodecay rates of known elements are regular and inalterable by any conceivable environmental force. Lately, this view has been challenged.

Once the quantavolutionary hypothesis is substituted for the evolutionary hypothesis of uniform and gradual changes based upon the change rates of recent centuries, the majority of tests simply is nullified. The reason is that the constituents of time-measurement are nature-dependent -the time-makers are, like undisciplined and free workers, able to speed up or slow down and hence cannot be counted upon for an indefinitely long series of regular movements or changes.

If there are 59 different measures of time, say, each one will have to know enough about a certain changing phenomenon of nature to guarantee that it has given off a set of signs or signals throughout a specified period, that these signals composed intervals translatable into current understanding such as solar years or millennia or some usable sequential juxtaposition, and these signals that were once given off can be reliably reproduced, observed, or inferred when recently or currently the signals were registered and/ or interpreted.

Considering the prevalence of scientific opinion on the side of a universe, solar system, Earth biosphere, and hominoidal presence, each of long duration - say, of 6 gigayears, 5 gigayears, 3 gigayears, and five million years - the challenge which short-time chronologists present to the time-keepers of science should be easily disposed of: these need only provide one incontrovertible proof of long duration where short duration is claimed. Should it be demanded that the short-time advocate offer his proofs first, one may plead that the long-time chronometrician is rich in experimental resources, hence *noblesse oblige*.

The stakes in radiochronometry are very high: all of the natural sciences have a stake in the game, plus ancient history, pre-history, anthropology, archaeology, indeed all of the humanities and, in the end, philosophy, theology, cosmology.

At Valsequillo (Mexico) human occupation is evidenced by sophisticated stone tools but the horizons occupied have been dated by the fission-track method on volcanic material and by uranium dating of a camel's pelvis at 250,000 years of age.[3] At least one of the team believes the age to be "essentially impossible."

North of the border, at the Calico site, California, early humans occupied premises and employed several categories of tools. Uranium-thorium tests yielded a date of 200,000 +/- 20,000 years for the artifacts.[4] Meanwhile, in Israel, at the 'Ubeidiya site, previously dated to 700,000 years, fossil mammals were redated to a human site containing Acheulian artifacts at two million years, "500,000 years older than any record of Early Acheulian artefacts or *Homo Erectus* in Africa."[5]

These claims support my attack in *Homo Schizo I* upon the hominid chronology asserted in such studies as those of R. Leakey and Johanson in East Africa. That is, all datings of hominids and early man are far too old, and the so-called hominids were probably human. They also support the thesis of *Chaos and Creation* that assigns an ecumenical culture, worldwide, to Pangea, prior to the breakup of the continents.

In the realm of legend, challenges to radiochronometry emerge as well. The following abstract from *Catastrophist Geology* may be quoted in its entirety:[6]

---

[3] Virginia Steen-McIntyre *et al.*, "Geologic Evidence for age of Deposits at Hueyatlaco Archaeological Site, Valsequillo, Mexico," 16 *Quaternary Res.* (1982), 1-17.

[4] Ruth D. Simpson, "Updating Early Man, Calico Site, California," 20 *Anthro. J. Canada* 2 (1982), 8.

[5] C. A. Repenning and O. Fejfar," Evidence for Earlier Date of *'Ubeidiya*, Israel, Hominid Site," 299 *Nature* (1982), 344.

[6] E. Guerrier, "Le Forgeron Venu du Ciel," 17 *Kadath* (1976), 30-6.

Lake Bosumtwi (diameter 8 km) in Ghana is by geologists generally interpreted as the impact scar of an extraterrestrial body, and the Ivory Coast tektite field has been correlated with it on chemical and geochronological grounds. The Dogons, who live 800 km away in Mali, preserve an ancient tradition attributing the Lake to the fall of a fiery metallic mass of unusual dimensions. This legend is also an integral part of the cosmogony of many other West African peoples, such as Mandingoes and Bambaras. Many priests make a pilgrimage to the Lake or to the nearby town of Kumassi, and also many blacksmiths visit the Lake before initiation to their sacred profession. Glass from the impact rim around the Lake has been radiometrically dated at 1.3 to 1.6 million years, a period when Africa was inhabited by Australopithecines.

The moment is opportune for some scholar to compile such victories of oral traditions. No less than eight hypotheses of this book are combined in and supported by this single story. And who dates the Australopithecines and how? The problem is global.

Every proposition that supports exoterrestrial influence on Earth threatens radiochronometry. Radiochronometry has meanwhile thrown biostratigraphical chronometry into disrepute. Vita-Finzi, for example, places his hopes for quaternary geochronology on radiochronometry.[7] Richer in his turn writes:

> Radiochronometric dating thus laid to rest once and for all the idea that rocks can be dated, even in a gross way, by their lithology or by the extent of their deformation and metamorphism. Radiometric dating also revealed that Precambrian time was far greater than anyone previously imagined."[8]

(Precambrian time is accorded 80% of all rock time and Precambrian rock by one estimate surface over 17% of the Earth.)

---

[7] *Op. cit.*

[8] *Op. cit.,* 65.

Fossil-time is heavily theory-dependent. Alter the assumed speed of evolution and one alters fossil-time, and the dating of its associated sediments. Evolution-time, once we dismiss the pretensions of natural selection (adaptation and survival of the fittest), and microevolution (neo-darwinism) and introduce quantavolution, can be calibrated on practically any time-scale, allowing only a perceptible succession and superposition of species.

The boundary times between the Cretaceous and Tertiary periods are increasingly recognized to have been catastrophic. From tall mountains to the deep abyss, notable turbulence occurs. Asteroids or comets have been called forth to explain the phenomena, which are holospheric. One study[9] concentrated upon a single core drilled at 4805 meters of ocean depth off Africa into a fan of a submarine canyon cut into the Walvis Ridge; at about 205 meters below the bottom the C/T boundary was ascertained and its materials analyzed. Numerous anomalous chemical conditions were discovered, leading the 20 authors to support conclusions, some suggested elsewhere, that the state of carbon dioxide, oxygen, iridium, platinum, cyanide, osmium, arsenic, calcium carbonates, terrestrial ejecta dust, and exoterrestrial dust indicated and/ or caused general decimation of marine invertebrata, and, by extension, as suggested elsewhere, insufferable conditions for flora and fauna of the continents, with magnetic disturbances, a rise in temperature of 8 degrees centigrade, flash-heating of the atmosphere at the explosive moment, difficulty in photogenesis, and starvation.

The mixing of C/T fossils above the boundary for two meters led to an unresolved question as to whether bioturbation or a prolonged extinction process was proceeding after the extincting event. There was only a film of sedimentary clay to

---

[9] K. J. Hau *et al.,* "Mass Mortality and Its Environmental and Evolutionary Consequences," 216 *Science* (16 April 1982), 249-56. (20 authors, now at 13 different institution, 2 funding organization, sponsoring center, and a number of readers were involved.)

work with at the boundary. Above lies core material ending with lower Eocene fossilized ooze at the surface: thus, most of the Cenozoic or recent period is unrepresented. Basalt is first struck at 280 m subbottom depth, below which it alternates to 340 meters with volcanoclastics, clay and sand. Above the latest basalt occur the same, with fossils at intervals intermingled with a sandstone marl, and, toward the present, chalks, cherts, limestones, and ooze. A layer of ash is found at -200 meters just above the C/ T boundary transition and another at -60 meters.

There is an obvious sequence from older to newer nannofossils, but there is also a gnawing doubt as to the length of time which the total deposition, even in its presumably truncated form, actually required. If, for instance, in the 280 meters of postbasalt deposits, some 7 meters consist of ashes, which must fall rapidly, then ashes amount to about one-fortieth of the column, but they must have dropped in a matter of days.

If, too, the submarine fan was laid down turbulently from its parental canyon, and, all the while, heavy volcanic fall-out was occurring, one might conjecture again in terms of days, or months, or years, but hardly in millions of years. The sudden cessation of deposition at the Lower Eocene of 50 million years ago suggests a bottom of prolonged stillness, but then what comes before as here must suggest a brief turbulence. The sequence of fossils could extinct and proliferate in centuries or millennia, or, less likely, occur by instant turbulent crossbedding from different sources.

The authors and others are looking for a medium-sized astrobleme that would have been the disastrous Intruder of the C/T boundary; a 25 km/diameter crater at Kamensk (S. Russia) is alluded to. By our theory the Earth may have suffered numerous meteoroid explosions at this time. In earlier pages, the exponential rate of astrobleme discoveries was noted. There is no chance of finding a solitary culprit. Cretaceous craters will be numerous, and if time is compressed, distinction among the ages of most astroblemes may be vitiated.

All of this is ominous. If geology and geophysics are so ready to sell out biostratigraphical chronology, on which natural history has depended almost entirely from the beginning of its modern phase 150 years ago, then those disciplines, if not bankrupt, are poor. One cannot be blamed for addressing them with alternatives.

Moreover, one must consider whether radiochronometry would ever had developed if geochronology had not already felt the need to posit macrochronism. The presuppositions of radiochronometry are such that it would have had hard going against a microchronism. Basic among these uncertainties of radiochronometry are, first, the setting of zero time for the start-up of radioactive decay of the measuring elements such as 238-uranium, second, the need to assume a constant intake of exoterrestrially produced elements during a long Earth history, and third, the belief that electric charges within the crust and their magnetic fields are either constant or do not affect rates of radioactive decay of the elements whose decay is used as a measuring rod.

As Cook has argued, the early state of the Earth is hardly empirically known or deducible. Yet radiochronometry must proceed as if it were, and, furthermore, somehow, whatever is found now as the result of decay was not present in the beginning but finds its only source in the decay process.[10]

> There is a basic weakness of all radioactive decay methods of chronometry that is too frequently ignored. All these methods must *assume* a given composition of species at zero time. For example, in the original 'lead method' it was assumed that the total 'chemical' lead was zero in uranium-thorium minerals at their time of origin. In later lead 'isotope' methods the decay isotopes were assumed to be absent in the original sample. Later work showed that such assumptions were very doubtful if, indeed, not untenable. Any such method would seem on its surface to be invalidated as soon as one obtains evidence

---

[10] *Prehistory and Earth Models*, loc. cit., 24.

regarding an appreciable abundance of decay products at zero
time unless some means were available to determine the zero-
time concentration of the radioactive decay products.
Unfortunately, one may only guess these concentrations, and
the age results thus obtained can be no better than this guess.
The apparent hopelessness of this situation is exemplified by
relative lead isotope abundance data presented in extensive
tables by Faul and Kulp (Landsberg, 1955).

Cook proceeds to the second problem, that of cosmically
produced nuclear transformation of the isotopes being used to
measure time.

A few years ago radioactive decay processes were the only
natural ones known. Perhaps all of the nuclear reactions
previously described as 'artificial' as well as many others
involving energies quite outside the range of artificial
transmutations actually occur probably at appreciable rates in
the earth. Puppi and Dallaporta (Landsberg, 1955) showed that
the average star (cosmic ray-promoted nuclear explosion) rate
is about $2/$ cm^2 /s or 10^{19} /s in the atmosphere alone. George
(Landsberg, 1955) gave star count data which would suggest
possibly about another 10^{25} inside the earth. Moreover,
spontaneous uranium fission alone should produce 10^{26} stars/
year inside the lithosphere. Since a particles emitted from
radioactive elements have enough energy to penetrate the
coulomb barrier in nuclei of atomic number Z up to at least 20,
perhaps upwards of 10^{-4} of these particles (geometrical cross-
section about 0.02 barns) should produce secondary nuclear
transmutations. If this is the case, natural decay processes
should effect at least $10^{29} - 10^{30}$ secondary transmutations in
the earth's crust each year.

This would be enough to disjoint the radio clocks.
Jean Perrin, as noted by Baranov,[11] has gone farther than
Cook to argue that radioactive decay is not spontaneous, but is

---

[11] In Interaction of sciences in the Study of the Earth, loc. cit., 221-2.

caused by ultrahard radiation coming in from exoterrestrial sources. That is why "natural" radioactivity is concentrated within the crust of the Earth.

We have stressed that exoterrestrial bombardments of the Earth by particles from nova explosions and other sources of hard radiation have been repeatedly experienced by the crustal rocks of the Earth. The present state of the Earth must be receiving a small fraction of its historical radiation. Yet scientists who have provided some of the chemical proof of these catastrophes have been, inconsistently, strong advocates of timing their own disproofs of cosmic particle equilibrium by the very radioactive levels being simultaneously disproved. Like the proverbial military headquarters, they issue bulletins that "the situation is developing well; our troops are withdrawing on all fronts."

A similar problem is to be seen in the separation of electricity from radioactivity. Ignoring the electrical state (or, better, the electrical history) of the Earth may foreclose alternative life-experiences of radioactive materials. But we have intimated earlier that the Earth has had heavy periodic electrical transactions with exoterrestrial bodies and plasmas. Further, the Earth has had electric potentials differing from its potential today.

Sykes placed a standard radioactive cobalt-60 specimen between the poles of a magnet with an estimated flux-density of 0.1 Tesla, positioned a gamma radiation detector in proximity, and took readings of the emissions when the magnet was on and when it was off. The "decay constant," which is supposed to be invariable if it is to be used to clock geological time, speeded up about 2% when the magnetic field was applied. He concluded that "the thesis of decay constancy under all environmental conditions cannot be maintained."[12]

---

[12] N. J. G. Sykes, "A Simple Investigation of the Thesis of Isotope Decay Constancy," III S. I. S Rev. (Aut. 1978), 43-5, 45; cf Don Robins, "Isotopic Anomalies in Chronometry Science," II *S. I. S. Rev.* 4 (1978), 108-10.

These experimental results move in the direction theorized by Juergens and experimentally indicated by Anderson and Spangler.[13] The half-life of radioactive isotopes appears vulnerable to external electromagnetic influences. Since the strength of the Earth's magnetic field has been diminishing, along with that of magnetized rocks, the radio clocks within the rocks will have been slowing down. Further, it is not alone a matter of a long-term trend. In any quantavolution, strong electromagnetic forces are likely to be applied to crustal rocks causing sharp increases in the speed of passage of "radio-time."

Furthermore "electric discharges of cosmic proportions should be capable of creating new elements; even atmospheric lightning is credited with producing radionuclides, and all artificial element-creation starting with the first fusion reaction ever achieved in the laboratory - producing technetium from molybdenum, in 1937 - has involved harnessing the forces of the electric discharge." So writes Juergens.[14] Tesla, his biographers recall, once began experiments to make of the whole Earth an accumulator of induced atmospheric charge; in 1982 an immense electrical current was traced from its North Pacific origins through the Strait of Georgia behind Vancouver Island past Tacoma (Wash.), into Oregon, paralleling a fault line.[15]

"What role," Juergens goes on to say, in passages cited briefly in our chapter on lightning, "might environmental electrification play in setting the rules for nuclear stability, radioactive-decay rates, and energies of particle-emissions in decay processes?" The ambient electrical stress would be different, whether continuously or for short periods of time. "It would seem to follow that decay rates for radionuclides might well differ radically from today's norms. Polonium isotopes now exhibiting very little stability [referring to Gentry's experiments] might then

---

[13] 77 *J. Phys. Chem.* (1973), 3114.

[14] III *Kronos* (all, 1977), 3-17, 11.

[15] J. R. Booker and G. Heusel performed the work; see "Nature's Hidden Power Line," 90 *Sci. Dig.* (Oct. 1982), 18.

acquire -briefly, but long enough, half-lives in keeping with the evidence of the Earth's crustal rocks." Gentry had shown the existence of short-lived polonium without evidence of association with uranium-decay, whereas polonium has been considered an essential link in the chain of decay that ends in 206 lead. Critics of Gentry objected that his findings would cause "apparently insuperable geological problems."

Juergens proceeds farther. Following experiments by Gamow in wave mechanics, he describes the nucleus as having a well-potential or "' potential-well" out of which alpha particles must climb to "decay," mustering sufficient energy to escape. He regards the Earth's electric charge as a principal "well-builder." "The Earth appears to be strongly charged with negative electricity, so that its surface potential is low, which is to say, highly negative."

Suppose, then, that Earth potential is suddenly lowered by just 1 million volts - this, in all likelihood, is an almost negligibly small fraction of the planet's 'normal' negative electric potential. Alpha particles could, so to speak, climb out of the well readily. "Any abrupt lowering of Earth potential by a mere million volts could be expected to produce rampant radioactivity, with consequent lethal or at least strongly mutational effects on all forms of life."

Even presently, under quiet cosmic conditions, the possibility of electrical intervention in radioactivity is not to be ignored. Radioactive radon is released from rocks in earthquakes.[16] This is revealed by a sudden decrease, followed by a sharp increase, in the radon content of the water table just prior to an earthquake. The mechanism is obscure, but it can be conjectured that the electrical fields being generated in the area of the faulting play a role in the phenomenon. When these occur under conditions of a largely quiet exosphere (though we bear solar-storms correlations with seismism in mind) piezoelectricity

---

[16] Hiroshi Wakita *et al.*, "Radon Anomaly..." 207 *Science* (22 Feb. 1980), 882-3.

is to be suggested, as rock is being recrystallized under pressure and heat.

What happens to cause a radon deficiency in the subsurface rock may be happening to other radioactive elements as well, including uranium and potassium isotopes. If so, such rocks may be incapacitated to serve as radiometric clocks, supposing, for example, that potassium 40 is under the same stress. It will either leak out of the rocks, or decay rapidly into the more stable form of Argon 40. If it leaks, and Argon 40 remains, the rock will become promptly "older" in K/A testing. If the Argon 40 leaks disproportionately from the rock, the rocks will become "younger." More likely, the ratio of the two will change and establish itself in a false gradation within the local geological column that will, upon testing, confirm relative age differences with perhaps little more chronological information than is supplied by simple superpositioning of the strata. But what is "local" is probably large-scale, inasmuch as rocks everywhere have been involved in seismic disturbances.

A treatise or symposium negatively critical of the macrochronal pretensions of radiochronometry would be welcome and is overdue. The objections raised here cannot be sustained without much more elaborate treatment. Nor can we more than mention the problems of radiocarbon dating, so important to holocene and Pleistocene geology with which we deal heavily in these pages. As I have written elsewhere, the fragility of this index of time is such as to make it less useful beyond 2,500 years ago.[17]

As with every radiochronometric process, various fluxes of cosmic and terrestrial electricity, large fluctuations of the gaseous and radiation intake of the atmosphere, and biospheric conflagrations all contribute to radiocarbon disequilibrium. Given, for instance, that a solar magnetic storm of the 1950's was observed to add 1% to Carbon 14 of the atmosphere, hence the intake of the biosphere, the probably much heavier solar storms

---

[17] A. de Grazia, *Chaos and creation, loc. cit.,* 51, and Chapter 3 generally.

associated with several kinds of atmospheric turbulence of antiquity might seriously affect dating, which, indeed the studies of H.E. Seuss have proven.[18] We bear in mind, too, the calculations of Cook, which, retrojecting the small but perceptible increase in Carbon 14 in the atmosphere under uniformitarian conditions today, come out with a figure of zero-carbon in the air some 13,000 year ago.[19]

Like every radiochronometric process, with its half-life calculations, radiocarbon decay is figured at a declining exponential rate. The mathematics of exponentialism subjects the process to time collapse; exponential rates in chronology are an unreliable ally of uniformitarian rates in biostratigraphical measures of time and of macrochronism generally. In the clamor of debate over the significance of the multitudinous mammoth (and antelope, rhinoceros, and other) fossils of recent times, the long spread of Carbon 14 dates assigned to the finds has attracted attention, but their meaning for carbondating has been ignored.

If frozen mammoth finds are dated from 44,000 years ago at one extreme to 2,500 years ago at the other extreme, an impossible pattern of climatic changes has to be developed, all allowing some of the cadavers to persist unthawed during the whole period, while letting others cadavers give all signs of eating warm-weather plants just before death.[20] The Carbon 14 dates must be invalid. The same dates, if collapsed and rendered simultaneous, then support an abrupt event, as opposed to an event occupying many thousands of years. The same reasoning would apply to other Carbon 14 problems of the end of the ice ages. Here then, one would refer back to the last chapter and its stress upon the abruptness of biological and geological change.

Is there then nothing whose history when retraced on an exponential rate of development must still have been of long

---

[18] 4 *Radiocarbon Geophysics* 3(1980), 113-7, 117.

[19] "The Radio Carbon Method," 39 *Utah Acad. Sci. Arts Letters, Proc.* (1961-2). 11-5.

[20] Cardona, I *Kronos* (Winter 1976) ' 77-85; Ellenberger, *op. cit.*

duration? Most likely to limit the microchronic concept of quantavolution are certain biological phenomena. Thus, if a living bristlecone pine tree shows annual growth rings now, and if these go back in time for hundreds of years on the same tree, and these live trees are positioned above a locality of fossil trees, which exhibit the same many rings, and in turn connect with the rings of other trees obviously buried continuously below them, a lengthy period of time begins to develop which, founded upon the need for the species to have evolved beforehand, would begin to push time back by thousands, if not many thousands, of years. Proof of such retrogression is not quite satisfactory yet.

With an enthusiasm born of religious convictions and impelled by many years of frustration at playing the other fellow's game, a group of creationist geologists, without spending much time at the task, can readily explain the history of the world's landforms in terms that allow only a few thousand years. That they can do so constitutes in itself a formidable challenge to conventional geology. Still, even granted that they can do so, are they correct?

If they pursued the line of thought that I follow in my books, they might first dispose of the missing half of the Earth's crust by removing it, in a major incident, from the Pacific Ocean hemisphere. They can place the removed crust on the Moon and in planetary space. Then the minor oceans of the world open up to let the continents raft into their present position. Practically all of the ocean bottoms are of recent lava.

Next, they tackle the waters, which descend largely from the heavens, and from boiling metamorphosizing basalt foundations. Next they fashion the rivers from the world's infinite cracks and faults, big rivers from big faults. The mountains are folded and thrust up forward and aft of rafting continents. Huge tides create deserts and fill some lakes. Precipitation fills others. The ice comes from precipitation in darkness, and from exoterrestrial falls.

The sedimentary rocks are ground up from the turbulence of winds, tides, and the friction of moving land masses. Their fossils, when they occur, denote rapid deposition. Volcanoes spurt up along the forward edges of movement in vast numbers and volcanic fissures vent even more than cones. All of the sea and some of the land is lava-covered, an igneous composition. Another large part of the land and ocean shelves is of the original basalt base of the earlier all-land system and is called shield rock or Precambrian exposures. The biosphere that has been destroyed by drowning, burning, burial, poisoning, and freezing exhibits itself largely in a few assemblages, as fossils, coal, fusain, and some types of oil. The metals coming from earlier explosions among the planets fall in dust or globules, mixing with the turbulence as deposits. Repeated falls of dust, terrestrial and exoterrestrial, mingle with the slowing floods to give the Earth its patina of soils in favored places. Here, and also at one time in the drowned slopes of debris off the shores of continents and around submerged volcanic heights, most of the surviving and adapting biosphere found its home.

Who needs more time than several thousand years to explain all this, they may well say? Only a small fraction of the operations and product of the earth sciences and biology depends directly upon the chronologies that have been developed in natural history. Determining whether the dinosaurs were exterminated five thousand or fifty million years ago may have little to do with deciding whether the mammals had reptilian ancestors. King Kong may still be alive in some jungle for all the difference it would make to primate zoology. The protozoans are alive and studied without reference to the discovery of similar Precambrian species.

Even the science of radiology is independent of its use to measure time; geophysicist Melvin Cook, following upon his trenchant criticism of radiochronometry, is prompt to praise other uses of radiation physics in geology. Similarly, Dudley attacked vigorously the idea of stability of radioactive decay measures even though he was a professor of radiation physics in medicine and quite aware of the value of radiation science.[21] When asked to comment on tests by Anderson indicating the non-random and unreliable decay of C14, "scientists said that it could be possible to accelerate or control the release of energy from decaying nuclei... This could lead to..."[22] It's an ill wind indeed, that blows no good.

When a group of scientists and philosophers, perhaps the most notable of them being Albert Einstein, radically criticized the notion of time, the progress of physics is said to have been assisted. Even when time is conceived to run backwards in certain physical, chemical and astronomical theories, the idea is treated as possibly a positive contribution to the solution of perplexing issues.

Nor does the radical alteration of other hard-shelled concepts throw the sciences into unhealthy turmoil. For some time now, the gravitational constant has been assailed as an inconstant, possibly diminishing on the Earth and in the cosmos, following the work of Dirac, Dicke, and others. In an accompanying volume, Earl Milton and the present author, in a history of the solar system, seek to dispense with the concept of gravitation entirely, save for the notion of inertia. At the same time, we seek to work with the concept of a single charge in electricity, endeavoring to solve cosmogonical problems without the two-century- old idea of positive and negative charges.

Why, then, does it matter at all when, in looking upon a mountain or dealing with a human being, one person says he is looking at a historical creation of a great many millions of years while another person says he is observing the creations of a few thousand years? Each sees beauty in the sight, let us grant; each

---

[21] See *Chem. and Engin. News,* Apr. 7, 1975, "Comment."

[22] Interview *NY Times* (30 Mar. 1971), Following presentation of paper, see IX *Pensée* 4 (Fall, 1974).

understands the morphology; each commands techniques for mastering problems that arise in connection with the mountain and the human. Indeed, each may exclaim, "What wonders hath God wrought!" - God taking much time to the first observer, little time to the other.

But, now the second person adds that he believes in the validity of certain scriptures that purport to convey the word of God, among which are some sentences that describe how God made the world, including a time-schedule of the construction. The first person has no interest in these same scriptures except as possible scientific testimony, and as such he finds them almost totally incorrect, and says so.

Now an issue is joined. But note that the issue concerns time only incidentally. The issue is whether a body of writings can be the words of God; many other parts of the scriptures are at issue which do not concern time at all, such as, for example, a statement forbidding the eating of pork and shellfish.

The issue of sacred authority is beyond the method of the present work (and is treated in my book, *The Divine Succession*) unless, as a consequence of this book, either the one or the other person derives support from it, which he can then use in proving that the alleged words of God do or do not conform to a historical reality, proved by other means.

However, it is deemed permissible to employ the scriptures in a secular sense here, as a source of facts, allegations, and hypotheses about natural history; in so doing, we submit the scriptures to the same respectful treatment we give to all the rare ancient documents treating in their own way of scientific subject-matter, such as Hesiod's Greek *Theogony* and the Hindus' *Rig-Vedas*.

Therefore, questions of the elapsed time for accomplishing the present surface of the Earth have to be answered with a set of intellectual instruments called the scientific method, which are presumed useful to all persons engaged in seeking such answers. The primary tools are the empirical proposition, the testing of this by factual evidence, and some control of reality under the government of the propositions -that is, hypothesis, proof, and application (prediction being one form of such).

This is all elementary, but leads us to ask about time. If the duration of historical time is unimportant and inconsequential in most of the work of the earth sciences, why should it be important in natural history? If it can be shown that natural forces could have provided all of natural history through the agency of hundreds of millions of years, why trouble oneself with showing that they could provide the same in a few thousand years?

There are two answers, not identical even though usually correlated: one set of solutions may be more consonant with reality; further, one set may be more useful. A satisfactory explanation of those answers (apart from the problem itself) would require a volume of philosophy on the true and the useful. We might, for instance, find ourselves concluding that the short and long chronologies are both equally true, but the short answer is useful for people who wish to correlate perfectly their natural philosophy about the empirical world with their beliefs in the words of their sacred scriptures.

Alternatively, we might discover that the long-term view is really true and we might as well accept the reality principle as our guide, instead of the sacred doctrine. Since this is a fairly weak view (why hold to reality if it doesn't make pay-offs?), it is often strengthened by a historical, acquired fear of negative experiences in treating with persons holding to the scriptural text. Taken together with various sociological forces - such as professionalism and bureaucracy - truth *per se* and historical fear can generate a strong sense of the utility of the truth. The stage is then set for an enduring struggle between creationists and gradualists.

Here we end up in a distinctly different position. We wish no quarrel with anyone; yet, in a sense, we have to quarrel with everybody. We say that, properly understood, natural forces can have created the present world in a vastly compressed span of time. Too, they may have done so. In arguing that they may have done so, we probably lend a hand to creationists; we do so, too, by according respect to ancient holy writ, as we find this source of evidence shabbily treated in both scientific and humanistic circles.

On the other hand, we see no divine miracles in a nature operating by quantavolutions over a short time. Nor do our time schedules and calendar of events correlate fully with the sacred ones that we know. Nor, finally, unless I underestimate my work, do our explanations facilitate the introduction of an animate divine intelligence into natural history.

Indeed, "creation science," as is called the systematic effort to validate the natural history of the Bible, may be self-defeating. It lets a holy statement, which might better be believed as a different kind of truth-telling and saving instrument, enter into competition in the contests of science, where the rules, the umpires, and the rewards are greatly different. I say this while expressing appreciation of the distinctive contributions that creationists have continuously made to the earth sciences, and realizing that, were it not for their religious zeal, their scientific interests alone would not have given birth to their hypotheses and research.

Supposing that a respectable case has been made for its actuality, what utility does mini-temporal natural history possess? It displaces time as dictator of events. Although it does not abolish historical time, it allows natural forces to play flexibly with time in history.

It lends historical stimulus to inventive ideas that would be hopeless if time were by its very slackness a limiting factor. We see this kind of idea now seeking realization in such fields as elemental physics and genetic engineering. Third, it permits the amalgamation of the earliest records of mankind into the natural sciences, makes man a creature and creator of nature in a holistic sense, helps understand the human story and uses that story to help explain nature.

Here arises the theory of which Velikovsky was the leading exponent, that the morale and behavior of the human race would be improved if humans would appreciate their catastrophic history. Once recalled and realized, the catastrophic record would keep mankind alerted to its compulsion to repeat its past. The death-wish of the human race could better be kept under control, especially now that the suicide of the human race is facilitated by nuclear armaments.

Since there exists a high correlation between millennialist attitudes (the expectation that world-destruction is imminent) and support for catastrophist scientific theories, I doubt that a therapy for the unconscious compulsion to destroy the world is to be found so easily. "If one is going to go to heaven, the sooner the better." More complicated solutions will be addressed in another work concerning religion. It is conceivable that quantavolution offers possibilities of a new effective synthesis of religion and science, which existing creationism and evolutionism cannot afford.

Beyond such utilities rest the several advantages that a microchronic model provides in association with the other elements of the theory of quantavolution: such as the negative exponential principle, the holistic principle, and the transactions of exoterrestrial and terrestrial forces. For example, moon-eruption theory (G. Darwin, Fisher, Pickering *et al.*) was first posited as occurring in early stages of the Earth's formation by macrochronic reckoning.

When Wegener advanced his continental drift theory, he was impelled by paleontology to place the rifting continents in the Cretaceous period. An opportunity to join lunar outbursting and continental drift was lost because of vast differences in timing the two events. Both theories, moon-eruption and continental drift, were placed in abeyance for many years.

Then, when Wegener's theory was revived, an elaborate mechanism of tectonic plates moving by convection currents was devised (Hess *et al.*). Again an opportunity was lost. But microchronism, together with its allied quantavolutionary principles, brings all three events together: paleontological ecumenicalism, the moon-eruption, and continental cleavage and rafting.

For propagandistic purposes, one might take advantage of the credibility that attends long time scales: granting quantavolution, may I not still allow a few millions of years for the resurfacing of the Earth, or only a million, or even a hundred thousand? Or use the Pleistocene, that period of "ice ages" which can be stretched from 100,000 to 2,000,000 and has as many climates and ice advances as we have fingers and toes, thus to avoid a furor of reproaches? Why do I crowd the Holocene so?

I reject this admittedly tempting idea for one large reason alone. As is demonstrable fully in my books on *Chaos and Creation* and the rise of *Homo Schizo,* I find evidence in the earliest behavior and beliefs of mankind that I cannot dismiss, which attests to human experience with every form and scale of quantavolution. At this point in the study of quantavolution, I would lengthen the time scales only if some incontrovertible proof of a relevant far-distant event were offered, or if it were to be discovered that the earliest humans whom we know about were survivors of earlier advanced civilizations whose true long natural historiography was handed down in garbled form. Neither seems likely.

EPILOGUE

This book will conclude without a chapter given over to the explosion of the Moon from Earth. In *Chaos and Creation* and *Solaria Binaria* lunagenesis is treated more directly, whereas here we have mentioned at many points its relevance to geological processes. Lunagenesis was the paramount holospheric event. No major geological process can be understood without a theory of the origins of the Pacific Basin. The reader can, if so minded, judge the plausibility and the consistency of the theory by tracing it with the help of the Index.

Geology has not been able fully to confront lunar fission because of its notions of time. Nearly all studies favoring the idea have placed the event in the most remote eras, because to place it later would require the reconstruction of later natural history, including that of the biosphere, Furthermore, recent explorations of the ocean bottoms have revealed their astonishing "youth." This finding has been thought to disprove even the earliest fission of the Moon, since lunar fission theory without the Pacific Basin as its point of departure would be unappealing.

But the new evidence piles up in favor of lunar fission from Earth. The physical calculations of mass fit are plausible; the Moon fits its hole. The Indo-Pacific Basin is there; the ocean bottoms are all freshly paved. The land has been cleaved into great and small chunks and directed at the source of the eruption. The cleavages have occurred at a negative exponential rate down to the very present. The only force capable of such large interlocked effects would be the passby of a gigantic exoterrestrial body interacting electrogravitationally with the Earth. Such evidence is resisted because it is felt that the atmosphere, lithosphere, hydrosphere, and biosphere would be totally destroyed. This is not a challenge to be met by theory alone. If the facts occur to demonstrate the prior existence of a totally encrusted and thriving world surface and, then, after an epic quantavolution, continuation of the same processes, greatly altered, lunar fission has to be believed.

Yet theoretical logic - call it speculation - has a large role to play, not the least in calculating whether the biosphere would survive. A review of all that has been written on this subject allows an affirmative. The extinction of a species is difficult; the extinction of tens of thousands of species is more difficult; the extinction of nearly all species requires the total explosion of the globe. Exponential reproduction over a few years can hide the most drastic reductions of population by fire, flood, thrusting, explosion, fall-out, radiation, de-oxygenating, and de-photosynthesizing conditions.

The very excesses of blast may harbor the secret of survival. Cyclonic action fashions its own boundaries. The cyclonic form is applicable to water, heat, dust, debris, electrical charge, radiation -to all that in a spread-out form would tend to exterminate life. An atmosphere permitting survival, by the theory of *solaria binaria*, would have been present in a huge plenum or sac surrounding the planets; in addition, atmospheric gases in close encounters can be exchanged, possibly even created under extreme conditions out of water and other compounds.

Surely survival would not be guaranteed. It might even be considered miraculous. Yet there is enough plausibility in survival so that extinction should not be assumed; what is perhaps the most useful and credible theory to explain the tortured Earth should not be passed over. If the Moon was assembled out of a blasted Earth in a highly developed and recent epoch, then the origins and behavior of continental drift are explained, world geography and physiography are explained, the oceans are explained, and the present state and distribution of the biosphere are explained.

It is astonishing and dismaying to consider the huge differences in time allowances between evolutionary and revolutionary morphology. The Grand Canyon has been a showpiece of geology as well as American tourism. Its accepted history is in the range of one to two billion years for the walls and 10 millions and more for the gorge. M. Cook's explanation calls for only 10,000 years to develop the whole complex. The whole world is implicated in such discrepancies, for the types of geological structures of the Earth are limited to a couple of dozens and they are nowhere unique.

But beyond these considerations goes the nature of the field. Geology operates upon a few basic concepts, among them superposition, erosion, heat and pressure. And these are commonsense to begin with. When one rock rests upon another, it is younger, unless some force has intervened; erosion is the effects of wind and water upon landscape; heat and pressure can transform and transmute a substance.

Catastrophists do not deny these ideas; in fact, they invented them. Geology also has a large dictionary of names that are given to things large and small, representing infinite combinations of substances, heat, pressure, erosion, and position. The genius of geology is to bring order to this immense variety and to use this knowledge to practical ends like making cement and finding oil. To all of which the quantavolutionist says "amen."

Neither geology, nor any other science in its historical aspect, has to fear the idea of collapsed time, but can derive theoretical benefits from it. Let us speak for a moment of chemical evolution. Should it be as well termed quantavolution? I have here above spoken of the Miller-Urey experiments on the initiation of primitive life processes, and have generally considered the possible derivation of earthly existence from exoterrestrial and atmospheric sources. In *Solaria Binaria* we go farther into the matter, elaborating the life-creating and sustaining plenum of primeval Earth.

In 1983 C. Ponnamperuna reported the discovery of all five of the so-called "precursors of life" in the Murchison meteorite that fell in Australia in 1959.[1] The compounds are adenine, guanine, cytosine, thymine, and uracil, which are key molecules in DNA and RNA. He subsequently created all five bases "in one fell swoop" by subjecting a mixture of methane, nitrogen, and water to electrical discharges. This, he said, evidenced that chemical evolution could have been accomplished in a single pool of liquid (or dense atmosphere?) in primitive times. The process might have occurred exoterrestrially as well as on Earth, commented Melvin Calvin who had also studied chemical evolution and won a Nobel prize.

"In one fell swoop:" what, if anything, is this expression but a way of saying collapsing time and quantavolution? Nor can one arrogate to man alone the ability to compress time. Nature may be blind, but she is infinitely large, powerful, and busy.

Therefore, collapsing time may boggle the mind but does not destroy geology. Collapsing time introduces the need for high energy forces than can do in weeks what erosion can do in millions of years. The forces - wind, water, heat, pressure - are already present; it is a question of their organization and intensity. The more intense the forces, the more they depart from our experiences, and resemble the catastrophic recitals of the earliest humans.

Also, the more intense the forces, the more likely that they originate exoterrestrially. There appear to be no means whereby the scientific ideology pervading the earth sciences for the past century and a half can continue legitimately to ignore exoterrestrial causes and exoterrestrial effects in explaining our lately tortured Earth

---

[1] P. M. Boffey, in the *New York Times,* 30 Aug. 1983.

INDEX

www.ingramcontent.com/pod-product-compliance
Lightning Source LLC
Chambersburg PA
CBHW031804190326
41518CB00006B/196